ABSTRACTS OF PAPERS PRESENTED IN SCIENTIFIC SECTIONS

XVIIth International Congress of History of Science
University of California, Berkeley
31 July - 8 August 1985

Acts, Volume 1

Office for History of Science and Technology
University of California, Berkeley
1985

Copyright © 1985 by the Regents of the
University of California

ISBN # 0-918102-13-8

Library of Congress catalog
card no. # 85-62072

PREFACE

This volume contains abstracts of papers presented to the Scientific Sections of the XVIIth International Congress of History of Science. All abstracts received by 15 June 1985 are included.

The abstracts are arranged by topic and within each major topic by session. An alphabetical index by author at the end of the volume designates the session in which the author's paper is scheduled. The designations, in the form "Xy", are also entered at the foot of each abstract to serve as pagination.

CONTENTS

Aa ASTRONOMY AND ANCIENT CULTURES IN THE EAST
1 August Thursday 9:30-12:30

Ab ARABIC AND EUROPEAN ASTRONOMY IN THE MIDDLE AGES
2 August Friday 9:30-12:30

Ac ASTRONOMY AND COSMOLOGY FROM COPERNICUS TO GALILEO
2 August Friday 14:30-17:30

Ad ASTRONOMY IN THE ENLIGHTENMENT
3 August Saturday 9:30-12:30

Ae MODERN ASTRONOMY
8 August Thursday 9:30-12:30

Ba TRADITIONAL MEDICINE AND BIOLOGY: EAST AND WEST
2 August Friday 9:30-12:30

Bb BIOLOGY AND PHILOSOPHY IN THE 17TH AND 18TH CENTURY
3 August Saturday 9:30-12:30

Bc EARLY MODERN PHYSIOLOGY AND ANATOMY
2 August Friday 14:30-17:30

Bd NATURAL HISTORY AND SYSTEMATICS
1 August Thursday 14:30-17:30

Be BIOLOGY: 19TH AND 20TH CENTURIES
5 August Monday 9:30-12:30

Bf MODERN GENETICS: CONCEPTS AND TECHNIQUES
5 August Monday 14:30-17:30

Bg BIOMEDICAL SCIENCE IN THE 20TH CENTURY
6 August Tuesday 9:30-12:30

Bh GENETICS AND IDEOLOGY
  6 August Tuesday 14:30-17:30

Bi DARWIN
  5 August Monday 9:30-12:30

Bj EVOLUTION: THEORY AND PHILOSOPHY
  5 August Monday 14:30-17:30

Bk IMPACT OF DARWINISM
  6 August Tuesday 9:30-12:30

Bl MEDICINE
  6 August Tuesday 14:30-17:30

Bg BIOMEDICAL SCIENCE IN THE 20TH CENTURY
  6 August Tuesday 9:30-12:30

Ca EARLY CONCEPTS OF CHEMISTRY AND ALCHEMY
  1 August Thursday 9:30-12:30

Cb 18th CENTURY CHEMISTRY: IN HONOR OF HENRY GUERLAC
  1 August Thursday 14:30-17:30

Cc ANALYSIS AND THEORY IN EARLY 19TH CENTURY CHEMISTRY AND GEOLOGY
  2 August Friday 9:30-12:30

Cd USES OF MODERN CHEMISTRY
  7 August Wednesday 14:30-17:30

Ce CHEMICAL THEORY IN MODERN EUROPE
  5 August Monday 14:30-17:30

Ga GEOLOGY AND GEOPHYSICS
  3 August Saturday 9:30-12:30

Gb ORIGINS OF MODERN EARTH SCIENCES
  7 August Wednesday 14:30-17:30

Ha SOCIAL SCIENCE BEFORE 1800
  3 August Saturday 9:30-12:30

Hb MODERN SOCIAL SCIENCES: DISCIPLINES
  6 August Tuesday 14:30-17:30

Hc MODERN SOCIAL SCIENCES: FOUNDATIONS
  7 August Wednesday 9:30-12:30

Hd PSYCHOLOGY
  7 August Wednesday 14:30-17:30

He MODERN SOCIAL SCIENCES: METHODS AND CONCEPTS
8 August Thursday 9:30-12:30

Ma ANCIENT MATHEMATICS AND CALCULATION
1 August Thursday 14:30-17:30

Mb NEAR AND FAR EASTERN ANALYSIS IN THE MIDDLE AGES
2 August Friday 9:30-12:30

Mc GEOMETRY AND ALGEBRA IN EARLY MODERN EUROPE
1 August Thursday 9:30-12:30

Md EUROPEAN MATHEMATICAL ANALYSIS IN THE 16TH AND 17TH CENTURIES
1 August Thursday 14:30-17:30

Me PROBABILITY AND STATISTICS
7 August Wednesday 14:30-17:30

Mf MATHEMATICAL INFLUENCES IN THE 19TH CENTURY
6 August Tuesday 9:30-12:30

Mg MATHEMATICAL ANALYSIS IN THE 19TH CENTURY
8 August Thursday 9:30-12:30

Mh MATHEMATICS IN THE FAR EAST
5 August Monday 9:30-12:30

Pa MATHEMATICAL PHYSICS IN THE 17TH CENTURY: DESCARTES AND NEWTON
2 August Friday 9:30-12:30

Pb MATHEMATICAL NATURAL PHILOSOPHY IN THE 18TH CENTURY
1 August Thursday 9:30-12:30

Pc THEORY AND PRACTICE IN THE PHYSICAL SCIENCES: LIGHT, HEAT AND ELECTRICITY
2 August Friday 14:30-17:30

Pd NUCLEAR TECHNOLOGY IN WAR AND PEACE
6 August Tuesday 9:30-12:30

Pe LITERARY INFLUENCE ON PHYSICAL SCIENCE
7 August Wednesday 14:30-17:30

Pf MECHANICS, ELECTRODYNAMICS AND MAXWELL'S THEORY
5 August Monday 9:30-12:30

Pg HISTORY OF SOLID STATE PHYSICS
5 August Monday 14:30-17:30

Ph ORIGINS OF RELATIVITY THEORY
   6 August Tuesday 9:30-12:30

Pi REACTIONS TO RELATIVITY THEORY
   6 August Tuesday 14:30-17:30

Pj STATISTICAL MECHANICS AND THE EMERGENCE OF QUANTUM THEORY
   7 August Wednesday 9:30-12:30

Pk EARLY QUANTUM THEORY
   7 August Wednesday 14:30-17:30

Pl NIELS BOHR'S LEGACY AND ATOMIC STRUCTURE CIRCA 1925
   8 August Thursday 9:30-12:30

Pm EARLY QUANTUM FIELD THEORY
   5 August Monday 9:30-12:30

Pn EXPERIMENT AND THEORY IN MODERN PHYSICS
   5 August Monday 14:30-17:30

Po NATIONAL STYLES IN MODERN PHYSICS
   6 August Tuesday 9:30-12:30

Pp PHYSICAL CHEMISTRY AND PHYSICS
   6 August Tuesday 14:30-17:30

Qa PRACTICE OF SCIENCE IN THE MIDDLE AGES
   3 August Saturday 9:30-12:30

Qb NATURAL PHILOSOPHY IN THE EARLY 17TH CENTURY
   1 August Thursday 9:30-12:30

Qc EPISTEMOLOGY OF THE SCIENTIFIC REVOLUTION
   1 August Thursday 14:30-17:30

Qd LATE 19TH CENTURY VIEWS ON THE UNIFICATION OF SCIENCE
   6 August Tuesday 9:30-12:30

Qe SCIENCE, LOGIC, AND COGNITION
   6 August Tuesday 14:30-17:30

Qf THEORY AND REALITY
   7 August Wednesday 9:30-12:30

Qg REVOLUTIONS IN SCIENCE
   7 August Wednesday 14:30-17:30

Sa SCIENTIFIC DEVELOPMENT IN LATIN AMERICA
   8 August Thursday 9:30-12:30

Sb SCIENCE AND TECHNOLOGY IN COLONIAL INDIA
5 August Monday 9:30-12:30

Sc ISSUES IN NATIONAL SCIENTIFIC DEVELOPMENT
7 August Wednesday 9:30-12:30

Sd SCIENCE, TECHNOLOGY, AND MEDICINE IN NAZI GERMANY AND IMPERIAL JAPAN
7 August Wednesday 14:30-17:30

Se 19TH CENTURY AMERICAN SCIENTISTS
8 August Thursday 9:30-12:30

Ta INVENTION AND TECHNOLOGY IN MEDIEVAL CHINA
2 August Friday 14:30-17:30

Tb METALLURGY IN ANCIENT CHINA AND INDIA
3 August Saturday 9:30-12:30

Tc WATERWORKS AND ENGINEERING
2 August Friday 14:30-17:30

Td APPLIED MECHANICS AND ORIGINS OF COMPUTERS
7 August Wednesday 9:30-12:30

Te HIGH TECHNOLOGICAL INNOVATION
5 August Monday 9:30-12:30

Tf INVENTION AND INNOVATION IN THE 19TH CENTURY
5 August Monday 14:30-17:30

Tg MODERN TECHNOLOGY
6 August Tuesday 9:30-12:30

Xa SCIENCE, TECHNOLOGY, AND ENGINEERING AFTER THE 17TH CENTURY
3 August Saturday 9:30-12:30

Xb ACADEMIC ORGANIZATION IN THE 19TH AND 20TH CENTURIES
6 August Tuesday 14:30-17:30

Xc SCIENCE AND ITS INSTITUTIONAL SETTING
7 August Wednesday 9:30-12:30

Xd SCIENCE AND RELIGIOUS BELIEF
7 August Wednesday 14:30-17:30

Xe IMMIGRATION AND INTERNATIONAL EXCHANGE IN PHYSICAL SCIENCE
5 August Monday 14:30-17:30

Xf WOMEN IN MODERN SCIENCE
  6 August Tuesday 14:30-17:30

Xg MISCELLANEOUS PAPERS
  8 August Thursday 9:30-12:30

Xh DEMONSTRATION OF CNRS DATABASE FRANCIS
  5 August Monday 10:00 Main Library Room 322

CODE

A Astronomy

B Biology, including medicine

C Chemistry

G Geology

H Human and social sciences

M Mathematics

P Physics

Q Philosophy

S Science by regions

T Technology

X Miscellaneous

The second letter indicates the session sequence.

Ronald A. Wells

Dept. Near Eastern Studies; University of Calif., Berkeley

SOTHIS AND THE SATET TEMPLE ON ELEPHANTINE: AN EGYPTIAN "STONEHENGE"

Observations *in situ* during January 1984 indicated that the temple of Satet built by Hatshepsut/Tuthmosis III faced the rising star Sirius, the namesake of the goddess. The orientation of the temple axis, however, corresponds very closely with the azimuths of midwinter sunrise (entrance) and midsummer sunset (rear). The heliacal rising of Sirius was New Year's Day of the Egyptian civil year which consisted of 360 days (twelve 30-day months) plus 5 epagomenal intercalary days. Computations have shown that the time interval between the day of summer solstice and the heliacal rising of Sirius was 5 days only at Elephantine during a short period of about 105 years at the time the temple was constructed in the 18th Dynasty. Further north, this interval increased by 1 day per degree of latitude. It also increased by 1 day per 105 years after this era. Consequently, the design of the temple's portico, containing a surrounding colonnade of 30 portals, could have been used as a calendar to mark the civil year by daily priestly offerings in each portal. The end of 12 such offering circuits of the temple, which began on the heliacal rising of Sirius, would occur on the summer solstice. The 5-day interval until the next Sothic rising and New Year was thus a natural representation of the intercalary Epacts unique to Elephantine at this period in history.

Scenes and inscriptions on the temple walls support the calendarial nature of the temple. Groups of deities are arranged on the walls in the sacristy in numbers representing the lunar month and year; the interval from new to full Moon; the phase relationships of the first and last quarter Moons with respect to full Moon; and the fact that full Moon can occur either on the 15th or 16th days of the lunar month. Moreover, an inscription on the temple entrance portal specifically states that night offerings were to be made on the 1st, 2nd, and 6th lunar days and on the days of first quarter, full, and last quarter Moons!

Computations have further indicated that the temple portals were so arranged that special phases of the rising and setting Moon could be seen through them. When this observation and the scenes and inscription of the previous paragraph are combined with the suggestion that the temple could have been used to mark the days of the civil year, a fundamental conclusion is reached: *the principal function of the 18th Dynasty Satet temple on Elephantine was therefore to permit priests to relate the dates of feasts in the lunar calendar to dates in the civil calendar by a simple counting scheme based on a daily ritual of offerings that could be regularly checked by observation.*

Further details of this study will be available in the summer 1985 issue of the journal *Studien zur altägyptischen Kultur* (University of Hamburg).

Supported by National Endowment for the Humanities Grant RO-20538-83 and by a Senior Fellowship from the Fulbright Commission in Cairo.

David Ulansey

Visiting Assistant Professor, University of California at Berkeley

THE ASTRONOMICAL SYMBOLISM OF THE MITHRAIC "BULL-SLAYING" ICON

The "mystery religion" of Mithraism flourished in the Roman empire from the first through the fourth centuries A.D., and was one of the major competitors to Christianity. The cult centered around an icon depicting the god Mithras in the act of slaying a bull, surrounded by a number of subsidiary figures including a scorpion, a dog, a snake, a raven, and a cup. Recently, a number of scholars have noted the similarity between this group of figures and a series of zodiacal and extra-zodiacal constellations, their conclusion being that the Mithraic icon is actually a star-map of some sort.

The current paper argues that the Mithraic icon is indeed a star-map. In particular, it is argued that the constellations represented are those which lie on the celestial equator when the equinoxes are placed in Taurus and Scorpius (the figure of Mithras himself is connected with the constellation of Perseus which is directly above Taurus the Bull).

The paper goes on to demonstrate how Mithraism could have originated among Stoicizing intellectuals in the city of Tarsus in the second century B.C., as a religious response to the discovery by Hipparchus of the precession of the equinoxes. From a geocentric perspective the precession must have appeared as a previously unknown movement of the <u>entire cosmos</u>-- a phenomenon which, to an age steeped in astrology and astral religion, could obviously have had profound religious significance. The image of Mithras slaying the bull, then, would represent the personification of the newly discovered deity believed to be responsible for this new phenomenon. His slaying of the bull would symbolize that cosmic force which had, in primordial times (reconstructed by the originators of the cult on the basis of Hipparchus' discovery) destroyed the power of the bull by moving the entire cosmos in such a way that the spring equinox moved out of the constellation of the Bull and into its then current position in Aries. Clearly, a deity capable of moving the entire cosmos could be seen as being supremely powerful. In particular, such a deity could be seen as being able to assist his devotees in conquering the astrological powers of fate, and also as assuring a safe passage through the planetary spheres after death.

Vladimir S. Tuman

Professor of Physics, Calif. State College, Stanislaus, Turlock, CA

APPLICATION OF COMPUTERS TO THE ASTRONOMICAL DATING OF THE ANCIENT
BABYLONIAN BOUNDARY STONES, KNOWN AS "KUDURRUS"

"Kudurru" is a Babylonian term used for boundary stones. These are the documents of ownership of land and property that Babylonian kings gave to their sons, faithful officers, ministers, or others. There are also documents in which kings offered lands and property to a temple and its priest. All these events are recorded in cuneiform script on Kudurrus.
There are about 110 Kudurrus scattered around the world, housed in different museums. Many are in the Louvre Museum, Paris; the British Museum, London; and the Staatliche Museen, Berlin. Numerous Kudurrus are carved from the black limestone, usually from large river boulders, and they are roughly dated from 1450 B.C. to about 700 B.C.
The documents often contain the size of the land, its geographical location, the name of the king, the names of the recipients, and the names of witnesses (legal records). They invoke deities, to protect the owner and his family and curse his enemies (religious practices of the time).
Aside from the aforementioned categories, the Kudurrus also contain an astral portion, in which iconographic symbols represented the abode of the deities and their attributes. On some Kudurrus, such as Susa #1, the names of the astral symbols are inscribed. For the last hundred years or so, these symbols and their astronomical representations have been deciphered. It is now evident that through iconographic symbols, the scribes have recorded the position of the planets with respect to the known constellations, primarily in the vicinity of the plane of the ecliptic.
In this paper, we concern ourselves only with the astronomical interpretation of the symbols carved on the Kudurrus. Such a knowledge is used to date three different Kudurrus. We shall briefly discuss the iconography of the astral symbol, and present a sample of the iconography representing the identified planets and constellations. The list will include only those discussed in this paper.
The results obtained for Nazi Marutash Kudurru Susa #2 will be discussed in detail; this Kudurru is dated as 1293 B.C. It is important to realize that the concepts of simultaneous rising and setting as well as simultaneous rising-culmination and setting-culmination were used. The constellations and bright stars used were known as "Ziqpu stars". These are discussed in the mul Apin tablets, and they are found among these Kudurrus. From the mul Apin tablets, primarily about the time of Asurbanipal, we know that these concepts were practised; now we know that these astronomical techniques were used back in 1293 B.C.

Aa

HO, Peng-Yoke

Head, Department of Chinese, University of Hong Kong

TOWARDS A MORE COMPLETE CHINESE ASTRONOMICAL RECORDS: HALLEY'S COMET IN MING CHINA

Efforts have been made recently to collect ancient and medieval Chinese astronomical records from sources outside the traditional Official Dynastic Histories. Three examples of the appearances of Halley's comets in the years 1378, 1456 and 1531 are quoted from the Veritable Histories of the Ming emperors to illustrate their superiority over the Ming Official Dynastic History both in detail and in accuracy. Progress in recent efforts to collect Chinese astronomical records from the various gazeteers and from the Veritable Histories is reported in this paper.

S.D. Sharma

Reader in Physics, Department of Physics, Punjabi University

Halley's Comet in Indian Tradition

Old apparitions of Halley's comet since 240 B.C. up to Halley's time are authentically decoded in Chinese tradition by T.Kian and reported in Memoirs of Royal Astronomical Society (1971) 76, 27-66. A similar preliminary report of European records up to back date 2647 B.C. is prepared by J.L. Brady (Pre-print: Levermore Observatory UCRL-74776). In Indian records there are some similar reports which are being analysed by the author the this paper. From Brhar-Samhita of Varahamihira (5th A.D.) and commentary on the same by Bhatotpala (10th A.D.) and other old encyclopaedic works (Samhitas) and Sanskfit, it is very well confirmed that there were some sages whe studied Comets and believed in their periodicity (Ref. Report by the author in A.S.I. Meeting at Bombay, Nov. 1984) and thus were keeping records on observations of Comets. The Samhita texts reveal that some records certainly belong to Halley's Comet. Physical characterisctics, formation of a long tail ($\sim 90°$ or even more), the reported directions of its appearance, combustion and disappearance, its locus among asterisms (Naksatras) confirm, at least one observation recorded at the time of Varahamihira or earlier. Computaions are being done for its loci and directions of appearance etc. in the past, in order to identify this Comet reported by ancient Indian Skygazers. This paper will expose old records of Halley's Comet in Indian tradition which are scattered over vast literature in Sanskrit.

Keizo Hashimoto, Kyoto, Japan

Professor, Kansai University, Osaka, Japan

ASTRONOMIA DANICA IN CHINA

Although the Tychonic world system and planetary theory were adopted at the astronomical reform in China in the last decades of the Ming period (1368 - 1644), we cannot find any positive evidence that the reform was accomplished on the basis of the translation of Tycho's works as far as the treatises on lunar and planetary motions are concerned. Instead it was the Astronomia Danica by Longomontanus in 1622 that played essential part at the compilation of the Treatise on the Motions of the Five Planets (Wu Wei Li Chih, a) in the mid-1630's. We have found 1) that his name has been transcribed as So-Wu-Li-No (b) after his middle name Severini in Latin (his full name was Christianus Severini F. Longomontanus). 2)His version of the Tychonic system and exposition of the theory of planetary motions as a matter of fact provided substantial material for the theoretical construction throughout the Treatise although it should not be overlooked that extensive observational confirmantion was carried out even by making use of the telescope during the process of the reform. Indeed the chief compiler of the Treatise, J. Rho (Lo Ya-ku, c), and his colleagues have alluded to Kepler's work on planetary motions, i.e., the Astronomia Nova in 1609. 3)But they were not ready at all to make use of it. Apart from the cosmological argument which was a very sensitive problem in those days, Rho has apologized for it, saying that any tables based on it were not yet made available for them. 4)Nevertheless there is even an implicit evidence that Adam Schall (T'ang Jo-wang, d) partly made use of the Rudolphine Tables in 1627 when he compiled the Tables of Eclipses (Chiao Shih Piao, e).5)

(a) 五緯曆指　(b) 色物利諾　(c) 羅雅谷　(d) 湯若望
(e) 交食表

1) Wu Wei Li Chih, ch. 3 (Hsin Fa Suan Shu, ch. 38), 20a and WWLC, ch. 4 (HFSS, ch. 39), 1b.
2) Astronomia Danica (Amsterdam, 1622), Appendix, pp. ( ) 1 & 3.
3) WWLC, ch. 4 (HFSS, ch. 39), 1b, 20a and 25a-b.
4) Ibid., 1b.
5) Chiao Shih Piao, ch. 9 (HFSS, ch. 80), 1a.

Juan Carandell

Universidad de Barcelona (SPAIN)

The Analemmas of the Risāla fī ʿilm al-ẓilāl of ibn al-Raqqām.

The manuscript 918,11 of the library of El Escorial (fols.68v-82v) is a tratise on the construction of sundials by the late thirteenth century Tunisian astronomer Ibn al-Raqqām (d.1315) who worked in Granada for the second king of the Banū Naṣr dinasty.

The manuscript (Risāla fī ʿilm al-ẓilāl) is a treatise on the construction of sundials for religiosus purposes. It consists of a discussion, arranged in 44 chapters, on the mathematical and astronomical principles relevant to sundials, such as the determination of the hour lines, the curves of the prayers of the ẓuhr and the ʿaṣr, a section (chapter 43) on the determination of the azimuth of the qibla and on the calculus of the geodesic distance between a locality and Mecca.

Exact solutions of the problems are found by using a graphic and descriptive method of hellenistic tradition known as Analemma. This method is used for every demostration and resolution of problems in spherical geometry, such as the determination of the azimuth and altitude, transformation of astronomical coordinates and for the resolution of more general problems of space geometry which are not specific to spherical trigonometry, such as the determination of the incident point of a sunbeam in an inclining and declining plane.

Ab

Mercè Viladrich

Universitat de Barcelona, Barcelona, Spain.

"ḤABASH AL-ḤĀSIB´S CHAPTERS IN   IBN AL-SAMḤ´S BOOK
ON THE USE OF THE PLANE ASTROLABE"

The aim of this abstract is to make known that five of the chapters of the Abū-l-Qāsim b. al-Samḥ´s (a disciple of the Andalusian astronomer Maslama al-Majrīṭī, d. 426/1035) book on the use of the astrolabe provide some methods that we may attribute to the astronomer Ḥabash al-Ḥāsib (fl. 835). Even though the text ascribes this chapters to a certain "Ḥanash", they should be considered as quotations of an unknown work on the astrolabe by this famous astronomer. This arabic treatise can be found in Ms. British Library Add. 9,602 (fols. 25 b - 55 b). The chapters dealing with Ḥabash al-Ḥāsib´s methods (fols 50 a - 55 a) are the following ones:

Chap. 109: An approximative determination of the latitude of the Moon knowing the true longitude transferred to the first quadrant In fact, it identifies: Lunar ecliptic with Solar ecliptic projected on the astrolabe; true longitude of the Moon with the ecliptical point having the same longitude; the lunar latitude with the solar declination.

Chap. 110: Determination of the true longitude of the Moon inserting an adjusted lunar parallax as a function of apparent altitude to obtain the true altitude. The adjustements concern the Ḥabash al-Ḥāsib´s Table in Ms. Berlin (Ahlward) 5750, fol. 153.

Chap. 111: Determination of the ecliptical point that sets -in the first half of the month- or rises -in the second half- simultaneously with the Moon. It introduces two adjustments to the observed meridian longitude of the Moon and another one, as a function of the corresponding latitudes to obtain the rectified longitude on the ecliptic of the astrolabe.

Chap. 112: One method for the prediction of the first visibility of the new Moon in the rays of the setting Sun, using 12° and 4/5 of a night hour as parameters of the difference in setting.

Chap. 113: Determination of the night hour knowing the lunar altitude and the latitude and longitude components of an ecliptical point that has the same altitude at the moment of measurement.

I think there could be a scribe´s confusion between Ḥabash al-Ḥāsib and Ḥanash b. ʿAbd Allāh al-Ṣaʿnanī (from Sanʿāʾ, near Damasc), a $tabi^c$ who lived in Zaragoza just after the islamic conquest of al-Andalus and who had a reputation as an astronomer. He was celebrated as the builder of the $miḥrāb$ in the $jāmi^c$ mosques of Zaragoza and Córdoba after having determinated the azimuth of the $qibla$ in both sites.
Ibn al-Samḥ seems to be the first Andalusian astronomer who quotes Habash al-Ḥāsib. On the other hand, I do not know any other andalusian or hispanic treatise on the use of the astrolabe that discusses this methods.

Ab

José Martínez Gázquez, Universidad Autónoma de Barcelona. España.

Profesor Titular, Universidad Autónoma de Barcelona. España.

LE DE ASTRONOMIA DE PETRUS GALLECUS ET LE LIBER DE AGGREGATIONIBUS SCIENTIE STELLARUM DE AL - FARGĀNĪ.

L'étude et édition des folios 49 rb à 56 v du ms. 8918 de la Bibliothèque Nationale de Madrid qui contiennent la Summa de astronomia de Fray Pedro Gallego nous permettent d'élargir notre connaissance du contexte scientifique de l'époque de Alfonso X el Sabio dont ont vient de commémorer récemment le VIIe centenaire de la mort (1284 a. D.)

L'auteur de la Summa de astronomia, religieux franciscain que fut confesseur du Rey Sabio, fut consacré évêque de l'évêché de Cartagène qui venait d'être réinstauré, et c'est durant cette ètape qu'il sut trouver le temps, parmi les dures tâches de consolidation du royaume de Murcia rècemment conquis, de cultiver la science de son temps, en particulier le domaine des traductions de l'arabe au latin, comme l'ont bien montré le travail de Pelzer ou les travaux de divulgation de la science astronomique tels que ceux qui apparaissent dans le texte que nous venous de retrouver. Dans l'ensemble la Summa de astronomia ne semble pas être une oeuvre d'une grande originalité mais il ne s'agit pas non plus d'un ouvrage traduit directement d'un original arabe.

Dans l'état actuel de nos connaissances sur le degré d'approfondissement atteint par la science astronomique à l'époque et dans l'entourage de la cour du Rey Sabio, le travail de P. Gallego atteint à peine le niveau d'una récopilation des concepts les plus traditionnels de cette science et apparaît plutôt comme un simple résumé et une paraphrase de l'oeuvre d'astronomie bien connue d'Alfragano d'aprés la traduction de G. de Cremone.

Si l'on en croit l'index de l'introduction, l'oeuvre comprend 9 chapitres. Le premier nous présente un exposé plus personnel de P. Gallego qui reprend certaines des idées de Ptolémée, de Al-Bitrūjī et une synthèse de sources latines. Au cours des chapitres 2 à 7, l'indication du tître ab Alfragano composita, semble pleinement justifiée; il s'agit, en effet d'une copie de l'oeuvre de Al-Fargānī, réalisée d'une manière très littérale, avec des recoupes quant à l'information mais aussi en y ajoutant quelques brèves réélaborations adaptées à son entourage culturel. Les chapitres de la fin sont une élaboration plus personnelle de l'auteur à partir de la doctrine commune héritée du monde latin, en particulier de M. Capella à qui il fait directement allusion.

Ab

Dr. K. Maghout und Dr. S. Chalhoub
I.H.A.S. Aleppo Direktor , Bereichsleiter der Naturwiss.

Die sphärische Trigonometrie
bei Abū - ᶜAbdllāh Muh. b. Gāber b. Sinān al-Battānī

Al-Battānī lebte in der 2. Hälfte des 3. Jhd. H.
(o. 9.Jhd. u.z.) Die ersten Beobachtungen machte er in
AR-Raqqa 264H. (o. 877 u.z.) er hat auch in Bagdad
gearbeiter und ist zwischen Ar-Raqqa und Bagdad 317 H.
(o. 929 u.z.) gestorben. Er ist einer der bedeutensten
Astronomen des 10. Jhd., er erreichte ausgezeichnete
Ergebnisse in seiner Zeit mit einfachen Instrumenten. Er
machte viele Korrekturen zu Ptolemäus in seinem Buch
"Al-Zig-As-Şabi'ī". Aber dieses sollte man mehr und
gründlicher untersuchen. Besonders hat er sich mit der
Zeitrechnung und dem Azimut der Kibla auf der Grundlage
der Cosinusregel beschaftigt.

Im.11. Kapitel seines Buches benutzt er den Cosinus α

$$\cos \alpha = \frac{\sin \delta - \sin h \sin \varphi}{\cos h \cos \varphi}$$

δ ist die Deklination der Ekliptik
h ist die Höhe der Sonne
φ ist die Breite des Landes

das entspricht etwa der heutigen Umsetzung

$$\cos a = \cos b \cos c + \sin b \sin c \cos A$$

Wobei $a = 90 - \delta$ $b = 90 - h$ $c = 90 - \varphi$ $A = \alpha$

Im 56.Kapitel benutzt er einen Naherungswert für den sin

Azimut der Kibla = $\dfrac{\sin \text{ der Differenz der Breite des Landes}}{\sin \text{ der Distanz des Ortes und Mekka}}$

So ähnlich finden wir es später bei Abū-l-Wafā'
Al-Būzagānī. Beides auf der Grundlage der spährischen
Trigonometrie. Bei meinen Studien stellte ich fest, dass
die spharische Trigonometrie bei al-Battānī bis heute noch
nicht gründlich untersucht worden ist, ebenso ist es mit
der Rechnung der Datierung im 10.Jhd. und es bleibt auch
zu vergleichen mit späteren Datierungen z.B. im 14.Jhd.
mit Ibn -Al-Shaṭer

Ab

Roser PUIG
Dpto. Arabe/Filología. Universidad de Barcelona. SPAIN

HISPANO-ARAB ASTRONOMICAL INSTRUMENTS. THE SAFĪHA
SHAKKĀZIYYA OF AL-ZARQĀLLUH (c.XI).

The safīha is an astronomical instrument devised in the eleventh century by the Toledan astronomer al-Zarqālluh. Its chief characteristic is that it offers on its front a stereographic meridian projection which has as its plane of projection that of the solstitial colure, superimposing the projection from Aries on the projection from Libra. The celestial and terrestrial spheres are thereby totally projected on the instrument which consequently may be used in any latitude.

There exist two obvious varieties of safīha, both of which are attributable to al-Zarqālluh: the zarqāliyya and the shakkāziyya.

Concerning the zarqāliyya which is the best known safīha there are two treatises, on construction and usage respectively, and which may be readily consulted in the Alphonsine versions contained in the Libros del Saber. It is distinguished by having on its front the aforementioned projection, whilst showing on its back, in addition to the lines usually drawn on an astrolabe, an orthographic projection.

The other variety of safīha, little known until only recently, is the shakkāziyya. Concerning the use of this type of safīha I have gathered a considerable number of treatises -in Arabic- without so far having been able to discover any treatise on its construction. By examing the manuscripts I have succeeded in classifying them into two groups, according to their number of chapters: those which consist of 60 and those which consist of 23. They all have an introductory chapter describing the instrument which is followed by an explanation of its usage. I have observed that the difference in the number of chapters corresponds not only to an abridgement of the text; it also points to the existence of modifications which in the course of time made changes in the instrument itself.

In any case, all the treatises on the shakkāziyya agree in ascribing to the latter a series of characteristics which contrast it with the zarqāliyya, and these are as follows: a simplification in the lines of the coordinates projected on the front, and the complete disappearance of the projection on the back.

Therefore, the safīha was an instrument constantly evolving from the time it was invented, which had a perceptible influence on subsequent instruments.

Ab

Celina Ana LERTORA MENDOZA

Chercheur du CONICET, Buenos Aires, Argentina

L'OEUVRE ASTRONOMIQUE DE ROBERT GROSSETESTE

Nous pouvons synthétiser l'apport de Robert Grosseteste à l'astronomie en trois points:
1. Méthodologie: Il a formulé une série de principes généraux de méthodologie, qui constituent un cadre systématique des disciplines particulières (spécialement dans ses Comm. in VIII Phys. et in Post. Analy.). Les éléments fondamentaux de la méthodologie proposée sont: la méthode analytique-synthétique avec possibilité de vérification empirique et la formulation exacte (mathématique) des résultats. Ceci doit s'appliquer à l'astronomie en tant que science descriptive, en constituant son critère de scientificité. On sépare ainsi le domaine scientifique de la pseudo-science. Par ex. dans son analyse des phases lunaires, le trajet des comètes (De sphaera, De cometis), la formation des étoiles (De generatione stellarum), etc., il utilise ces principes méthodologiques. Il accepte l'influence naturelle des astres sur les objets inanimés, mais pas sur les événements humains spirituels, bien que l'on puisse accepter une quasi-prédiction dérivée d'un raisonnement sur la nature et la propriété spécifique de cha que planète ou corps céleste.
2. Réception de contenus nouveaux: La première moitié du XIIIe. s. a incorporé beaucoup de contenus nouveaux, parfois de manière désordonnée. Par contre, Grosseteste s'est efforcé de systématiser et approfondir dans les sources; c'est lui qui a commencé le traitement théorique des thèmes qui auparavant n'avaient fait l'objet que de descriptions et mesurages isolés. Il s'occupe donc de l'étude causale des phénomènes célestes, veut établir des hypothèses générales à confirmer (p.ex. la nature des étoiles dans De generatione stellarum) et l'influence de la lune dans les marées dans De accesu et recessu maris.
3. Apport empirique: Il a permis le progrès car il a parfait les calculs et les observations. On sait que la crise et le dépassement des anciennes théories astronomiques ne sont pas dus à la confrontation parmi des théories contraires, mais au contraste entre la prédiction déduite de la théorie et un mesurage empirique correct et minutieux. En ce sens, Grosseteste a fait un apport personnel qui a été repris par ceux qui l'ont suivi (p.ex. les observations et mesurages pour adapter les Tabulae de Toledo au méridien local).

Robert S. Westman and Owen Gingerich

Professor of History, University of California, Los Angeles, USA
Professor of Astronomy and of the History of Science
Harvard-Smithsonian Center for Astrophysics, Cambridge, MA, USA

PAUL WITTICH AND LATE-SIXTEENTH-CENTURY COSMOLOGY

J.L.E. Dreyer, the great historian of Tycho Brahe´s life and work, wrote that "The idea of the Tychonic system was so obvious a corollary to the Copernican system that it almost of necessity must have occurred independently to several people..."

In this paper, we argue that the Tychonic system was not quite so obvious to sixteenth-century astronomers as it seemed to Dreyer. The history of the construction of geoheliocentric cosmologies actually emerged through a complex web of personal relationships that we have patiently recovered by studying a remarkable family of annotated copies of Copernicus´ De revolutionibus (1543).

At the center of this network was the now virtually unknown mathematical astronomer Paul Wittich. Four copies of Copernicus´ book annotated by Wittich are today found in Prague, Liège, Wrocław, and the Vatican, and important related copies are in Aberdeen, Edinburgh, Wrocław, Wolfenbüttel, and New Haven. We consider Wittich´s relations to Erasmus Reinhold, John Craig, Duncan Liddell, Andreas Dudith, Wilhelm of Hesse, and Johannes Praetorius, and his very problematic friendship with Tycho Brahe. We suggest that Brahe´s famous world system may owe more to the obscure Wittich than either he or his later biographer allowed.

Wilbur Applebaum

Assoc. Professor - Illinois Institute of Technology - Chicago, IL - U.S.A.

"Realism" and Empiricism in the Reception of Keplerian Astronomy

Several factors have been considered as having had a bearing on the evaluation of Kepler's theories by 17th-century astronomers. One is the observational evidence of various kinds that began to emerge after Kepler's death in 1630 and the results of the systematic efforts to compare Kepler's Rudolphine Tables with others during the middle years of the 17th century. Another factor is the role of Kepler's belief that the astronomer could determine the real and not just the apparent orbits and positions of the celestial bodies. Little attention has been given to this latter factor as one which may have had persuasive force for astronomers in opting for or against the Keplerian system.

The proliferation of "empty-focus" equant theories by those who may have accepted elliptical orbits on empirical grounds or from empirical and extra-empirical considerations raises questions about the role of "realism" in the transformation of astronomy in the post-Keplerian period. The post-Keplerian equant allowed the preservation of the ellipse and non-uniform motion without the difficulties of the area rule. When and why do astronomers, having earlier embraced Copernicus's rejection of the Ptolemaic equant, become convinced that the apparent non-uniform motion is real? How could a "realist" follower of Kepler like Horrocks propose purely instrumental or non-realist mathematical models, notably for his lunar theory? Seventeenth-century astronomy can only be understood through a careful analysis of the interaction between empirical and non-empirical factors and their changing roles and functions in different contexts.

Michael Chriss

Professor of Astronomy and Humanities, College of San Mateo,

San Mateo, Calif.

THE GALILEAN GAMBIT OF 1619

In August 1612, Galileo was finishing his second letter on sunspots. Using the occasion to strike another blow in his battle with the Peripatic philosophers, Galileo pointed out that the sunspots, recently discovered by him, showed that the heavens were not immutable at all. And enlisting the aid of the respected Tycho Brahe, Galileo pushed on saying, "And as if to remove all doubt from our minds, a host of observations come to teach us that the comets are generated in celestial regions." This reference to Tycho's precision parallax measurements of the comet of 1577 seems to be Galileo's only recorded tribute to the Danish astronomer. It is quite striking hten, to compare this praise with the diatribe put into the mouth of Salviati, in Galileo's Dialogo, 20 years later:

> ...As far as the comets are concerned, I, for my part, care little whether they are generated below or above the moon, nor have I ever set much store in Tycho's verbosity ... I doubt whther the comets are subject to parallax ...

How are we to take these two sharply differing attitudes toward Tycho? Was Galileo insincere in 1612, or was it in 1632? Or did something happen between these two years to cause such a profound change of heart, from praise to bitterness? The answer appears to be a complex combination of these alternatives. The change of heart was occasioned by the appearance of three comets in 1618, which excited much interest and attention in the populace. At the time, Galileo was suffering after three years under a Church decree which forbade the teaching of the Copernican theory. Seizing the opportunity, Galileo in 1619 decided to take a calculated gamble: advance Copernicanism indirectly, by attacking Tycho and his system. And, since the issue at hand concerned comets, it was on this ground that the battle would be waged.

This paper will discuss the circumstance which led Galileo to take this gamble, which involved what was, after all, a false issue, the comets. The implications of Galileo's thoughts toward Tycho, and perhaps even Kepler, will also be examined.

Winifred Lovell Wisan

GALILEO ON GOD AND CREATION

Galileo on God and Creation

God's creation is a dominant theme in Galileo's writings. It is not a mere rhetorical device and it doesn't begin with his observations with the telescope. The creation theme first appears in Galileo's earliest manuscripts on motion, which were composed around 1590, and it steadily develops throughout his life. His earliest attempt to explain the arrangement of the world draws on Aristotle and on the atomists, but then the latter are dropped for a more Christian and pious account. Then, following Archimedes, Galileo discovers that Aristotle's theory that there is natural upward motion as well as natural downward motion must be wrong. At first, he tries to rescue Aristotle by making some distinctions. But soon he finds himself forced to abandon Aristotle's physics and is pushed yet further away when he begins to study mechanics. If motion toward the center of the earth is natural and motion away unnatural, as Galileo finds, then an ideal body on a smooth, round, ideal earth would have no tendency to move. It cannot move down and it resists being moved up. But there is no resistance to motion that is not away from the center. Given the least push, an ideal sphere, for example, would roll about the earth forever, again contrary to Aristotle.

Now, if Aristotle's earthly physics is wrong, what about his cosmology? Galileo begins cautiously to consider the Copernican theory. By time he hears of the telescope, he knows what to look for. Hastening to make one, he turns it to the heavens. He makes discoveries that lead him to see himself as the messenger of the stars, and he tries to convert the Church, from the lowly, recalcitrant _frati_ to sophisticated prelates. Then, silenced for a while by the Church, which would accept the Copernican system only if proved apodictically, Galileo gets another chance when Barberini becomes Pope Urban VIII. Urban believes Galileo's argument for the earth's motion (from the tides) to be apodictic, and he permits publication of the _Dialogo_, provided Galileo admits God might have caused the tides in another way. Galileo, however, in his _Dialogo_, refutes Urban on the ground that God acts only in the simplest, most economical way. Moreover, he puts Urbans's argument into the mouth of Simplicio, who admits he is not capable of fully understanding Salviati's (i.e., Galileo's) argument. Barbarini is affronted, embarrassed, and hurt. He speaks of being "deceived" and of a doctrine that is "perverse in the extreme degree." It is not disobeying an old injunction but the personal insult that causes Urban's fury and brings on the trial and punishment of his old friend. (This is clearly shown in a previously unnoticed letter written by Galileo after the trial.) For Galileo, understanding God's creation had been the deepest commitment of his life. As an obedient son of the Church, he could abjure. But he did really believe he should and must teach the Church (and its Pope) to understand better the created world.

Richard S. Westfall

Professor of History of Science, Indiana University

The Trial of Galileo

Galileo had many hesitations about his *Dialogue*, and the Church had even more reservations about the manuscript they saw. Why then was the book ever published? Without questioning Galileo's conviction that the book spoke the truth about the world system, my paper argues that motives inherent in the system of patronage help to explain this joint decision.

Galileo had remained virtually unknown until his revelations of new bodies in the heavens. The telescope projected him, not just into prominence, but to the very pinnacle of the Italian and indeed European intellectual scene. He became a most desirable client, and from his prominence he reaped the benefits that patronage offered. In addition to material reward, they included others which were less tangible but not less real, such as prestige and standing in the world. He could not maintain his position however, without continuing to demonstrate his excellence. Pressures generated by the system of patronage appear to have functioned as a major factor in his decision to take what he knew to be a very large gamble.

Cardinal Maffeo Barberini made Galileo's personal acquaintance soon after his move from Padua to Florence. Like virtually all Italian intellectuals, Galileo hailed the election of Barberini as Pope Urban VIII. He hastened to Rome to have the restrictions imposed on discussion of the Copernican system in 1616 lifted. Whatever the exact nature of the assurances Urban gave, there is no doubt about the extraordinary demonstrations of personal favor that the Pope extended toward Galileo. Urban saw himself as the patron of learning and culture. When Galileo brought the manuscript of the *Dialogue* to Rome in 1630, the authorities were very uneasy about the book. Urban himself appears to have made the final decision, encouraged by two partisans of Galileo, Castelli and Ciampoli, who belonged to his entourage. As the patron of learning, he could not bring himself to refuse the man recognized by him and by everyone as the symbol of learning in Italy.

When a crisis in the Thirty Years War cut the ground from under Urban, he found himself forced in new circumstances to turn upon Galileo. Even in the trial, however, Galileo's prestige protected him; harsh as his punishment was, the system of patronage offered him advantages that other unfortunates in the hands of the Inquisition did not receive.

Suzanne Débarbat, Solange Grillot, Jacques Lévy

Astronomes, Observatoire de Paris, France

LA MERIDIENNE DE L'OBSERVATOIRE DE PARIS

La méridienne est située dans la salle du 2ème étage du bâtiment, dont le roi Louis XIV et son ministre Colbert confièrent la construction à l'architecte Claude Perrault en 1667.

Cette salle, qui abritait un laboratoire d'optique depuis plus de 50 ans, a été récemment libérée, puis restaurée ; la méridienne est redevenue accessible, et les premières photographies en ont été présentées au Longitude Zero Symposium tenu à Greenwich, en juillet 1984.

La méridienne avait été inspirée à Cassini I (1625-1712) par celle qu'il avait établie dans la cathédrale de Bologne, et qui lui avait permis d'importants et célèbres travaux sur l'orbite du Soleil et sur la réfraction astronomique. Le projet de Cassini I ne fut qu'incomplètement réalisé ; son fils Jacques (1677-1756) entreprit de la faire reconstruire entièrement, en 1729-1730, et c'est celle-ci que nous voyons aujourd'hui.

Comme l'écrit Cassini IV en 1810, "le gnomon et la ligne méridienne... [ne sont pas] d'un grand usage dans l'astronomie moderne". Mais ils méritent d'être conservés "comme ancien monument et comme le premier de tous les instruments qui, dès les temps les plus reculés, a servi à déterminer les mouvements du Soleil".

Le bâtiment de l'Observatoire présente un plan de symétrie, exactement orienté. La méridienne est dans ce plan ; longue de 30 mètres, elle est formée par une suite de barres métalliques. Sur les plaques de marbre qui l'encadrent sont gravés les repères et les signes du zodiaque correspondant à la hauteur du Soleil au cours de l'année.

Le gnomon proprement dit était placé en arrière d'une fenêtre ménagée dans le mur de la façade sud, à 10 mètres de hauteur au-dessus du dallage de la salle. Cette ouverture est actuellement obturée.

La méridienne matérialise le méridien de Paris, méridien origine employé en France jusqu'à la création du méridien origine international, et figurant sur les cartes géographiques françaises jusqu'à une époque récente. Le rattachement de ce méridien au méridien international s'effectue par l'intermédiaire des observations faites à l'astrolabe à pleine pupille situé dans le jardin de l'Observatoire, à une vingtaine de mètres à l'est du méridien de Paris.

Derek Howse

National Maritime Museum, Greenwich, England

ANECDOTES OF NEVIL MASKELYNE, ASTRONOMER ROYAL

From 1765 to 1811, Nevil Maskelyne was Britain's Astronomer Royal at the Royal Observatory, Greenwich. He was the conceiver and first editor of *The Nautical Almanac and Astronomical Ephemeris*, still published today, and one of the main reasons why, in 1884, the International Meridian Conference at Washington, DC, chose the meridian of Greenwich to be prime meridian of the world. He was at the centre of the scientific establishment in England and corresponded with astronomers and mathematicians all over the world, particularly France despite the various wars.

The author will relate some of the anecdotes encountered during research for a forthcoming biography.

Craig B. Waff

THE FIRST INTERNATIONAL HALLEY WATCH: WORLDWIDE
ANTICIPATION AND OBSERVATION OF COMET HALLEY, 1755-59

Despite its importance as the first successfully predicted return to perihelion of a comet, the appearance of Comet Halley in 1758-59 has received less attention from historians and modern astronomers than the later appearances in 1835 and 1910. Most commentators have limited themselves to Alexis-Claude Clairaut's celebrated month-off prediction, based on perturbation theory, of the perihelion date; Johann George Palitzsch's charming Christmas-day recovery of the comet; and Charles Messier's persevering search for and independent recovery of the comet and his subsequent long series of observations. Their accomplishments, though certainly important, were, however, only a small part of the worldwide effort, described in this paper, that was exerted by a large number of individuals in anticipation and observation of the comet--an effort that might well be called the first International Halley Watch. The changing position and appearance of the comet, reconstructed from over 400 observations made by more than 50 individuals, is also described. While the 1759 return was not as spectacular visually as the later returns, it resembles, more than any other previous return, the forthcoming apparition in terms of observing conditions.

Edoardo Proverbio

Istituto di Astronomia e Fisica Superiore, Cagliari, Italy

The Boscovich's activity and the astronomical instrumentation at the Brera Observatory in Milan since 1765 to 1772.

The Brera Astronomical Observatory in Milan organized and managed by the Jesuit Fathers of the Brera College began its activity near the end of 1762.

In the summer of 1764 on the occasion of one his visit at the Brera Observatory R.G. Boscovich was invited to plan a plain of reconstru ction of the same Observatory. Upon the project of Boscovich the new Observatory was finished in the Summer of 1765.

There are very poor informations about the first activity and the instrumentation used in the Brera Observatory for the period 1765- 1772 in which Boscowich actively worked. On the basis of the analysis of the available original documentation including the whole corrispondence of Boscovich the history of this period making special attention to the role undertaken by R.G. Boscovich has been reconstructed. Boscovich developed an intense and original activity in the study of the new astronomical instrumentation. He gave an impor tant contributions to the theory of instrumental errors and obser- ving techniques. It is interesting to underline the remarkable simu larity between his method of investigating the instrumental errors and these discussed some years before indipendently by Tobias Mayer.

Ad

Seymour L. Chapin

Professor of History, California State University, Los Angeles, U.S.A.

LALANDE AND THE LENGTH OF THE YEAR; OR, HOW TO WIN A PRIZE AND DOUBLE PUBLISH

The exact determination of the duration of the solar year is one of the most important calculations that astronomers must undertake. Not only, of course, are the distances of all the planets to the sun determined by the relation of their periodic times to that of the earth, but the utilization of earlier observations - whether of eclipses for chronology, of planets for the establishment of their movements, or of stars to determine their variations - presupposes a knowledge of the positions of the sun at those earlier times, all of which depends upon the true length of the year.

Given its constant importance, it is not surprising to find that Lalande - an astronomer who concerned himself with all aspects of celestial phenomena, including periodic revisions of many astronomical constants - treated this particular subject on more than one occasion. The interest of this paper, however, will not be so much to survey his work on the sun from 1752 through his participation in the creation of a new calendar during the French Revolution, as it will be to concentrate on simply one episode therein. Specifically, it will focus primarily on a mémoire which was originally submitted in response to a prize question posed by the Royal Society of Sciences in Copenhagen but appeared in print both under its aegis and that of his native scientific Academy.

Karl F i s c h e r , Wissembourg, France.

ASTRONOMIE im ELSASS vor der GRÜNDUNG der UNIVERSITÄT

Die älteste astronomische Denkwürdigkeit des Elsaßes ist der Cod.88. der Burgerbibliothek Bern, laut einer Marginalie im Besitz des Straßburger Bischofs Werinharius (1001-1029). Der älteste Astronomie beinhaltende Kodex des Elsaßes, der "Hortus deliciarum" der Herrad von Landsberg verbrannte 1870 bei der Erroberung Straßburgs. Seine Ikonographie war einzigartig. Der heute älteste astronomische Kodex des Elsaßes zeichnete für die Stiftsdame Guta der Marbacher Chorherr Sintram. Seine Kalenderilluminationen weisen auf arabische Vorlage hin. Ein weiterer illuminierter Kalender stammt aus dem KlosterKönigsbruck. Er bildet mit dem Koppenhagener Cod.Thott 239b. und mit dem Kalender der Stadtbibliothek Straßburg Cod.258. eine ikonographische Reihe, dessen jünstes Glied der Cod.Ross.176. der BAV ist.
Selbständige Malerreihen bilden die Illuminationen von Caspar Engelsüß und Diebold Lauber. Da im Reichsgebiet im 15.Jht. keine Malerschulen existierten, kann man fast alle illustrierten HSS als Erzeugnisse Elsäßischer Werkstätten ansehen. Der Maler und der Schreiber waren nicht immer dieselbe Person, wie im östlichen Randgebiet der deutschen Kultur (Prag oder Krakau). Außer anderen stelle der Autor zwei Reihen fest, die in gewissen Merkmalen übereinstimmen : Pierpont Morgen Library Cod.384, Salzburg M-II.-180., Berlin SB MS germ.fol.244, Roma,BAV Cod.palat.lat.1370, Berlin SB MS germ.fol.1191, Darmstadt LB Cod.266., Donaueschingen Cod.494, Karlsruhe, St.Georgen 81. Diese Reihe ist wahrscheinlich dem C.Engelsüß aus Straßburg zuzuschreben. Zweite Reihe : Bonn UB Cod.S-498, Lexinghton Library, Kalender, einziges von D.Lauber aus Hagenau signiertes Exemplar, Burg Pürglitz bei Prag Cod.I-E-7, Salzburg M-I.-19., Heidelberg UB Cod.germ.palat.261,298, München Cgm 595., BAV Cod.palat.lat.1369, Berlin, SB MS germ.fol.557., München, Cgm 867.
Das astronomische Leben im Elsaß war rege. Es war die Zeit der Klosterschulen. Dominikaner hatten in Straßburg ihr "Studium generale", auf dem Albertus Magnus und Ulrich von Engelbert tätig waren. Mag.H.Collis OFM stellte Sonnenuhren her, ebenso wie Arnold Kunig aus Mergentheim. Joh.Sachs galt als ein Wanderstudent. Auf der Domschule war tätig Conradus Argentinensis um 1200, um 1500 Joannes Lichenberg, der wahrscheinlich Privatastrolge des Bischofes Albert von Sachsen war.
Als die Domschule um 1520 versagte, treten zwei Stiftsschulen in den Vordergrund, nähmlich die Thomasschule und die Schule "beim Jung und Alt St.Peter". In diesem kulturellen Niemandsland gewinn der Name des Jakob Sturm, Gründer des Straßburger Gymnasiums, immer mehr an Bedeutung. Mit ihm begann eine neue Epoche der Astronomie im Elsaß. Zu diesem Zeitpunkt ist die Lehre in Straßburg streng orthodox an den Ansichten von Melanchthon orientiert, also geozentrisch.

Michael J. Crowe

University of Notre Dame, Notre Dame, Indiana, USA

ASTRONOMY AND THE EXTRATERRESTRIAL LIFE DEBATE 1700-1910

This paper draws upon researches undertaken for the author's forthcoming book <u>The Extraterrestrial Life Debate 1750-1900: The Idea of a Plurality of Worlds from Kant to Lowell</u> to show that astronomy was in a number of ways significantly involved with ideas of extraterrestrial life. For example, about three fourths of the most prolific pre-1880 astronomers participated in the debate, most advocating the affirmative side. It is shown that some leading astronomers were drawn to their field by early interests in plurality of worlds speculations which also in a number of instances influenced various astronomical theories and even observations. Extraterrestrial life ideas played a major role in efforts to increase public interest in astronomy, leading in at least five cases to funding for the establishment or improvement of major observatories. Such fields as selenography and areography were stimulated by these ideas which also increased interest in sidereal astronomy and in theories of the origin of the solar system. In some cases authors opposed to extraterrestrial life claims formulated arguments that eventually led to new conceptions of the planets and of nebulae. The effects of the involvement of astronomers in this debate were not in all cases positive; for example, some speculations provoked bitter controversies in the astronomical community, diminished its credibility, and distracted practitioners from more tractable topics. Overall, it is suggested that the debate over extraterrestrial life deserves a place in histories of modern astronomy and that some of the themes developed in this paper shed light on the continuing controversy over the question of life elsewhere in the universe.

S.D. Sharma*& J.G. Chhabra

*Reader in Physics   Depart. of Physics, Punjabi University

Validity of Laplace's Laws for Prediction of Pluto

In 1910 AD, V.B. Ketakara used an extended version of Laplace's laws on consonance of orbits (derived for the cas of Jovian satellites) for prediction of Pluto. From our recent abinitio derivation of these laws for the triåd Uranus, Neptune, and Pluto (U-N-P system) it is now clear that the laws used by Ketakara belong to the case I of the three possibilities discussed by Laplace*. The differential equation for the consonance of orbits is

$$\frac{d^2 V}{dt^2} = \beta n^2 \sin V \quad \therefore \quad \frac{dV}{dt} = \sqrt{c - 2\beta n^2 \cos V}$$

where $V = (n-4n' + 3n")xt + \epsilon - 4\epsilon' + 3\epsilon"$, n's being daily mean motions and ϵ's the epoch longitudes.

Calculations show that $c \sim 1.12$ and $\beta \sim 6.09 \times 10^{-6}$. It is shown that the consonance results into the requisite opposition configuration periodically after about 1800 years, which last occurred around the year of prediction of Pluto by Ketakara. Probably, this is the first example available which corresponds to the case I of laws of consonance of orbits discussed by Laplace. The system of three Jovian satellites corresponds to the case II of Laplace's laws. Since Ketakara extended the laws of motions of Jovian satellites to eh U-N-P system at an appropriate time and that is why he was quite successful to predict the position of Pluto. All the details of these findings are elaborated in this expostion and compared with the works of Lowell and Pickering.

REFERENCES:

* " Celestial Mechanics" by Laplace, Book II, Vol. 1,
    p. 639-71. Translation and commentary by Nethaniel Bowditch
    (Ed. 1966).

Barbara L. Welther

Harvard-Smithsonian Center for Astrophysics

ELEVEN CEPHEIDS: SHAPLEY'S STANDARD CANDLES

When Harlow Shapley and Heber Curtis published their opposing views on the scale of the universe in the Bulletin of the National Research Council for May 1921, Curtis challenged Shapley's cluster distances by questioning the foundation on which they rested: Shapley's calibration and application of the Period-Luminosity (P-L) relation for Cepheid variables. Rather than defend his work on the Cepheids, Shapley strengthened his case for cluster distances based on other types of luminous stars. Nevertheless, the two disparate diagrams for the P-L relation that Curtis and Shapley each published need to be examined and explained.

On Shapley's diagram,[1] 57 points representing 11 Cepheids in the galaxy, 25 in the Small Magellanic Cloud (SMC), and the rest in five different globular clusters all line up like beads on a string, with little deviation from a well-defined curve. Shapley had deliberately patterned the curve after Leavitt's for the SMC Cepheids[2] and had calibrated it for absolute magnitude using Hertzsprung's method for galactic Cepheids.[3] In contrast, on Curtis' diagram[4] new data points for other Cepheids show little correlation with Shapley's defined curve. Among other problems, Curtis questioned Shapley's "elaborate system of weighting."

Shapley's first step had been to apply statistical methods to Boss' proper motions for only 11 galactic Cepheids. Thus, he calculated an average absolute visual magnitude of -2.35 at a mean period of 5.96 days. In today's terms this value was too faint by 1.35 magnitudes. Next Shapley derived the absolute magnitudes for the 11 stars individually. Plotted against their periods, these magnitudes formed a scatter diagram. However, by averaging the values in groups of 3, Shapley defined a curve that he could superimpose on Leavitt's. Then he plotted the smoothed values of the absolute magnitudes for the 11 Cepheids so that they also fell neatly on the defined curve.

This paper will examine some of the art, science, and fortunate coincidences that Shapley employed in calibrating the P-L relation and thus deriving a scale for the galaxy from only eleven Cepheids.

---

[1] Astrophysical Journal 48, 104, 1918.
[2] Circular 173 Harvard College Observatory, 3, 1912.
[3] Astronomische Nachrichten 196, 201, 1913.
[4] Bulletin 2 Nat. Res. Coun., 205, 1921.

Norman Sperling

Lecturer, Sonoma State University

The Four Great Questions of Astronomy

The failure of many developments in astronomy to follow directly
fr m the previous ones demonstrates that consecutive narrative is
not satisfactory to analyze this science. A revolution in one
specialty scarcely influences others. That is because astronomy
is not a single pursuit, but four intertwined lines of investigation.
The entire history of astronomy can be analyzed as the quest for
answers to four Great Questions. These overlap in time but rarely
in personnel, and they deal not with equipment but with the resarch-
es that equipment was set to. The investigation of each object
individually follows virtually lockstep in 2-3-4 order, but the
work of astronomers is typically based in a single question (or in
instrumentation).
    I. Terrestrial Mechanics: What time is it? Calendars and
clocks reckon the periodicities of Earth's motions. The critical
problems arise from those motions' irrational proportions.
This deals with the Celestial Sphere, so it also includes navigation
direction-finding, and practical astronomy in general. The calendar
aspect was essentially solved in the 1500s but time-keeping
refinements continue through the present. Ethnoastronomy in general
dwells almost entirely on this aspect.
    II. Celestial Mechanics: "Saving the phenomena". Kepler satis-
fied the problem basically by mid-1609, and Newton formalized it
in Principia Mathematica. 150 years or so ago, one's astronomical
worth was judged largely by mastery of Laplace's Mecanique Celeste.
The specialty is now called "Dynamical Astronomy."
    III. What is the physical nature of the things in space?
This question is now the mainstream; most astronomers are occupied
figuring out the pieces of this puzzle. It is variously
called "descriptive astronomy" and "astrophysics;" synonyms
at the turn of this century, the first now means "without math"
and the second "in the manner of physics". This question became
accessible only in late 1609, a few months after Kepler solved
Question II, when Galileo first applied his telescope astronomically.
The question really got rolling in the late 19th century when the
split of national and personal styles (traditional German dynamics
versus new English/American physical astronomy) signalled the emergence of Question III with spectroscopy and photography. Now
space probes, nonvisual observational means and computers drive this
aspect.
    IV. How do objects , and the universe, evolve? This question
has become accessible only in recent decades. The careers of stars
are now known in some detail. Theories of cosmology still vie with
one another, though a plurality opinion has developed for the
Big Bang. This question occupies the second-greatest number of
astronomers.

Ae

Armelle Debru

Maître de Conférences à l'Université de Lille, France.

PHYSIQUE ET MEDECINE : LES QUALITES DE L'AIR DANS LA PHYSIOLOGIE ANTIQUE.

Dans la physique antique, le monde se forme par altérations des éléments : changement l'un dans l'autre ou altération propre, notamment par densification ou raréfaction alliées à la chaleur et au froid. Les qualités et les propriétés physiques de l'air (vaporisation, condensation, dilatation) jouent un rôle important, à l'image de la météorologie, dans l'explication de nombreux phénomènes physiologiques, comme la sueur, le sommeil, les mouvements du coeur etc... Mais à partir de l'époque alexandrine, des conceptions mécanistes liées à l'utilisation technique de ces propriétés (pompes, expériences sur le vide, orgue) interviennent plus directement dans l'explication physiologique. Dans son <u>De Usu Respirationis</u>, Galien discute dans ces termes des mouvements du thorax et du blocage respiratoire. Dans la pathologie, l'explication physique par les propriétés de l'air (rareté, gravité) n'élimine pas un autre mode explicatif lié aux théories de la sensation et de la connaissance, comme en témoignent différentes conceptions de l'asphyxie et des maladies dues à la nocivité de l'air.

KRISHAN D. MATHUR

Professor, Department of Political Science
University of D.C., Washington, D.C. 20008

Some methods of inoculation in medieval India

Both smallpox and measles have been fatal diseases in the ancient times throughout the world. European travellers to Greece noted that either smallpox did not occur in Greece or, "there was hardly ever an instance of a native of the Island of St. Helena, man or woman, that was seized with the distemper escaped life." The same was true of Africa and the Middle East. The first accounts of smallpox were noted in Egypt during the times of Imam Omar, the successor to Prophet Muhammad and the first Imam Abu Bakr, but no records of providing a cure were found. It was surmised that these diseases were imported from distant lands, through the vast trade that existed from ancient times between Egypt and the East.

So far as India is concerned, these diseases have been present there from ancient times. However, very little has been written about it by the Indian writers themselves. We are indebted to the scholars sent by the East India Company during the Eighteenth century to study the religions, cultures, sciences and technologies prevalent in India at that time. The documents of British scholars are invaluable in the understanding of the history of Indian sciences, because many earlier records are either missing or were destroyed by religious fanatics.

The British scholars give a vivid description of the disease of smallpox as they witnessed, as well as the treatments offered by the natives. The treatments include the ceremonies of worship and prayer to the goddess, as well as scientific administration of inoculation, that was not found in other parts of the world.

One of the accounts of the inoculating process is contained in a report to the London College of Physicians by Dr. Holwell, F.R.S. This document is remarkable in furnishing details of the inoculation process as practised in India in 1744. It gives a graphic description of the instruments used in inoculation, and the successful results in the treatment of the disease. This paper deals with the nature of the disease and the process of inoculation as it evolved in India during the medieval times.

Guerrino, Antonio Alberto

Professor - History of Medicine - University of Buenos Aires

Teaching of Anathomy in the Middle Age.

In Bolonia who seemed entirely faithful to scolastic medicine, Guillermo Corvi, Dino del Garbo and Tomaso del Garbo obtained their greater triumphs. The work of Guiller-Corvi, Articella de Galeno was considered a classic book during two centuries. After 1300 a new tendency in the study of Anathomy and Surgery is observed. Anathomy up to then has been taught only in schools and with Galeno's book, which nobody dared to discuss. Anathomic dissections wery very seldom practised, and almost exclusively on animals, and never by doctors but only by surgeons or employees. Bartolome Varignana was known for having examined a corpse of a person who was thought of having been poisoned. Guillermo de Saliceto writes a book teaching surgeons anathomy of different parts of the body. Enrique de Mondeville was his disciple. But the first anathomist was Mondino de Luzzi, an anathomist of great success and also famous as politician and considered a wise man. The first dissection was done by Mondino in the year 1315. His book was used for teaching anathomy for many centuries.

Xie, Huan Zhang

Professor, Beijing Institute of Technology, Beijing, China

ANCIENT CHINESE QIGONG TECHNIQUES INSPIRE MODERN TREND OF SCIENCES

Qigong is a more than 3000 years old traditional Chinese transcendental meditation technique which can cultivate human body's innate bioenergy to improve health, consciousness and to cure diseases through relaxation, concentration, respiration and meditation. The history of Qigong is almost as long as the history of the written language of Chinese nation. The discovery and founding of Chinese Qigong was much earlier then the establishment of Chinese Taoism and Buddhism. Qigong played an important role in the creation and evolution of Chinese ancient culture and civilization. Ancient people practised Qigong, improved their cleverness and studying ability for the integration of Chinese civilization. It was a great help for ancient Chinese to cognize self body and outside nature through long time Qigong practices. For example, some of the Qigong practitioners could produce "internal vision" to cognize the organs inside the human body and to establish the famous Meridian Channels Theory which is the base of the Chinese traditional medicine.
Today, the traditional Qigong techniques have been spreading more popularly in China and have increasingly been proved as an effective method for preventive health practice and an encouraging potential therapy for control and cure of the cancer disease of mankind as well as other chronic diseases. Now a new scientific branch called "Qigongology" has been establishing in China. Surprising results and encouraging discoveries have lately been reported. such as: Various electromagnetic waves, infrasonic emissions, magnetic field and an unknown ultramicron particls flow have been monitored from the palms of Qigong practitioners; A stronger ultramicro radiations along the exact traditional meridian channels have been detected on human body; Various physical, biological and physiological effects of Qigong have been practised and confirmed by researchers from different part of China; many conventional physical and biological laws might be subjected to modification and new interpretation through the inspiration of the phenomena of Qigong. Therefore, the ancient Chinese Qigong techniques may inspire another direction of developing modern sciences in the fields of physics, medicine, biology, anthropic science, etc. For example, Qigong can produce more extraordinary phenomena then the Maxwell's Demon of 1883 which led to the establishment of many physical principles. (Some Qigong phenomena will be demonstrated at the meeting)

Fournier, Marian

Curator at the Museum Boerhaave, Leiden, the Netherlands

Mechanical philosophy and the fabric of life

Around 1660, nearly sixty years after its invention, the microscope became a scientific tool. As soon as the first observations were published it was believed that the microscope shed new light on the construction and functioning of the parts being studied. Thus microscopy flowered, its most famous representatives being Marcello Malpighi, Robert Hooke, Jan Swammerdam, Nehemiah Grew and Antoni van Leeuwenhoek. They studied the microscopic structure of animal and human organs, the anatomy of insects and plants, and minute organisms were found to populate the earth in the most unlikely places.

However, the heyday of microscopy lasted for only a relatively short period. By 1690 scientists no longer concerned themselves in a serious way with microscopy, except Van Leeuwenhoek.

It is well-known that the belated introduction of microscopy as a scientific discipline coincided with the rising influence of the mechanical philosophy in the life-sciences. As mechanical explanations of the phenomena of life presupposed structured matter the employment of the microscope for the elucidation of these phenomena constituted a perfectly pertinent course of action.

In this paper it will be argued that the subsequent decline of microscopy was to a great extent inherent in the application of the mechanical philosophy to physiological processes. Within a short time a gap in interpretation opened between microscopical observations and prevailing, mechanistic views of the basic structure of living matter.

Anto Leikola

Department of Zoology, University of Helsinki

FRANCESCO REDI AND THE GALL INSECTS

Francesco Redi (1626-1697) is widely known for his decisive contribution to the problem of spontaneous generation: he provided experimental evidence for the view that insects and other small animals are not generated spontaneously in mud and putrid matter. As to the insects of plant galls, he came to the conclusion that they are generated by the procreative virtue of the plant itself. This was not inconsistent with his experimental findings which denied spontaneous generation in lifeless matter but allowed an equivocal generation in living beings, such as the spontaneous production of intestinal worms in vertebrates. Redi adhered to this view even after Malpighi had brought forth evidence that gall insects are generated from parental seed, in the same way as other insects. The problem, however, puzzled Redi for the rest of his life, and this dissatisfaction was probably the main reason why he never published his treatise on different fruits and animals which are generated in oaks and other trees. Its unfinished manuscript, which has recently been found and published by Santini, Tongiorgi Tomasi, and Tongiorgi (Redia 64, 1981), contains only descriptions of galls and gall insects and lacks all theoretical discussion about their origin.

Virginia P. Dawson

NASA History Office, Washington, D.C.

Abraham Trembley and the Influence of Dutch Science

Abraham Trembley's classic work, _Mémoires, pour servir à l' Histoire d'un genre de Polypes d'eau douce, à bras en forme de cornes_ (Leiden, 1744), exerted an enormous influence on the intellectual life of the eighteenth century. In particular, it produced a new interest in the problems of generation and revived the debates over preformation and epigenesis. It gave impetus to the materialistic and vitalistic philosophies of La Mettrie, Diderot, Buffon and Bordeu. Trembley, however, refused to be drawn directly into the metaphysical debates of his time, though it is likely that he shared some of the philosophical notions of his close friend, Charles Bonnet.

Trembley's experimental scrutiny of the polyp revealed perplexing characteristics which seemed to belong to both the plant and animal realms. Its mode of locomotion was like that of an animal, yet its curious mode of reproduction through budding and cuttings was more like a plant. In fact, early in his study, he thought he might have discovered a "zoophyte," a being midway between the two realms. Trembley, however, as a careful observer, "suspended his judgment," until he had determined through a study of the structure of the polyp, how it took food. Because hydra can go many months without eating and it is very difficult to see the mouth when it is closed, it took nine months for Trembley to discover that the polyp was a carnivore which drew its food through its mouth into its central cavity. This left no doubt that the polyp was an animal.

Trembley's use of nutrition as the criterion on which to base this decision reflects the influence of Herman Boerhaave and the scientific community at the University of Leiden. Boerhaave had called animals "reversed plants" since they had "inner roots" or "lacteal vessels" for the absorption of food. Playing experimentally with Boerhaave's definition, Trembley reversed, or inverted, the polyp. He mistakenly concluded that it could draw nutritive substances equally well through its inner and outer surfaces, but only through whichever surface was structurally on the inside.

The experiment of reversal demonstrates how ideas, even one which is no longer valid, contribute to the creative exploration of nature. Trembley is justly regarded today as a model of the experimental approach, but it should not be forgotten that he was part of a scientific community to which he contributed and drew inspiration. Moreover, a study of the correspondence between Trembley and Bonnet reveals the strong links between the Leiden circle and the intellectual life of Geneva.

Shirley A. Roe

Harvard University, Cambridge, Mass., U.S.A.

MICROORGANISMS AND MATERIALISM: JOHN TURBERVILLE NEEDHAM VERSUS CHARLES BONNET

Prior to the mid-eighteenth century, microorganisms were considered to be miniature animals that arose directly from eggs. Yet after 1749, and the appearance of two new theories of generation from Georges Louis de Buffon and John Turberville Needham, the nature and status of microorganisms were called into question. As part of a broader challenge to preexistence theories, microorganisms played a key role in the theories of both Buffon and Needham by providing evidence for the powers of active matter in generation.

This paper focuses on two of the principal participants in the debate over active matter, generation, and materialism. Needham's microscopical work, and the theory of epigenesis he built upon it, became a focal point of controversy in the 1760s, when the preformation-epigenesis debate reached its height. One of the major proponents of preformation was Charles Bonnet, who sought throughout his career to oppose both epigenesis and materialism, and was especially critical of Buffon and Needham.

Microorganisms received a considerable amount of attention within this broader debate for several reasons. First, the observational evidence presented by Buffon and Needham seemed to demonstrate that generation in the microscopic world proceeded via material decomposition rather than through preexistent germs. Second, there was a real sense, in microscopical research in this period, of actually observing matter in action at the most fundamental level at which matter and life interact. Finally, the generation of microorganisms from lifeless matter became a favorite example and a frequent metaphor in the writings of materialists like Denis Diderot and the baron d'Holbach. Thus microorganisms became something of a test case, as it were, on which the success or failure of two widely diverging biological and philosophical positions seemed to depend.

My purpose in this paper is to illustrate, through two case studies, the ways in which the microorganism controversy became enmeshed in the wider debate over materialism and thereby to show why the subject of microorganisms assumed so much importance in the 1760s and 1770s. Bonnet and Needham were chosen not only because they represent the two opposing generation theories but also because they were in close contact with one another through a twenty-year correspondence. Their letters to one another and to other contemporaries reveal much of the motivations that underlay their biological work. Furthermore, although Needham and Bonnet disagreed over generation, they were united in their opposition to materialism and to irreligion. The differing ways in which they each attempted to save generation theory, metaphysics, religion, and morality from the rising tide of materialism reveal much about the broader context of the generation debate in the mid-eighteenth century.

Peter McLaughlin

Freie Universität Berlin

The Rational Core of Eighteenth-Century Theories of Spontaneous Generation, on the Example of Buffon

In 1775 in the second supplement volume of his *Histoire naturelle* the French natural historian Georges-Louis Leclerc Comte de Buffon published a table of data giving rather exact dates for the spontaneous generation of sea monsters on the various planets and moons of our solar system. Buffon believed he could determine that at a particular time (say 13,624 B.C.) at a particular place (near the poles on Jupiter's third moon) organisms of a particular species (now extinct on earth) arose spontaneously according to the universal laws of matter in motion. Besides the surprising audacity and superfluous exactness of the speculations, what is interesting about these assertions is that Buffon proposed them as empirical hypotheses and offered experimental proof for them. Close to two full volumes are devoted to the experiments and calculations that resulted in this table of dates. The proof Buffons offers consists in measurement data on the cooling rates of metal balls of various sizes and compositions, which he had heated up in his iron foundry in the village of Buffon.

I argue in this paper that Buffon is merely drawing the logical consequences of premises implicit in Enlightenment materialism, and that the belief in spontaneous generation was at the time the necessary consequence of some of the premises of scientific materialism. I pursue the question: on what assumptions and under what conditions is it rational, plausible, or even compelling to believe that the dates of the appearance of particular animals can be inferred from a knowledge of the surface temperatures of the various planets and that these temperatures can in turn be established by measurement in a laboratory on earth.

Buffon recommends himself for this kind of analysis first of all because he draws explicitly some of the conclusions merely implicit in much of Enlightenment thought and thus enables us to infer what the hidden premises must have been. Furthermore, he also explicates many of these premises and presuppositions in the course of his broad-ranging *Histoire naturelle*. Thus, most of the assumptions that must be articulated in order to "reconstruct" his theorem on the spontaneous generation of conspecifics of earth inhabitants on various planets can be found explicitly stated in some part of his works.

Henry Lowood

Bibliographer for History of Science Collections, Stanford University

HOW DOES YOUR FOREST GROW?: FORESTRY, MATHEMATICS, AND CAMERALISM IN GERMANY AROUND 1800.

During the last two decades of the 18th century, an alliance of naturalists, foresters, and government officials sought to reform the theory and practice of German forestry. Opposed to the negligence, inadequate education, and harmful practices of Jägermeister, farmers and woodcutters, these reformers hoped that the systematic study of all natural phenomena pertaining to the forest, along with better education and training for foresters, would bring improvements in forest management. In 1797, J.M. Bechstein summarized the accomplishments of this movement: "scattered pieces of knowledge have been fit into systems, and all sorts of activities, previously left to habit, have been transformed into a science." Today, historians of forestry would seem to agree; they refer to Bechstein's contemporaries as the Forstklassiker.

An active branch of German scientific forestry was Forstmathematik, an approach to sylvan science that figured prominently in the curricula, writings, and practices of enlightened foresters around 1800. Forestry mathematics was defined as the application of mathematical techniques to the management of forest resources, and included such diverse topics as forecasting planting and harvesting cycles, mapping the forest, determining growth rates of trees, estimating yields, taxing and pricing lumber, and even comparing the burning and heating value of different kinds of wood. In order to carry out these calculations, many and diverse variables had to be accurately estimated, a prerequisite of which were botanical and physical studies of trees, wood, soil, climate, and topography. The men who tackled these formidable problems were in the main enlightened forestry officials concentrated in Prussia, Hesse and Saxony-Thuringia, and instructors--at universities, cameral colleges, and forestry schools--trained in the cameral or natural sciences.

The ambitious program of forestry mathematics could not possibly be carried out quickly; critics jabbed at the new, scientific forestry, opposing the dominance of "academics and ancillary sciences" and calling for an end to distractions from the few simple maxims of careful forest management. While not abandoned, mathematical techniques occupied a backseat in the subsequent progress of forestry for many decades. This may have been due in part to the critics, and in part to the diverse paths followed by technical and university education during much of the 19th century, in sharp contrast to the synthesis sought by the cameralists. The story of forestry mathematics around 1800 exemplifies the role played by cameralist administrative practice, coupled to faith in mathematics and experiment, in promoting the application of science to a number of trades, techniques and professions during the later years of the German Enlightenment.

Anne Bäumer, Arbeitsgruppe für Geschichte der Naturwissenschaften, Universität Mainz, Federal Republic Germany

THE IMPORTANCE OF CHICKEN DEVELOPMENT FOR EMBRYOLOGY IN THE 16TH CENTURY.

In the 16th century, bird development was discussed, on the one hand in encyclopaedic literature, but, as was the case, in the equivalent compendia of Classical Antiquity and the Middle Ages, it was of minor importance. If described at all, the development of the bird was mentioned only as a part of the very detailed description of the entire animal kingdom. But the study of chicken development occupied a special position in encyclopaedic literature, in so far as up to the 16th century, the bird was the only creature whose development was described in detail. This changed when, in the 16th century, the embryos of other animals were also examined. On the other hand, within the study of embryology as a science in itself, the afore-mentioned research once again led to the study of chicken development as the main object of examination. Because the anatomists had not been interested in the particulars of animal development, but rather only in foetal anatomy, they were unable to answer the question which had been posed since Antiquity, namely which part of the body develops first. Only by the deliberate return to the knowledge and the methods of Classical Antiquity during the age of Renaissance-Humanism, could a new way be found to resolve this question. Ulisse Aldrovandi (1522-1605) was, without doubt, inspired by the proposal put forward by the author of the Hippocratic work "On the Nature of the Child", namely to examine systematically the course of development of one animal: he successively observed the different stages of chicken development in order to answer definitively the question, which part develops first. It is evident that Aldrovandi's basic conception was the same as that of the ancient authors: what is observed in one animal is representative for the development of all blood animals - the hypothesis of the comparability of animal development is the basis for all embryological research. The chicken embryo was chosen according to the method developed by the presocratic philosophers who used clearly observable phenomena to elucidate obscure ones. Inspired by his teacher Aldrovandi, Volcher Coiter (1534-1576) also proceeded from this principle, and additionally tried to solve the problem concerning the origin of veins; therefore his research was even more precise than that of his teacher. For embryology, which in the Renaissance period still belonged within the realm of anatomy, the examination of bird development played an important role, especially by virtue of its very decisive function.

Annette FELIX

Attachée scientifique, Centre d'Histoire des Sciences, Brussels, Belgium

THE CHICKEN HATCHERY OVEN ("FOURS A POULETS"), FORERUNNER OF ARTIFICIAL HATCHING ?

The conception of artificial hatching is not recent; it could already be found in Pliny and Diodorus Siculus' writings. One of the originalities of the 18th century is the determination to raise this technique to an industrial level and many trials have been performed. The correspondence between the French physicist Réaumur and "amateurs" from France and elsewhere among whom the French naturalist Jean-François Gaultier (living in Canada), shows evidence of the interest in the field.

The Royal Library Albert I of Brussels owns a very interesting manuscript describing a series of ovens ("fours à poulets") intended to hatch 4 000 eggs at once.

BUCHS Mina

Chargée d'enseignement, Faculté des Lettres, Université de Genève, Suisse.
L'histoire de la petite circulation et sa "redécouverte" au XVIe siècle.

La description de la circulation du sang par William Harvey au XVIIe siècle a indiscutablement permis à la physiologie de franchir une étape décisive. Cependant, ce pas a été précédé de la découverte de la petite circulation, ce qui constituait le début du rejet du dogme de Galien.
Plusieurs savants du XVIe siècle, dont Michel Servet, Realdo Colombo, Juan Valverde et André Vésale, ont contribué à cette "redécouverte". S'il est intéressant de rechercher les liens qu'ont pu avoir ces savants à un moment ou l'autre de leur vie, il est néanmoins inutile de rechercher lequel d'entre eux décrivit le premier la circulation pulmonaire. Il est en effet maintenant bien établi que ce mérite revient à un savant du XIIIe siècle, Ibn-an-Nafis. Sa description se trouve dans son commentaire du Canon d'Avicenne dans lequel an-Nafis réfute aussi bien Galien qu'Avicenne. Il rejette notamment à plusieurs reprises et sans ambiguïté le dogme galénique d'une communication entre les ventricules.
Ce n'est probablement pas par hasard que plusieurs médecins, tous formés à Padoue, ont donné presque en même temps une description acceptable de la circulation pulmonaire. La nomination d'Andrea Alpago à la chaire de médecine théorique de l'Université de Padoue en 1521 a fort probablement joué un rôle déterminant dans la "redécouverte" de cette théorie qui avait déjà fait l'objet d'une description trois siècles auparavant. En effet, ayant vécu plusieurs années au contact de la langue arabe, Alpago devient un traducteur reconnu des oeuvres des savants arabes, et notamment de celle d'Avicenne et de ses commentateurs, parmi lesquels on trouve justement Ibn-an-Nafis. Andrea Alpago n'enseigna que durant très peu de temps à Padoue où il mourut soudainement mais où son neveu Paolo, qui l'avait accompagné dans certains de ses voyages, hérita de ses manuscrits et poursuivit le travail de son oncle. Il est certain qu'il se trouvait à Padoue occupé à la traduction de textes arabes en latin, en même temps qu'y vivait Realdo Colombo. Finalement, une comparaison attentive du texte de Ibn-an-Nafis qui traite de la petite circulation avec ceux que nous venons de citer, montre indubitablement de nombreux détails similaires dans la manière de présenter des phénomènes.

Isaac Benguigui

L.D.E.S., Université de Genève, Genève, Suisse

UN PRECURSEUR DE L'ELECTROTHERAPIE :
LE PHYSICIEN GENEVOIS JEAN JALLABERT (1712 -1768)

Le développement des travaux sur l'électricité au XVIIIe siècle a permis d'envisager l'application de l'électricité aux traitements de certaines maladies.
Parmi ces travaux nous mentionnerons en particulier:
1) les expériences récentes de l'électricité par frottement.
2) les observations sur l'analogie entre la lumière, le feu, la chaleur et l'électricité.
3) la preuve de l'électricité de l'air.
4) l'expérience de Leyde (1745) et la constatation que, par des stimulations électriques, les muscles et les nerfs peuvent être facilement excités.
Parti de ces observations, Jean Jallabert considère que la paralysie est souvent produite par l'interruption du cours du fluide nerveux et il pense que les secousses violentes qu'excite, tout à coup, dans les nerfs la commotion, pourraient en certains cas, dissiper les obstacles qui embarrassent le cours du fluide, et rendre aux nerfs la liberté de leurs mouvements.
Pour expliquer les mouvements-convulsifs que cause le fluide électrique, il précise que le gonflement et la contraction des muscles sont produits par les écoulements d'un fluide "très subtil" dans les fibrilles nerveuses. Jallabert signale un certain nombre d'effets tirés des expériences méthodiquement menées.
a) accéleration du pouls.
b) amplitude croissante du jet dans l'expérience de la saignée.
c) accroissement de la chaleur du corps.
Mais l'effet le plus important et le plus réel qui ait donné des résultats thérapeutiques, c'est que l'électricité agite les muscles de mouvements convulsifs.
Jallabert mettra à l'épreuve cette dernière observation, pour tenter de voir quel effet l'électricité produit sur un paralytique.
L'oeuvre de Jallabert se répandit très vite en Europe, procura à son auteur une grande renommée et donna une impulsion puissante à l'électrothérapie après que celle-ci avait été discréditée pendant un certain temps. Considéré par ses contemporains comme "le père de l'électrothérapie," Jallabert a été le premier à noter le phénomène de la contraction musculaire comme le résultat d'une stimulation électrique, ouvrant ainsi la voie aux recherches de Duchenne, un siècle plus tard. Le paralytique dont Jallabert s'est occupé demeura au centre de toutes les recherches en électrothérapie, et fit de Genève le berceau de l'èlectricité médicale.

Naum Kipnis, Fellow, Bakken Library of Electricity in Life and University of Minnesota, U. S. A.

THE DISPUTE ON THE "ANIMAL ELECTRICITY," 1792-1804

In 1791 Galvani announced that a muscle contracted, when connected to a corresponding nerve by two different metals. He explained this by the action of animal electricity. In Volta's view, electricity was involved, but it originated in the contact of different metals.

It is believed that Galvani's theory was rejected as erroneous. In fact, both theories were equally right and wrong, for both kinds of electricity acted in those experiments. Neither side could account for all the experimental results, and it was Newton's requirement of not multiplying the number of causes that finally won: Volta claimed (A. Volta, Opere, 1918, v.1, 268) that since common electricity could explain most of galvanic phenomena, there was no need for a special "animal electricity." Later the experiments neglected by Volta gave life to a new science - bioelectricity.

Another fact neglected by historians is that some of Galvani's opponents stated that his phenomenon had nothing at all to do with electricity and was rather of a chemical origin. They referred to the fact that galvanic fluid affected no electrometer and produced neither shock nor sparks. Besides, Humboldt demonstrated (Versuche uber die gereizte Muskel- und Nervenfaser..,1797, 42-48, 82-88,) that the muscular contractions occurred even when the nerve-muscle circuit was open, which was impossible to reconcile with the electrical theory. No one could refute Humboldt, but Volta's 1799 discovery of the electric pile, which exhibited all of the wanting properties of electricity, gave to his theory a new support.

Unable to recognize that all of these effects appeared only in circuits devoid of any animal organs, scientists abandoned the "animal electricity" and began exploring the physico-chemical properties of the pile. The pile replaced a muscle as the object of investigation, and, accordingly, the term "galvanic phenomena" changed its meaning.

Thus, leaving aside Galvani's problem of existence of animal electricity, Volta substituted for it another one, that of the existence of a new sort of inorganic electricity. However, no one at the time recognized this substitution. It took more than forty years to prove that bioelectricity exists, and much more to grasp that explanation of muscular contractions requires a synthesis of both electrical and chemical models.

KEEL, Othmar F.

Professeur, Institut d'Histoire et de Sociopolitique des Sciences
Université de Montréal

Remarques sur l'interaction entre
chirurgie et médecine dans l'école de William et John Hunter au
cours de la seconde moitié du XVIIIe siècle.

Cette communication analyse le type de formation dispensé par les Hunters et leurs collaborateurs dans les écoles d'anatomie et/ou à l'hôpital.
Nous montrons que cette formation se base sur des recherches nouvelles poursuivies non seulement en anatomie et en chirurgie, mais aussi en anatomie pathologique, en médecine et en physiologie.
Ceci s'explique, selon nous, par la formation polyvalente, que nous mettons ici en évidence, des Hunters et de leurs collaborateurs. Autour des Hunters, s'est développé un réseau d'élèves et de collaborateurs scientifiques qui regroupait non seulement des chirurgiens, mais aussi des médecins (dont un nombre important avaient des fonctions hospitalières). Les uns comme les autres communiquaient aux Hunters les résultats de leurs observations et expériences, voire des pièces pour leurs musées.
Nous montrons aussi que J. Hunter a animé, avec certains médecins, plusieurs sociétés et publications qui avaient pour objectif de promouvoir l'articulation entre les connaissances chirurgicales et le savoir médical.
Le fait que, dans l'école huntérienne, on ait pu cultiver et enseigner simultanément l'anatomie, la chirurgie, la physiologie, la médecine, l'anatomie pathologique, etc... a eu pour effet de rendre possible une forme développée d'interaction et d'intégration entre ces différents domaines. Cette interaction a permis à son tour que se systématise dans cette école un point de vue chirurgical (anatomique, localiste) en médecine clinique, et, réciproquement, un point de vue physiologique et pathologique en chirurgie.
C'est ce qui explique, sans doute, que l'on ait vu sortir de cette école des figures comme celles de M. Baillie. Son traité d'anatomie pathologique est une production typique de l'école huntérienne.
Comme déjà les travaux des Hunters sur les corrélations anatomo-cliniques et la pathologie des organes et tissus, ce traité montre à quel point la clinique avait intégré, dans cette école, l'approche anatomique et localiste de la chirurgie.
Bien avant l'Ecole Clinique de Paris, l'école huntérienne a donc fait fonctionner, sur le plan scientifique, une forme d'unification entre médecine et chirurgie, nonobstant le fait que, sur le plan social et professionel, il subsistait, en Angleterre, une séparation et une inégalité entre les deux branches de l'art de guérir.

Shamin, Alexei, Prof., Dr.

Institute of the History of Science and Technology, USSR Academy of Sciences, Moscow, USSR

NATURAL HISTORY INCUNABULA AND THE PROBLEMS OF THE SCIENTIFIC CULTURE OF AN EPOCH

The history of the first century of printing (incunabula and paleotypes) considered in the context of the first "information revolution" permits to determine the limits of the first rather homogenous massif of sources and to introduce the scale into the contextual historiographic tradition. Studies of the incunabula and paleotypes on natural history, medicine, and mathematics commenced in the XIXth century. Several Soviet scientists worked in this field. Among those are V.S. Ljublinsky, V.P.Zubov, A.P.Markushevitch, N.P.Kiselyov. These works however make only the first attempts to study individual texts or editions and to compare them to manuscripts. Thus important problems needing the complex study of the vast integrated material remain untouched, such as the problem of the connections and relations of science and culture in the historical development. To identify the sources for such studies, the accounting and examination of incunabula on natural history, medicine and mathematics within Soviet collection were undertaken. It was found that Soviet collection comprises somewhat 1000 titles of such incunabula. They are characterized and catalogued. This material is sufficient for the studies of the principles of an epoch scientific culture "field" formation, and for more definite conclusions about the role of the "transmitters" (basic antique or medieval works or texts which influenced the formation of sc'ence of the Modern Ages).

Peter K. Knoefel

Emeritus Professor, University of Louisville, KY, USA

NICANDER VINDICATED

The THERIACA of Nicander of Colophon (2nd century B.C.) is the earliest surviving book to be devoted to poisonous animals. It has had many editions, translations, and commentaries. However, since the NICANDREA of O. Schneider (1856) it has often been said to be merely a versification of an earlier work by Apollodorus which is known only through citations by Aelian, Athanaeus, Pliny, and the scholia to the Theriaca. Nicander's poetry has been described as pedestrian, his vocabulary as bizarre, and his style as repulsive. Close study of Schneider's criticism of the matter of the Theriaca reveals many of his arguments as specious and his conclusions as conjectural. He arbitrarily substituted names, Apollodorus for Apollonius and Apollophanes. In general Johnsonian criticisms of the Theriaca have dealt with the antidotal use of plants, which is a large part of the book, but today is only historical interest, of factual value only as indicating what plants grew around the Mediterranean then. Nicander did borrow from many poets before him, but so did they. His real contribution, description of reptiles, their appearance, their habitat, habits, and the effects in man of their bites, is not found in quoted Apollodorus, nor is his distinction of the three dangerous types of reptiles, the neurotoxic elapids, the vasculotoxic viperids, and the constrictors.

L.C. Palm

Institute for the History of Science, University of Utrecht, Utrecht, the Netherlands

VAN LEEUWENHOEK AND SELLIUS ON MOLLUSCS: A COMPARISON OF TWO EARLY EIGHTEENTH CENTURY METHODS OF MALACOLOGICAL RESEARCH

Eighteenth century malacology can be characterized by the growing need for a systematic classification of the many varieties of molluscs. In this paper two types of malacological research, carried out in the Netherlands, will be explored.

The Delft microscopist Antoni van Leeuwenhoek (1632-1723) constructed a great many small simple microscopes of extremely high quality. With the aid of these tiny instuments he studied numerous objects of mineral, vegetable, animal and human origin, in order to try and thus to reveal the mysteries of nature. Although his results were published scattered over many letters in a rather chaotic manner, there were nevertheless a few general themes that he studied, e.g. the structure of matter and the phenomena concerning growth and reproduction. Van Leeuwenhoeck's malacological research mainly concerns the reproduction of mussels, oysters and swan mussels, apart from a few isolated discoveries like cilia. He did not carry out this research for its own sake. The motives which underlay his studies in this field were primarily the corroboration of his rejection of the theory of spontaneous generation.

For quite different reasons the Danzig lawyer Gottfried Sellius (1704?-1767) entered in the malacological field. Still living in Holland after finishing his studies in Leiden he was a witness of the shipworm plague that from 1730 to 1732 terrified the Dutch people and authorities, because of the damage inflicted upon the sheltering dikes. Apart from juridical and physical studies Sellius published one monograph in malacology only. His research is marked by a systematic survey of facts known from literature combined with original observation of living shipworms and the anatomizing of them. This method, in its systematical approach so completely different from Leeuwenhoek's, was derived from the German philosopher Christian Wolff. Sellius came to a correct classification and made some remarkable observations on shipworms.

These differences in the methods and the results of Van Leeuwenhoek's and Sellius's malacology count for the greater influence of the latter in eighteenth century malacology.

Brigitte Hoppe

Professor at the Institut f.Gesch.Naturw., Univ.München FRG

CONTRIBUTIONS OF PHYCOLOGY TO XIXTH CENTURY CYTOLOGY AND REPRODUCTION BIOLOGY

The researches on lower animals and plants including the Algae enlarged the empirical basis of biology during the 19th century. But deep historical studies on the new phycological researches and on their contributions improving the development of modern biology are wanting. In order to remove the deficiency the paper presents my recent results of historical studies on original phycological works and manuscripts of 19th century biologists.

The paper will explain that following the empirical researches on the morphology and the developmental physiology of the Algae the earlier doctrines on the main morphological structures and the main living processes of plants in general were changed. While the botanists of the 18th century tried to interpret the morphological structures of lower green plants by comparing them with seemingly equivalent parts of the higher plants, they learned from the 19th century researches the specific nature of the organs and anatomical parts of the pigmented Thallophyta. Furthermore they found out many unknown cellular structures and developmental processes of the microscopical plants. The observations furthered the origin and development of the new founded fields of modern cytology and reproduction biology.

Shamin Alexei, Prof., Dr.

Institute of the History of Science and Technology, USSR Academy of Sciences, Moscow, USSR

INTEGRATING ROLE OF THE PHYSICOCHEMICAL BIOLOGY

The analysis of the development of modern biological sciences shows that classic biology has been succeeded by nonclassic biology with its strong tendency to the formation of the new theoretical base. The main role in this process is played by the complex of the fields of science, formed in the region of interactions between biology, chemistry, physics and mathematics. This complex has been named the physicochemical biology and it caused the revolutionary changes in biology, which brought the appearance of the new structure of the new structure of biological sciences. The new "general biology" is being formed focusing its attention on the cell and the individual organism as the system of cells maintaining itself by means of the molecular mechanisms. Thus the whole hierarchy of systems from the molecular, supermolecular, subcellular and cellular to the individual systems of the organisms and the organism as a whole becomes the object of the new "general biology". The objects of the other complex of biological sciences are the communities of organisms. These two complexes can be easily differentiated, but they have several connecting "bridges" such as genetics by all means, chemical ecology, insect biochemistry and others. The process of the formation of the new structure in biology is accompanied by the much stronger tendencies to the changes in the relation of theoretic and empiric elements.

Stephen F. Mason

King's College, University of London, WC2R 2LS, U.K.

UNIVERSAL DISSYMMETRY FROM PASTEUR TO PARITY NON-CONSERVATION
AND THE ORIGIN OF BIOMOLECULAR HANDEDNESS

    The conjecture that a dissymmetric force pervades the physical world was derived by Louis Pasteur from his discovery of the spontaneous breaking of chiral symmetry by the crystallisation of a racemate (1848) and the surmise of nineteenth century French stereochemists that a morphological analogy obtains between a crystal form and the shape of the constituent molecules. Faraday's discovery of magnetically-induced optical activity (1846) led Pasteur to suppose that magnetism and other polar forces are manifestations of the universal dissymmetry, but Curie (1894) demonstrated that a chiral field is necessarily composite with parallel or antiparallel polar and axial elements, as in the case of left- or right-handed circularly polarized electromagnetic radiation.
    The parity-principle of Wigner (1927), requiring reflection symmetry in the forces of nature, contrary to Pasteur's conjecture, is found to be violated in the weak nuclear interaction, giving asymmetric β-decay of radionucleides. The weak interaction, unified with electromagnetism as the electroweak interaction, emerges as the universal chiral force of nature, for which Pasteur and his followers sought to account for biomolecular handedness and the homochiral biochemistry of living organisms.

    While Pasteur had taken the then-uncharacterised dissymmetric force as universal, extending even to the crystallization dish employed for the separation of enantiomers, Japp in the presidential address to the chemistry section of the British Association for the Advancement of Science (1898) on <u>Stereochemistry and Vitalism</u> proposed that the force is confined to the organic world as the <u>vis vitalis</u>. A grand debate on the cosmological origin of biomolecular handedness ensued, filling the correspondence columns of <u>Nature</u> during 1898. The alternatives to Japp's postulate, suggested by his respondents, involved either the intervention of external agencies, as by panspermatic celestial seeding, or heterogeneous mechanisms, such as a racemate crystallization, followed by the chance natural selection of a single enantiomer.

    Surprized to learn, in 1953, that the origin of biomolecular handedness still appeared to be problematic, Frank proposed a homogeneous kinetic mechanism in which the dynamic generation of a racemic product is metastable, and subject to a bifurcation catastrophe into either a L or a D reaction channel. Coupled to the universal chiral electroweak perturbation, the Frank mechanism came to account for the established dominance of the L-amino acids and the D-sugars among the natural products.

S.F. Mason, <u>Nature</u>, 311 (1984) 19-23.

Elizabeth Lomax

Adjunct Lecturer, Medical History Division, Dept. of Anatomy, UCLA

WHITEHEAD AND MEREI'S MID-19TH CENTURY PROTOCOL FOR THE
QUANTITATIVE ASSESSMENT OF INFANT DEVELOPMENT

At the beginning of the 19th century, the French physiologist, François Chaussier, devised an instrument for measuring fetuses in an attempt to correlate length with age and viability (J.M. Tanner, A History of the Study of Human Growth, 1981, p. 478). Reference to his studies of fetal growth were made by Albert Quetelet and Charles M. Billard, who by 1830 were both measuring newborn and older babies (A. Quetelet, Sur l'Homme, Paris, 1835; C.M. Billard, Traité des Maladies des Enfans, Bruxelles, 1835). At the same time, Eduard von Siebold was reintroducing the practice of weighing newborn infants at Götingen, and Joseph Clarke was measuring weight and head circumference at the Dublin Lying-In Hospital. In the 1840's, James Young Simpson of Edinburgh measured height, weight, and head circumference at birth to determine whether larger size could account for higher male than female perinatal mortality (J.Y. Simpson, "Memoir on the sex of a child...," Edinb. Med. Surg. J., 1884, 62, 387-439).
In 1856, James Whitehead and A. Schoepf Merei began a project for assessing the development of infants who were patients at the newly opened Manchester Clinical Hospital for Women and Children. Whitehead described the findings in 1859, by which time Merei had died (James Whitehead, Third Report of the Clinical Hospital, Manchester, London, 1859). The series included 2,584 patients under three years old, but 939 were excluded from analysis for various reasons. The developmental parameters chosen were head circumference, progress of dentition, and age at which walking began. Height, weight, and chest measurements were apparently taken but not used in evaluating the state of development. The reasons for the choice of criteria will be discussed, and this study, and Whitehead's conclusions, will be compared with contemporaneous ones.

Zeno G. Swijtink

Assistant Professor, State University of New York at Buffalo

From Error to Variation: the Reinterpretation of a Formalism in Agricultural Trials in 19th Century Germany

When the Method of Least Squares was developed at the beginning of the 19th century, it was just one more calculus for the reduction and combination of observations. But through the authority and results of Gauss, Least Squares, and the probabilistic justification he gave of it, became orthodoxy. It was used not only to determine the best reduction, but also to measure the uncertainty that is attached to it.

At the end of the 19th century Least Squares is applied in a completely different context. It is used to analyze the confusing variation in outcomes of comparative trials. The same syntax has obtained a different semantics.

To apply the analytical tools of the Theory of Errors to the data of comparative trials a series of conceptual hurdles had to be overcome. Two of these are analysed in this paper. They derive from two of the assumptions Gauss made in his development and justification of Least Squares as the calculus to reduce and combine observations that are subject to error: the assumptions that there exist true underlying quantities that are measured with error, and that direct measurements of the same quantity are probabilistically independent. One hurdle is to find some reinterpretation of the notion of true quantity. This has sometimes been called "the problem of the meaning or value of the mean." The other is what to make of the notion of probabilistically independent measurements when there is no real true quantity and the distinction between variable error and constant error breaks down.

In my paper I discuss these matters with respect to German agricultural trials in the 19th and early 20th century. The issue of independence attracts only attention at the end of this period when a probabilistic interpretation was explicitly sought after. Taking averages in combining observations of comparative trials can make sense without such an interpretation, as also the history of the Theory of Errors before Gauss testifies.

Richard M. Burian

Virginia Polytechnic Inst. & State University, Blacksburg, USA

MODEL ORGANISMS AND RESEARCH STRATEGIES IN MENDELIAN GENETICS:
THE CASE OF DEMEREC AND MULLER*

There are striking parallels between the careers of Milislav Demerec (1895-1966) and Hermann Muller (1890-1967). Each was a highly influential geneticist who devoted his career to the question, "What is the gene?". Each developed novel techniques (including chemical and radiological ones) designed to help characterize genes more narrowly. Each perceived himself as something of an outsider, on the edge of a number of different traditions that needed to be blended within genetics. The theoretical considerations deployed by the two of them were strikingly similar, as were their persistence and ingenuity in obtaining experimental support for their views and experimental resolution of the questions they posed. Furthermore, each was thoroughly integrated into an international network; there were very few major developments in genetics of which the two were not comparably aware.

In light of these parallels and of their keen desire to adjudicate discrepancies between their views, it is particularly interesting to examine their differences -- differences in experimental problems, theoretical commitments, research strategies, arguments deployed and interpretations of data. The present talk will show how some of these differences are connected to the choice of experimental organisms. The argument will explore the reasons for -- and consequences of -- Muller's abiding commitment to Drosophila and Demerec's contrasting switching of model organisms -- from maize and delphinium to Drosophila to molds to bacteria. One result of this exploration is to focus attention on the importance of so-called model organisms in the development of research strategies and the influence of the choice of organism on the results obtained. Some (though by no means all) of the differences between Demerec and Muller are traceable to this source. Another result is to highlight the mixture of advantages and disadvantages in each of the strategies employed, yielding an argument for pluralism with respect to model organisms and research strategies.

*Research supported in part by the National Science Foundation and the National Endowment for the Humanities.

Lindley Darden

Associate Professor, Committee on the History & Philosophy
University of Maryland of Science
College Park, Maryland, 20742 USA

THE DEVELOPMENT OF THE THEORY OF THE GENE, 1900-1926

This paper traces the development of the theory of the gene from its beginnings in 1900 to the version provided by T.H. Morgan in 1926. Reconstructions of the theory in 1900 and in 1926 provide the beginning point and end point for the analysis of theory change. Various factors involved in the development are identified: discovery of new data and expansion of the scope of the theory; anomalous data which necessitated a change in theory; conceptual clarification and development that contributed to the concept of the gene; interfield connections to cytology that provide insights into strategies for theory construction and alternative criteria for theory choice.

Bf

Margaret Somosi SAHA

University of Virginia, (Scholar-in-Residence), Dept. of History

The Wissenschaft Ideal as a Research Tradition in Early 20th-Century German Genetics

The growing body of literature which deals with the history of German genetics has consistently shown the approach of German geneticists to be quite different from that of their foreign counterparts. More physiological and holistic in nature, German genetics generally refused to isolate questions of gene transmission from gene expression, or even from evolutionary concerns.

There is little argument that the Wissenschaft ideal, with its powerful philosophical undercurrent and its emphasis on unifying the various branches of knowledge, served as a major source of inspiration for the course of 19th-century German biology. But it was equally responsible for imparting this unique character to early 20th-century genetics research in Germany.

Moreover, the Wissenschaft ideal, as applied to German research on heredity, displays all the major characteristics of a research tradition in Larry Laudan's (Progress and its Problems, Berkeley, 1977) sense of the term. It encompassed a number of specific theories. It contained a methodological as well as a metaphysical framework. Most importantly, this tradition, neither static nor rigid, underwent a natural evolution. This is particularly evident from the case of 20th-century genetics whereby this research tradition adapted to the demands of a highly experimental, often statistical science.

This tradition played a major role in the institutional sphere, by inhibiting the development of genetics' chairs. It likewise influenced the acceptance or rejection of various theories. Geneticist Paula Hertwig argued that the lukewarm acceptance accorded to Morgan's work was attributable to philosophical concerns associated with the Wissenschaft ideal. In a similar vein Richard Goldschmidt, Elisabeth Schiemann, Viktor Hamburger, and Carl Correns all stressed the importance of the Wissenschaft ideal in determining the approach of German geneticists and their espousal of such theories as phenogenetics, cytoplasmic and quantitative inheritance, all of which were more enthusiastically embraced in Germany than elsewhere. Finally, in Laudan's concept of a research tradition, a scientist can work in two competing traditions simultaneously, particularly in "troubled" times. Thus, by viewing the Wissenschaft ideal in terms of a research tradition, it also becomes possible to explain the divided loyalties of geneticists such as Erwin Baur whose work still reveals many aspects of the Wissenschaft ideal despite his infatuation with American genetics.

Ramunas Kondratas

Curator, National Museum of American History, Smithsonian Institution

COLLECTING AND INTERPRETING THE ARTIFACTS OF BIOTECHNOLOGY

Biotechnology, broadly defined, includes any technique that uses living organisms (or parts of organisms) to make or modify products, to improve plants or animals, or to develop micro-organisms for specific uses. Biological processes and organisms have been used with great success throughout history, for example, to bake bread or brew beer, and have become increasingly sophisticated over the years. More recently, the application of genetic engineering has enabled scientists to directly manipulate the genetic material itself and to accurately program the micro-organisms for their various tasks, such as making insulin, human growth hormone, or interferon.

The skill to manipulate the basic units of life and to identify and pick out single genes from the millions in the human gene set by means of recombinant DNA technology is a power unlike any other in biology. It parallels that of splitting the atom and has already allowed scientists to make remarkable advances in fields ranging from the study of the immune system to cancer research, to public health, to agriculture, to chemistry, to mining, to the study of evolution and human history. The potential for future scientific breakthroughs, novel and useful practical applications as well as commercial exploitation is tremendous.

Yet, there are some problems. Questions about potential effects on human health and the environment are major considerations, especially in the case of genetically engineered micro-organisms. Perhaps a more perplexing issue is the potential impact of the new genetic technologies on human values. As we gain the ability to manipulate life, we must face basic questions of just what life means and how far we can reasonably and safely allow ourselves to alter it.

In order to begin to grapple with these issues and to make socially responsible and prudent decisions as consumers, as administrators, or as regulators, we must raise the public knowledge concerning genetics and its potential. One way of doing this is through public exhibits.

In this paper I would like to briefly outline the conceptual script for an exhibit on the development of genetic engineering in the United States and discuss how artifacts of biotechnology can be used to document that story. The exhibit would inform the public on a fairly elementary level about genetic engineering -- what the technology is, where it came from, how it is being applied and exploited, how it has been received (by scientists and lay persons), and what are its implications. The exhibit would try to portray the enormous burst of scientific creativity and imagination that brought about this revolution in biology, and at the same time, raise the important social, political, economic, and ethical questions associated with it.

Claude Debru

CNRS Département de Médecine Expérimentale Lyon France.

L'EVOLUTION DE LA NEUROCHIMIE

La biochimie du système nerveux est longtemps restée "terra incognita". L'électrophysiologie, l'électroencéphalographie se sont développées aux dépens de la neurochimie. Les propriétés des courants et champs électriques semblaient fournir un équivalent acceptable de l'activité psychique. Cependant, à la fin du 19° siècle, l'oeuvre isolée de Ludwig Thudichum sur la chimie cérébrale trouve sa source dans l'influence de Liebig. Mais la chimie analytique des constituants cérébraux a cédé la place à la pharmacologie de la jonction neuromusculaire (Keith Lucas) et du système nerveux autonome (Elliott, Langley). L'étude du système autonome a imposé l'idée de la transmission neurohumorale (Dale, Loewi). La neurotransmission chimique, objet de longues controverses, n'a pas occulté l'aspect neurohumoral. L'étude de la fonction neuroendocrine de l'hypophyse et de ses relations avec l'hypothalamus (Harris, Roussy, Stutinsky, Scharrer), la mise en évidence de structures neurosécrétrices (Bargmann), la reconnaissance du milieu intérieur cérébral et de la "neurophysiologie humide" (F.O. Schmitt) ont imposé la généralité de mécanismes neurohumoraux à l'intérieur du système nerveux central ou reliant celui-ci au reste de l'organisme. En outre, la découverte récente de la pluripotentialité de certains neurones donne accès à d'importants phénomènes de régulation.

Rosaleen LOVE

Chairperson, Humanities Department, Swinburne Institute of Technology

WOMEN SCIENTISTS AND THE GREAT BARRIER REEF. A HISTORICAL PERSPECTIVE

The Great Barrier Reef Expedition of 1928/9 set up Australia's first marine biological station on Low Isles near Cairns, under the direction of Maurice Yonge. The laboratory was a workplace which allowed women full participation in scientific life. Two women scientists of the expedition, Sidnie Manton and Sheina Marshall, subsequently had distinguished careers in zoology in Britain. This paper gives an account of their work and places it in the context of early scientific exploration of the Great Barrier Reef.

The paper also reflects upon responses to this account of women's work in science in terms of the theoretical basis provided by Elizabeth Janeway in The Powers of the Weak.

Ruth Schwartz Cowan

Associate Professor, History, SUNY-Stony Brook

A HISTORY OF AMNIOCENTESIS AND ITS RISK ASSESSMENT

The technique which is known as "amniocentesis," and is currently the central element in prenatal genetic screening, actually consists of four separate techniques: (1) ultrasonography of the abdomen (to locate the position of the fetus and the amniotic sac); (2) the amniocentesis itself (a tap through the abdomen of the mother to obtain a sample of amniotic fluid for analysis); (3) karyotyping of the cells found in the fluid (the techniques of culturing the cells and preparing them for observation, as well as the observations themselves); and (4) chemical analysis of the amniotic fluid (to determine the presence or absence of substances which indicate developmental abnormalities in the fetus).
Each of these techniques has its own history. Each was developed by a separate community of scientists and in at least three of the four cases the original goal of the developmental work was something other than prenatal screening.
This paper will examine those four separate historical tracks, focussing on the manner in which they converged to create a new clinical stratagem just at the time that abortion laws were becoming liberalized. In addition the manner in which each community of scientists communicated the risks involved in the techniques (risks either of injury or of observational error) to the other communities and to clinicians will also be examined.

Anne Marie MOULIN, Institut d'Histoire des Sciences, C.N.R.S.

Paris

Tradition versus innovation in Medicine : the international debate on human leucocyte groups (1953-1973)

The history of Transplantation Immunology suggests a comprehensive model for medical progress in controversial issues. Too many contradictory data were collected on organ transplantation in the thirties ; they did not lead to any clearcut conclusion on the many biological factors involved in the usual failure of transplantation between individuals of the same species. Which usually happens in that case is that scientists seem to elude too specific experimental material and switch to a more comprehensive field when they do not feel so much committed to the biological framework of their time and become more sensitive to the general outlook of the question and to the formal and logical relationship between the elements. This second stage may be called a philosophical one.

In the thirties, merged the idea of an individual organism composed of " differentials " the combination of which was unique (but the differentials could be shared with other organisms). This type of approach suggested a marker molecule-oriented research and a combinatory strategy for the exchange of organs and the possibility of a social network to achieve such an exchange. This theoretical stage is usually preliminary to the shaping of a new autonomous science, namely Transplantation Immunology.

My claim is that this three-stage model could be useful to understand some discrepancies between European and American labs at the time of the description of leucocyte antigens as being transplantation antigens (1952-1963). The differences between experimental data is obviously irrelevant to the debate (1963-1973) on the crucial character of these antigens (HLA) for the selection of kidney donors. It might explain the persistence of conflicting views on the prevalence of kinship in the selection of donors .It emphasizes the involvement of biomedicine in theoretical assumptions, even when it copes with the emergencies and the entanglements of pathological situations.

Masayuki Ohbayashi, Sophia University, Tokyo, Japan

THE ORIGINS OF MOLECULAR BIOLOGY IN JAPAN

The purpose of this paper is to discuss the origins of molecular biology in Japan.

Recently the development of biotechnology in Japan has arrived at a high level, compared with that of other countries. However, the japanese molecular biology on which biotechnology depends does not have a long history since it started only after World War II. Especially, molecular genetics which uses "bacteriophage" had hardly been studied before the war and started by a few scientists after the war. This is one origin of molecular biology in Japan. But there are other origins, one of which is the group formed by biologists, biochemists and physicists interested in nucleic acids. This group also started just after the war. Still another origin is the group of enzymologists. Enzymology was one of the main subjects of biochemistry from before the war. In Japan biochemistry developed in continuity with the medical and agricultural sciences from the pre-war era. These played an important role in introducing molecular biology from Europe and the United States after the war.

In this paper I will try to clarify the history of molecular biology in Japan and the effects it received from biochemistry. I discuss mainly the history from the end of the war to the early 60's in which the post-war research institutions for biological sciences were established. Such an historical study, when compared with the history of molecular biology in Europe and the United States, may contribute to elucidate the nature of molecular biology. Moreover, such an historical study might help us to think of the future development of biotechnology.

Michael Hubenstorf

assistant professor, Free University Berlin, West-Berlin, Germany

THE EUGENICS MOVEMENT IN AUSTRIA 1900 - 1945

Studies on eugenic ideas in german-speaking countries are restricted to a narrow spectrum of authors from the former German Empire. Very little is known about Austrian doctors and intellectuals advocating eugenic concepts, although the had a decisive influence on the German debate.
This is the more astonishing because there developed a close cooperation between the Universities of Vienna and Munich from 1900 up to World War I in the fields of eugenics and social hygiene. The focal points for this cooperation of predominantly Viennese scientists have been the Munich branch of the "Deutsche Gesellschaft für Rassenhygiene" (German society for race hygiene) and the section on race hygiene within the International Exhibition on hygiene, taking place in Dresden in 1911. The contributors of this circle included experts on hygiene (Max v.Gruber), Ignaz Kaup), on anatomy (Julius Tandler), on psychiatry (Rudolf Allers), on pathological anatomy (Anton Weichselbaum) and on botany (Erich von Tschermak--Seysenegg).
Both Kaup and Tandler served successive terms of office as minister of public health in the first government of the Austrian Republic -rom1918 to 1920. As a measure of public health policy they subsidized another expert on eugenics: Heinrich Reichel, who became professor of social hygiene in Vienna and lateron of hygiene in Graz in 1933. He became one of the integrating figures of Viennese race hygiene since 1921.
Establishing a Race Hygiene Society in Vienna under Reichel and a Eugenics Society in Graz under the dermatologist Rudolf Polland as well as several smaller branches in different towns such as Innsbruck and Linz, race hygienists built a coherent framework of institutions during the twenties. Among others they deeply influenced the Anti-Alcoholics-Movement, the Society for Population Policy and the organisations for the protection of the family.
Political fractions within the movement can be singled out only in a vague manner: there existed a german-nationalist Darwinian group (Reichel,Polland) increasingly changing over to a national-socialist organisation, a socialist Lamarckian wing (Tandler, Kammerer, Goldscheid) and a catholic group undecided between those two. Political cleavages in Austria hampered practical application of eugenic ideas, reservedness on the side of the Catholic Church added to the restriction of the eugenic impact. Already in 1933 eugenicists instigated Medical Students´and Doctors´organisations to demand the introduction of courses on race hygiene into the medical curriculum. After the "Anschluss" of 1938 the german-nationalist group comprising medical doctors, hygienists and anthropologists quickly took over the "Offices for race biology"of the NSDAP as well as similar university institutes recently founded.

Baader. Gerhard, Freie Universität Berlin, Germany
Institut für Geschichte der Medizin

SOCIAL HYGIENE AND EUGENICS IN GERMANY 1900 - 1933

Eugenics - as it had been coined by Francis Galton in 1883-,is nothing else than the application of the findings of Darwin to social conditions i. e. to ameliorate the race by developing the prolificacy of generation. Such considerations can be compared with ideas that could be already found at early social hygienists as Eduard Reich in 1870/71. Especially the idea of elimination of the inferior as formulated by race-hygienists had got also influence on parts of social hygiene. For instance Alfred Grotjahn's proposal of eliminating the hereditary inferiors in mad-houses as part of his so-called hygiene of sexual reproduction is to be seen in this respect,and neither Benno Chajes, one of the pioneeers for founding ambulatories of the health insurances in Berlin as centers of social medicine and socialdemocratic politician, nor socialist city-physicians like Georg Loewenstein or Richard Roeder are free from such tendencies. Nevertheless it is not possible - as it had been done recently again in the case of Grotjahn - to put these men without differentiation into the neighbourhood of the ideological basis of medicine in national-socialism in Germany as the race-hygienists themselves. This is forbidden by the different basis idea of social hygiene in comparison to race hygiene, although tendencies of fitting outcasts into the social context of the bourgeois society can be found also there. These outward similarities with the conception of race hygiene consist rather in the same roots, i. e. the ideology of the society of German imperialism of improving economic efficiency in general. They therefore converge with the tradition of tendencies to force public order in the Public Health System in Germany upon the whole. Social hygiene itself has in fact nothing to do with preparing the concepts or with the practice of murdering outcasts in national-socialism and only little with their discrimination in post-war Germany today. This is more or less continuity of concepts of medicine of national-socialism still nowadays.

Sheila Faith Weiss

Assistant Professor of History, Clarkson University

RACE AND EFFICIENCY IN FRITZ LENZ'S EUGENICS

Recent scholarship on the history of German race hygiene, scant as it is, has at least implicitly been concerned with the overall aims of the so-called new science. Was it first and foremost preoccupied with improving the "Nordic race" or, as in the case of eugenics movements elsewhere, did it have other goals in mind? While there has been some consensus in the newer historiography that race (in the anthropoligical sense of the term) was not the primary focus of attention during the pre-Nazi period, little has been done in the way of articulating the concrete aims of the movement prior to 1933 and in rectifying the seeming contradictions between racist and nonracist eugenicists.

This paper seeks to examine the complex nature of German race hygiene during the Republic and Nazi eras through a close examination of the writings of Fritz Lenz (1887-1976). Recognized as Weimar Germany's most outspoken eugenicist, Lenz was the leader of the influential Munich chapter of the German Society for Race Hygiene; he also coauthored the standard eugenics textbook throughout the Republic and into the Third Reich. More importantly, however, Lenz's numerous publications lay bare the sometimes disparate, sometimes compatible aims of Nordic supremacy and national efficiency that colored Weimar and Nazi eugenics. While accepting the basic tenets of those nonracist eugenicists who viewed the new discipline as a tool for boosting national efficiency by monitoring the realtive number of unproductive and highly productive groups in society, Lenz went beyond them in his insistence on linking cultural productivity to race. Aryan blood, according to Lenz, was a necessary if not sufficient condition for a vigorous, efficient, and culturally productive nation. Although his emphasis on good Nordic stock as a prerequisite of national efficiency met with strong opposition by many eugenicists during the Weimar years, the connection between race and efficiency forged by Lenz helped make race hygiene ideologically acceptable to Nazi racial theorists and policy makers.

Mauricio Schoijet, Universidad Autónoma Metropolitana-Xochimilco, México 04960, DF, México

THE IDEOLOGICAL CONJUNCTURE PREVIOUS TO LYSENKOISM

Contrary to Loren Graham's interpretation, that because many Marxists and Socialists supported eugenics there was "no preordained connection between eugenics and conservative, even fascist sentiments,"[1] we submit that eugenics was supported by the leading sectors of the bourgeoisie because it worked as an instrument for the disarmament of the antagonistic social forces. Either support of eugenics or acceptance of some of its claims by Anarchists, Socialdemocrats or Communists, should be seen within the framework of contamination of working class movements by bourgeois ideologies, that led to support for the imperialistic 1914 war, and to lack of resistance against racism and chauvinism.

A series of texts by Socialists and Communists, published between 1910 and 1934, from Karl Kautsky's Vermehrung und Entwicklung... to Hermann J.Muller's Lenin and Genetics should be seen as attempts of demarcation vis a vis the ideological exploitation of science represented by eugenics. But while the texts by Kautsky, Hugo Iltis, Vassili Slepkov and N.M.Volotskoi are counterideological, in that they are left-wing darwinisticists, who posit a connection between eugenics and class struggle but also swallow eugenists' claim on "degeneration" of the "race," Max Levien and Hermann J.Muller denied such claim. Levien dismissed eugenics as a sham and a tool of imperialism. Muller also denounced Lamarckism as idealistic, and suggested that Lenin's ideas on the struggle of idealism and materialism in the physical science were applicable to the case of genetics. Muller's paper should be seen as the first Marxist critique against both eugenics and Lysenkoism, and it was correct both in the political and in the philosophical sense. It was ignored because the political needs of the Stalinist apparatus favored the Lysenkoist counterrevolution.

References
1. Loren Graham, Between Science and Values, p. 228, 230 and 247.
2. Karl Kautsky, chapter on Rassenhygiene in Vermehrung und Entwicklung in Natur und Gesellschaft, J.Dietz Verlag, Stuttgart, (1910) p. 258-268.
3. Loren Graham, appendix with H.J.Muller's Lenin and Genetics, in Science and Philosophy in the Soviet Union, Knopf, New York, (1971), p. 451-469.

Iskren Azmanov

Academy of Science, Sofia, Bulgaria

Academician Prof. Dr. Doncho Kostov and his relation with American Scientists

The academician Prof. Dr Doncho Kostov is well known sdientist in the field of Genetics /1897 - 1949/. He is bulgarian, born in a village at the vicinity of Sofia. In 1924 he graduated Agronomy at the University of Halle/S, Germany. During the period of 1926 - 1929 he is a Rockefeller fellow specializing and doing research work at Harvard University, Cambridge, Mass. USA. Later on, we see him as assistent professor at the state University of Sofia - Faculty of Agriculture and Silviculture teaching Genetics /1925 -1932/; as a lecturer and extention worker in the Universities of Leningrad and Moscow /1932 -1939/; and again as a professor of Genetics in Bulgaria /University of "Kliment Ohridski", Sofia/, (1939 till his death - 1949).

Prof. Kostov become a regular member (academician) of the Bulgarian Academy of Science /1946/ and member of Yugoslav Academy of Science and Arts - Zagreb /1948/. In his scientific work, academician Kostov has been in close relation and colaboration with many scientific Institutions and leading scientists of USA, USSR, Germany, England, France, Netherlands, Belgium, Italy, Sweden, Japan, India... Thus up to the end of his life he had connection with prominent American geneticists as: Prof.E.M.East, Prof.James I.Kendall, Prof.Hermann J.Muller /Nobelist/, Prof.Leslie C.Dunn Prof.C.B.Hutchison, Prof.Mislav Demerec, Prof.A.F.Blakeslee, Prof.T.H.Goodspeed, Prof.Roy E.Clausen, Prof.Lester W.Sharp...

Academician Kostov was as well member of a number Internatio nal institutes of Biological Science as: Docktor of Science of Leningrad University /1935/; member of American Genetic Association /1927/; member of the American Association of Botanicists /1928/; member of American Association of Biologists /1929/; member of the American Association for the Advancement of Science /1935/...

The paper I am presenting to the International Congress of History of Science the folloing is prepared on the ground of diferent annals, materials from private Kostov's archives; informations from the Rockefeller Foundation (its Archiving Center - N.Y.); information from the American Philosofical Association; Bulgarian Academy of Science; Sofia University; Academy of Science of Souviet Union; Leningrad University and others.

Kostov,D.(1959): Biobibliograрhy. Sofia; Azmanov,I.(1978): Jizn vo imya nauki. Znan. sila 5, 33 - 34, Moscow; — (1983): The Ame rican dossier of acad.D.Kostov. Vekove 1, 63 - 66, Sofia; — (1984): Structural criteria in the investigations of acad.D.Kostov. Philosofska misal 8, 94 - 96, Sofia; Atabekova,A.&Ustinova,E.(1980): Cytologia rastenii, Moscow; Carlson,E.A.(1980): Genes, Radiation and Society - The life and work of H.J.Muller, Corn. Press N.Y.; Tzikov,D.(1969): Doncho Kostov, Sofia;

Li Pei-shan

Deputy Director and Associate Professor of the Institute for
  History of Natural Science, Chinese Academy of Sciences

The Qindao Conference of 1956 on Genetics: The Historical
  Background and the Foundamental Experiences

The Qindao Genetics Conference of 1956 marked a turning point
in the studies of genetics in China by cutting loose from the
controlling Lysenko influence of that period. The Conference
was held in the light of the "Double-Hundred" policy for
flourishing culture and science: Let a hundred flowers blossom
and a hundred schools of thought contend. The then abnomality
in the studies of genetics was one of the principal factors
that resulted in the formulation of this important policy.
Soon after the establishment of the People's Republic of China
under the prevailing direction of "learning from the Soviet
Union", along with the advantages, some defects were also
brought in. The problem in biology, especially in genetics
was most conspicuous among others. Between 1949 to 1956, many
books written by Lysenko and his followers particularly his
report entitled "The Present State of Biology" written in 1948
were all translated into Chinese, and were published in large
amount. At that time, all the Chinese genetists who followed
the Morgan school were criticized for being pseudo-scientific,
idealistic and with bourgeois ideas; consequently, their lectures and research work were abruptly stopped. However, during
the 1956 Conference, the Morgan school was rehabilitated, with
all the political labels cast off; and this was officially an-
nouced by the leading cadres of the Communist Party and the
government of China. Thus the genetists of both schools participating the Conference each presented their scientific facts
principles and their own views, but in a friendly atmosphere.
Afterwards, the majority of the genetists gradually started
their own work again at ease. There was once a trace of
relapse around the year 1960, but was soon back to normal.
From then on, the study on genetics in China not only was on
the cellular level but also on the molecular one. During the
"culture revolution", both were criticized for "losing contact
with the reality", for "being followers of the foreigners", etc.
and works of both were stopped.
The fundamental experiences gained at the Qindao Conference
were: (1) Scientific arguments must be strictly distinguished
from political issues. Free discussion be fully encouraged in
settling differences in scientific arguments. (2) Freedom of
expression, especially exchanging ideas over scientific endeavors within the academic circles must be guaranteed; and demo-
cracy, the prerequisite of such freedom, must be exercised.
(3) Academic arguments must be based on the results of resear
-ch works. (4) Conclusions on scientific pursuits can only be
reached by scientists themselves.

William B. N. Berry

Professor, University of California, Berkeley

ON THE RELEVANCE OF CHARLES DARWIN'S GEOLOGICAL OBSERVATIONS IN SOUTH
AMERICA TO MODERN PLATE TECTONIC THEORY

Modern plate tectonic theory indicates that the crust of the
earth is composed of a number of plates, some of which bear continents
and oceans and others bear oceans only. The plates appear to slide
on a viscous, sometimes molten rock material. Plates appear to be
continually and always to have been continually in motion. New
crustal or plate material is formed at oceanic ridge/rise (oceanic
mountain chain) systems and is consumed at plate collisions. A
consequence of plate collision seems to be mountain-making. When a
continent-bearing plate collides with an ocean-bearing plate, as is
the situation off the west coast of South America, the ocean-bearing
plate is thought to slide beneath the continent-bearing plate. At
the collision boundary, the ocean-bearing plate buckles downward. As
it descends, the plate ruptures to produce earthquakes, and it heats
slowly as well. As heating continues, a melt or magma is formed that
commences to rise. Some of this rising magma may crystallize below
the surface to form granitic rocks, and some may erupt in the form
of violent vulcanism. Such eruptions appear to be the cause of the
large composite volcanos that occur along the crest of the Andes.

In the course of his geological observations in the Andes as
well as along the west coast of South America, Charles Darwin witnessed the close correspondence between strong earthquakes and violent
vulcanism. Darwin also noted that areas under the ocean prior to an
earthquake had been shoved out of the sea to form land as a result of
the earthquake. Darwin argued that the formation of mountains such as
the Andes and certain volcanic activity observed in such mountains
are related to the same cause. He went on to suggest that earthquakes and related mountain-making activity and vulcanism result
from some great change beneath the crust of the earth. He stated
that this change was a process presently in progress and that the
formation of mountains was a continuing, enduring process. To Darwin,
mountain chains were formed by a long series of slow movements. In
regard to the Andes, Darwin stated that "the crust of the globe in
Chili rests on a lake of molten stone, undergoing some slow but
great change." Furthermore, he drew note of the great length of
this lake for the related volcanic-earthquake-mountain-making
activity extended for hundreds of miles in a linear belt parallel
with the west coast of South America.

Darwin's observations and ideas on the relationship between
mountain-making and certain earthquakes and volcanic activity read
to the Geological Society of London in 1838 have received scant
attention from tectonics-oriented geologists. His observations and
conclusions have proven most perspicaceous in the light of the
development of plate tectonic theory. His observations are consistent with that theory, and his ideas on the presence of a vast
quantity of molten material beneath the crust most forward-looking.

Matthew J. James

Lecturer, Department of Paleontology, Univ. of California, Berkeley

CHARLES DARWIN'S CONTRIBUTIONS TO THE GEOLOGY OF THE GALAPAGOS ISLANDS

The accurate and meticulous geological observations made by Charles Darwin during his five-week stay in the Galapagos Islands in 1835 have generally been eclipsed by his better-known contributions to evolutionary theory. These geological observations closely correspond to the modern view of the volcanic history of the island group and to the plate tectonic setting of the eastern Pacific Ocean. Subsequent geological observations verified and expanded Darwin's original conclusions. Prior to the Beagle voyage, Darwin received informal geological training under Adam Sedgwick at Cambridge University. During the voyage itself, Darwin's practical knowledge of geology was increased by reading Charles Lyell's Principles of Geology and by careful application of Lyellian principles to observed geological phenomena in South America.

The Galapagos Archipelago consists of fourteen main islands and numerous rocky islets that sit athwart the equator about 900 km west of Ecuador. Darwin personally visited four of the principal islands and received rock specimens from most of the others in conjunction with the hydrographic surveying activity of the Beagle. In the Galapagos, Darwin made observations about: 1) the formation and physiography of volcanic craters, 2) the physical properties of lava flows, 3) the geomorphology of tuff cones, and 4) the linearity of island formation.

After geologizing in the islands, Darwin was probably the first to suggest the role of sea water in the formation of palagonite tuffs. Darwin's observations on the viscosity of lavas contradicted current opinion and resulted in an alternative interpretation. Without the aid of a petrographic microscope (which had not been invented) Darwin made noteworthy observations about presence-absence trends of mineral composition (e.g., plagioclase and hornblende) throughout the islands. Ocean waves as errosive agents on tuff cones were proposed by Darwin as a formation mechanism for spectacular volcanic remnants and pinnacles.

Modern geological investigatons have established that the Galapagos Islands are volcanic edifices sitting on the submarine Galapagos Platform, located at the northern-most extreme of the Nazca Plate. The location of the Galapagos Islands near the intersection of three plates, or triple junction, has great plate tectonic significance (unknown to Darwin but perspicaciously anticipated by him and published in 1844 in his Geological Observations). The historical perspective usually provided by geology also affords, through forward projection of plate tectonic theory, the prediciton that the Galapagos Islands will disappear in twenty million years when subducted under the South American Plate by way of the Peru-Chile Trench.

Dr. A. B. Millman and Dr. C. L. Smith
University of Massachusetts/Boston

THE STRUCTURE OF DARWIN'S THINKING IN THE B NOTEBOOK

The authors have developed an interpretation of Darwin's thinking in the B Notebook as exploratory thinking guided by a central concept cluster. At the start of the notebook, Darwin is committed to a central core of concepts consisting of <u>adaptive variation</u>, <u>reproduction</u>, <u>life span</u>, and <u>changing conditions of life</u> which he believes may provide a way into the theory he needs. The early part of the B Notebook can be seen as centered on the exploration of the relationships among the partially specified concepts in the cluster. Darwin is committed to holding that there are some illuminating relationships among these concepts much more than he is to any specific relationship.

We give several lines of argument in support of our identification of the concept cluster. We also discuss several examples of Darwin's play with concepts involving different specifications of the concepts and different relationships among them. For example, Darwin entertained both the hypothesis that more reproduction is associated with shorter life (B22-23) and the hypothesis that more reproduction is associated with longer life (B61). As Darwin considers these different relationships he fills out the specifications of the concepts in different ways. (1) He varies the entity whose death or reproduction is being considered. In the above example, Darwin first assumes that monads have a fixed life span and that species which reproduce the most have the shortest life span. Darwin then tries a variant of his original idea, but with a twist. He now supposes that the entity with a fixed life span is the species (rather than the monad); the entity with a variable life span is a superspecies unit, namely a branch of the monad. Therefore, those species which reproduce have longer life. (2) Another way Darwin fills out the concepts is by picking alternative specifications for a component dimension. In the example, reproduction changes from being seen as exhausting to being seen as revivifying. Species life span moves from being variable to being fixed.

We consider several further examples and draw some conclusions about Darwin's style of exploratory thinking.

Giuliano Pancaldi

Associate Professor, University of Bologna, Italy

DARWIN'S STRATEGY FOR THE DARWINIANS: FERTILIZATION AND TELEOLOGY

Darwin's attitude towards final causes is known to have been interpreted in a variety of ways by his contemporaries. Recent commentators seem to agree that he substantially rejected teleological arguments. Considering a work like <u>Fertilisation of Orchids</u> (1862), where Darwin dealt with the beautiful 'contrivances' by which insects pollinate flowers, it has however been admitted that this rejection had many facets.

Darwin's book on orchids was the starting point of a new specialty, cultivated by a small international group of botanists engaged in research on fertilization. These included outspoken adversaries of teleology, like Hermann and Fritz Müller, and supporters of design in nature, like Asa Gray and Federico Delpino. One possible, and so far unexplored way to map out Darwin's attitude towards teleology is to analyse the role he played within this group.

Such an analysis throws light on Darwin's public and private attitudes to finalism. It also reveals his efforts to orient the public reception of his views in the changing intellectual climate of the Sixties and Seventies, without however becoming involved in open discussions of a controversial issue.

It appears that the complex strategy adopted by Darwin played a significant part in generating the variety of interpretations his ideas underwent. Moreover this strategy did not help him to attain an objective which he otherwise pursued : to accommodate his language, rich in teleological expressions, to his many observations and experiments aimed at eliminating teleology from the life sciences.

Desiderio Papp- Dr. Phil.(BUDAPEST) & Dr. Honoris Causa

Membre effectif Academie Internationale d'Histoire des Sciences(Paris)
Membre Honoraire Academie de Medecine d'Argentine
Membre Honoraire Academie de Medecine du Chile, etc.

SI DARWIN AVAIT CONNU LES LOIS DE MENDEL

Lorsque les vagues succeseves de critiques qui suiverent l'apparition de L'ORIGINE DES ESPECES se furent apaisees, surgit soudain, en 1867, unargument d'un tel poids qu'il semblait menacer, encore que de facon passagere, l'idee maitresse meme de la Theorie de la Selection naturell. Cet argument imprevu, un prodigieux defi, ne venait point d'un biologiste, mais d'un ingenieur de grande erudition, Fleming Jenkin, professeur de technologie à Edinbourg, qui anticipa certains aspects de l'analyse sagace a laquelle LUDWIG JOHANNSEN, une trentaine d'annees plus tard, allait soumettre les variations fluctuantes de la genetique darwinienn.
L'argument de JENKIN, "l'effet de submersion"(swamping effect) signalait que l'heredite fusionnee reduirait avec chaque generation la variabilite, detruisant l'efficacite de la selection naturelle; l'inferiorite numerique des porteurs de caracteres avantageux finirait par eliminer completement la selection.
DARWIN ne repondit jamais au defi de JENKIN, ni ne le mantionna dans aucun de ses levres, main ses lettres ne laissent aucun doute quand a sa profonde preoccupation in face de l'objection de son adversaire.
L'auteur demontre comment DARWIN, pour anuler l'argument de JENKIN se vit oblige a admettre la direction ectogenetique des variation, orientees par l'adatation mesologique, for differentes des variations fortuites postulees originairement par DARWIN. Son compromis toutefois n'etait que partiel, car il continuait a maintenir l'hypothese de l'heredite fusionnee des "gemules", particules mosses, divisibles et miscibles, porteuses des caracteres hereditaires, postulees par sa theorie de la "Pangenese".
Pourtant le problems qui avait perturbe la quietude du plus grand biologiste de son seiecle se trouvait resolu, des 1865, dans la dense dissertation de MENDEL, ensevelie dans l'annuaire de la societe de Brun. Les facteurs mendeliens, particules dures qui ne se melaient point, conservant, inalteres les caracteres hereditaire, annulaient le "swamping effect" de JENKIN. Mais DARWIN aurait-il admis les decouvertes de MENDEL? D'eminents geneticiens ant manifeste des opinions divergentes. S'appuyant sur l'accueil fait par DARWIN a certaines propositions de CHARLES NAUDIN, precurseur de MENDEL, l'auteur demontre que DARWIN, tres probablement, aurait ecarte la these de MENDEL.

Joan Leopold

University of California, Los Angeles

CHARLES DARWIN AND THE ORIGIN OF LANGUAGE

Charles Darwin incorporated the question of the origin of language into a biological framework in <u>The Descent of Man</u> (1871) and <u>The Expression of the Emotions in Man and Animals</u> (1872). But long before, he had been issuing "Queries on Expression" to far-flung ethnographical informants and recording the acquisition of gesture and speech by his infant son.

This paper will outline Darwin's linguistic researches and opinions. It will attempt to situate him within an English linguistic context, delineating his reliance upon the work and support of his relatives, notably Hensleigh Wedgwood and George Darwin, and his use of the work of others, such as Edward B. Tylor and F. Max Müller, in a confirmatory or combative manner.

MARTINEZ - CONTRERAS Jorge
PRESIDENT. UNIVERSIDAD AUTONOMA METROPOLITANA-IZTAPALAPA

FREDERIC CUVIER AND THE ORIGINS OF MODERN PRIMATOLOGY

Frédéric Cuvier, younger son of the famous and powerful Georges Cuvier, founder of paleontology, has probably been the father of modern primatology.

F. Cuvier was a careful and keen observer who published many works on mammalian systematics and behavior - - - ("moeurs des animaux") of many different species, wild and captive in the "Ménagerie du Muséum d'Histoire Naturelle", institution he directed.

He made the first scientific description of a primate's behaviour by observing and classifying the acts, reactions, volitions, etc., of a juvenile female orang-outan brought to Paris at the beginning of the nineteenth century.

F. Cuvier, as his brother, was a creationist and a -- "fixist" (he did not believe in the transformation of species) violently opposed to the current transformationist of his time: Lamarck and Geoffroy-St.-Hilaire, both forerunners of the theory of evolution. Nevertheless, what I intent to prove in this paper, is that his observations were accurate because he was a fixist and not a transformationist. If Geoffroy-St-Hilaire had observed the same animal, he would have probably arrived to different -and inaccurate- conclusions. The fixist theory was necessary to the teory of evolution.

Kevin Padian, Paleontology, University of California, Berkeley

ON RICHARD OWEN'S ARCHETYPE, HOMOLOGY, AND THE VERTEBRAL THEORY: INTERRELATIONS AND IMPLICATIONS

The principle of the Archetype, the concept of homology (special, general, and serial), and the vertebral theory of the skull were central to Owen's teachings on morphology. None of these concepts was original to Owen; he adapted them from very different schools of 18th-Century French and German schools of thought. In his hands they became powerful and logically consistent tools of a classic transcendental philosophy of morphology (Report on the Archetype and Homologies of the Vertebrate Skeleton, 1848).

E.S. Russell (Form and Function, 1916) explained these three concepts in the sequence that Owen used in Archetypes and Homologies. Owen idealized the anatomy of the vertebra to extend the comparison of vertebral arches and processes to ribs, costal cartilages, the sternum, and even the skull, which he divided into four vertebrae. The vertebral theory, however, required grounding in a theory of homology. Owen emphasized positional criteria of homology (after Geoffroy St.-Hilaire) at the expense of developmental or functional criteria; this enabled him to meet objections that he was unjustified in homologizing certain dermal bones of the skull (e.g., the parietal) with others of endochondral origin (e.g., the supraoccipital). He did use ontogeny, but selectively: if, for example, at some point in their development two structures assumed a corresponding position in different animals (general homology) or in different segments of the same animal (serial homology). In his scheme, general homologues were those that "stand to" the archetypal plan of construction (= Goethe's Typus), regardless of their modification through development or adaptation (On the Nature of Limbs, 1849; Lectures on Invertebrate Animals, 1843). (So, for example, as he once noted [Monogr. Fossil Reptilia of the Liassic Formations, 1870, p. 73], "we may with certainty infer that Archaeopteryx was hot-blooded, because it had feathers [=Bird], not because it could fly.")

The argumentation used by Owen and reviewed by Russell proceeds from the specific to the general. I would argue the reverse: that the Archetypal Theory was the philosophical basis and rationale for disregarding modifications of function and development in favor of the immutable positional criteria, which assert the primacy of form over function. Owen (On the Nature of Limbs, pp. 2-3) related the concept of homology to the German word Bedeutung, translating it as "signification" or "nature", connoting predetermined patterns "answering to the 'idea' of the Archetypal World in the Platonic Cosmogony ... to which archetypal form we come, in the course of our comparison of these modifications, finally to reduce their subject." Owen thus replaced the 'final cause' of Cuvier (the need to perform similar functions) with the 'formal cause' (the Archetypal plan) in establishing the basis of homology, from which the vertebral theory of the skull proceeded as a consequence.

Bj

James A. Secord,

Churchill College, Cambridge, England

NEW LIGHT ON THE GENESIS OF ROBERT CHAMBERS' <u>VESTIGES OF THE NATURAL HISTORY OF CREATION</u> (1844)

The publication of Robert Chambers' <u>Vestiges of Creation</u> in 1844 sparked off one of the most violent episodes in the Victorian controversies about man's place in the natural world. But because of the secrecy that surrounded the authorship for forty years, remarkably little is known of the early development of the ideas contained in the <u>Vestiges</u>. Equally unclear is its precise place in the author's prolific output as a leading popular essayist and publisher of improving literature for the middle classes. Using a variety of new sources, this paper reconstructs the history of Chambers' thought from the early 1830s to the publication of the <u>Vestiges</u>. As a working journalist, Chambers developed many of his scientific ideas in highly public forums, especially in popular treatises (including a best-selling children's book) and articles in <u>Chambers' Edinburgh Journal</u>. These works indicate his attitudes towards contemporary developments in astronomy, natural history, and geology; they also shed light on his reformist views on religion, science education, and political economy. Together with manuscript letters, they make it possible to understand how the <u>Vestiges</u> came to be written.

Particularly significant is the emergence of Chambers as a force within the Edinburgh circle of George Combe, whose <u>Natural Constitution of Man</u> provided the single most important model for the discussion of creation by natural law in the <u>Vestiges</u>. Chambers' book is in part an expression of his growing hostility to organized religion ("priestcraft", not religion <u>per se</u>), an attitude that is most clearly evidenced by his rejection of the strict Presbyterianism of the Church of Scotland. Equally apparent is his abjuring of the conservative antiquarian themes so prominent in his early publications. In their stead, Chambers embraced a populist vision of science and expounded it in the <u>Vestiges</u>, seeing that work as an instrument of popular enlightenment and gradual reform.

Mary P. Winsor

Associate Professor, University of Toronto, Toronto, Canada

Vacariance Zoogeography versus Calculation of Dispersal: the Barbour-Matthew Debate of 1915

Shortly before his death in 1946, Thomas Barbour thought that the pendulum of scientific thought in biogeography would one day vindicate his opposition to "flotsam and jetsam" as a significant force in the geographical distribution of animals. The contrary school of thought, initiated by Darwin and sustained by Darlington, Simpson, and Mayr, continued to dominate until recently. Now the "vicariance" school, inspired by Croizat, claims to be the first to bring a truly scientific methodology to biogeography. Barbour's approach foreshadowed this approach insofar as he discounted rafts and windstorms as reasonable means of dispersal for many species, except for "waif" species not relevant to geographic history, and insofar as he saw patterns of distribution as evidence of former land connections between islands. Croizat is correct in saying that "Barbour did not have himself a critically elaborated plan of integration of what he otherwise knew, and so while he could tellingly criticize the Matthewian 'zoogeography' in detail, he had nothing with which to replace it in general". Without claiming for Barbour the status of a precursor, one can nevertheless better evaluate his work now that the pendulum is in some motion.

A review of the early work of Barbour, from his collecting in Malaysia in 1907 to his 1915 analysis of West Indian lizards, exposes the disturbed state of zoogeography in that period. Because of the great diversity of opinion among geologists, Barbour, inspired by A.R. Wallace, looked to the fauna to determine whether an island was "continental" or "oceanic". Experience with delicate creatures like Peripatus and Amphisbaena inclined him to sympathize with the view recently proposed by several specialists, including Baur, Pilsbry, and Scharff, that islands with similar faunas had not been colonized across water, but were the remnants of a former landmass. Reviewing with G.K. Noble the species of the genus Ameiva, Barbour insisted that the pattern resulting from accidental transport should be haphazard, whereas in fact the closest relatives were found on adjacent islands.

When the paleontologist W.D. Matthew published his landmark "Climate and Evolution" in 1915, Barbour replied in the same journal with a vigorous objection to the claim that geological time could convert a highly unlikely accident, such as a pair of animals arriving by natural raft and populating an island, into a reasonable supposition. Neither man convinced the other, but Matthew's view became orthodoxy. The affair makes clear that zoogeography in that period was not stalled in the mere description of faunal regions. It is instructive to ask whether either contemporary or modern definitions of scientific method could have broken their stalemate.

John C. Greene, University of Connecticut, USA

Professor of History

SCIENCE, PHILOSOPHY, AND METAPHOR IN ERNST MAYR'S WRITINGS

Ernst Mayr has played a major role in the rise and development of the modern synthesis in evolutionary biology. As a systematist he helped to develop the biological species concept, to bring out the importance of geographic and reproductive isolation in the formation of species, to distinguish species multiplication from evolutionary development, and to clarify and defend the role of systematics in biology. As a philosopher of science, he has expounded the differences between typological and population thinking, between proximate and ultimate causes, and between the conceptions of science and nature typical of the physical sciences and those appropriate to the study of biology. As an historian of science, he has presented his scientific and philosophical views in historical dimension, seeking to show how divergent ideas and researches in the various fields of biology were eventually brought together in the synthetic theory of evolution. As a polemicist, he has defended the synthetic theory from attacks on all sides.

In meeting these attacks Mayr has often been forced into seemingly contradictory positions and forms of linguistic expression. Against "finalists" and "vitalists" he falls back on concepts of causation and mechanism derived from 19th century physics and argues that the task of biology is "to explain the perfection of adaptation mechanistically." In answer to experimental biologists who would reduce biology to physics and chemistry and to philosophers of physics who are sceptical of the efficacy of natural selection, however, Mayr stresses the uniqueness of organic phenomena, the goal-directed character and hierarchical structure of living systems, and the distinction between proximate and ultimate causes. The activities and structures of organisms are viewed as goal-directed, but the evolutionary process is conceived "mechanistically" as the outcome of the interaction of random genetic processes and the "anti-chance" action of selection. Yet, actually, survival and reproduction seem to depend on the intersection of two series of events, genetic and environmental, both of which are random with respect to the needs of organisms. Linguistically, however, the process of evolution is described in language suggestive of design and purpose. The "function" of mutation and recombination, says Mayr, is to provide maximum variability. Living systems evolve "in order to meet the challenge of the environment." Like Julian Huxley and G. G. Simpson, Mayr, having rejected traditional religious and philosophical ways of finding meaning and value in the evolutionary process, uses teleological figures of speech to accomplish that end. This language is in open contradiction to his formal philosophical position denying teleology in nature. In other essays Mayr seeks to derive ethics from evolutionary biology.

JOHN LAURENT

Post-Doctoral Fellow, University of Wollongong

Alternative Social Darwinism, 1890-1914

Turn of the century Social Darwinism is usually taken to refer to the individualistic 'survival of the fittest' social theories of Herbert Spencer and others. But Darwin was interpreted in other ways as well. Late nineteenth and early twentieth century socialists in Britain, Australia, and the U.S. often more frequently cited Darwin in support of their arguments than Marx.

My paper will look at some representative lecturers and writers from the period, and these include a number of self-educated working people who were able to view Darwin from a different perspective, and to emphasize aspects of his writing apparently overlooked by the more familiar Social Darwinists.

David K. van Keuren

Mellon Fellow, American Philosophical Society

Darwin, Evolution, and Museums: Curators and Biological Evolution in the Natural History Museum, 1859-1884.

Rejection or acceptance of Darwinian evolution by 19th century scientists has been traced to a mixture of intellectual, professional, and social factors. Research specialty, academic training, and institutional affiliation are among those professional factors thought to have dictated whether a scientist accepted Darwinism, opposed it, or withheld judgment. One institutional center of opposition to Darwin in the 1860s and 1870s was the natural history museum, particularly the departments of zoology. In the British Museum Owen's opposition to Darwin's theory is well-known. He was joined in opposition by John Edward Gray, Keeper of Zoology, and Gray's assistant and successor, Albert Guenther. Not until William Flower succeeded Owen in 1884 was there an effective spokesman for Darwin within the Museum. Collectors and systematizers elsewhere, such as T.C. Eyton, also kept evolution at arms length.

Many curators had intellectual ties to Cuvier and accepted his division of the natural world into four branches sub-divided into morphologically stable categories, the most basic of which was the species. Taxonomic divisions were real entities which helped organize nature and could be duplicated within museum galleries. Daily experience gained from the identification and organization of rapidly growing collections showed that species were real and did not merge into one another. Museum work entailed a research program based on species identification and classification, the results of which were then published in the form of catalogs and guides. Both daily work experience and research orientation opposed a theory which appeared to undercut the validity of museum science, which seemed excessively theoretical, and which was divorced from the hands-on experience which curatorial work provided.

Glenna Matthews

Associate Professor of History, Oklahoma State University

DARWINISM AND DOMESTICITY: THE IMPACT OF
EVOLUTIONARY THEORY ON THE STATUS OF THE HOME IN AMERICAN CULTURE

"The circle was therefore complete: Darwin consciously borrowed from social theorists such as Malthus and Spencer some of the basic concepts of evolutionary theory. Spencer and others promptly used Darwinism to reinforce these very social theories and in the process bestowed upon them the force of natural law....The recent resurrection of the theory of sexual selection and the ascription of asymmetry to the 'parental investments' of males and females are probably not unrelated to the rebirth of the women's movement. We should remember that Darwin's theory of sexual selection was put forward in the midst of the first wave of feminism. It seems that when women threaten to enter as equals into the world of affairs, androcentric scientists rally to point out that our natural place is in the home."
Ruth Hubbard, 1979

Although this is an incisive and useful statement of the function of Darwin's theory of sexual selection with respect to women, it does not go far enough. In the early nineteenth century, a cult of domesticity had grown up in the Anglo-American world. The American variant saw the home as responsible for a great deal of historical progress, a view that gave the home a role at the center of the culture. By making competition among males the dynamic that leads to human progress, the theory of sexual selection, in effect, divorced the home from history, making the home a retrograde drag on human progress in the eyes of many educated Americans. Even those who tried to defend women often accepted the terms of Darwin's argument, an argument that assumed that woman's concerns were inherently trivial. Rather than ascribing her "triviality" to nature a la Darwin, they--I am thinking primarily of Lester Frank Ward and Charlotte Perkins Gilman--ascribed it to culture, i.e., the home. Given the widespread diffusion of Darwinian ideas in American society, the net result was that although the overwhelming majority of women were still housewives in the late nineteenth century, they were much more likely to define themselves as "just a housewife" than would have been the case a few decades earlier.

Rudolf Schmitz, Prof. Dr.

Director of the Institut für Geschichte der Pharmazie,

Marburg,

Fed. Rep. of Germany

ZUM ANTHROPOSOPHISCH-HOMÖOPATHISCHEN ARZNEIMITTEL

Ein spezifisch mitteleuropäisches - besser: deutsches Phänomen sind Homöopathie und anthroposophische Heilmittellehre. Folgt man der öffentlichen Meinung, so besteht ein objektiver Bedarf an einem solchen anthroposophisch-homöopathischen System zwar nicht, wohl aber ein subjektives Bedürfnis der Bevölkerung. Dabei wird von Seiten der Verantwortlichen in Staat und Gesellschaft die Frage, ob eine Wirkung von homöopathischen und anthroposophischen Heilmitteln überhaupt nachweisbar ist und worin diese besteht, meist umgangen.

Das Referat beschäftigt sich zunächst mit dem anthroposophischen Krankheitsbegriff. Dann gilt das Interesse sowohl der Findung als auch der Qualifizierung anthroposophischer Heilmittel. Die für den Außenstehenden kaum verständliche Dualität eines einfachen Begriffs im anthroposophischen System ist aufzulösen und zu definieren, da in diesem Zusammenhang immer wieder mit dem Hinweis operiert wird, daß ein anthroposophisches Heilmittel neben seiner Materialität gleichzeitig metaphysischen Charakter besitze. Sodann folgt ein Vergleich von Anthroposophie mit Homöopathie, wobei auf Komplexität und Kompliziertheit der Fachsprache und der Terminologie abgehoben wird. In diesen Rahmen soll auch die "Ganzheitsmedizin" gestellt werden, die sich angeblich in Widerspruch zur "naturwissenschaftlichen Medizin", und vor allem zum naturwissenschaftlich-synthetischen Arzneimittel befindet. Im Zuge der Rückbesinnung auf die sogenannten "natürlichen", d.h. weniger "künstlichen" Grundbedingungen des Lebens und der Krankheitsbekämpfung haben sich die als "unschädlich" apostrophierten homöopathischen und anthroposophischen Arzneimittel unvermutet schnell durchzusetzen vermocht.
Warum das der Fall ist, stellt die Schlußfrage dar.

R. Schmitz

François DELAPORTE

Professeur, Université Nationale du Mexique

COMMENT CARLOS J. FINLAY A DECOUVERT L'AGENT DE TRANSMISSION DE
LA FIEVRE JAUNE

Il s'agit, ici, de souligner ce fait ignoré, négligé ou occulté par les historiens de la médecine : l'élaboration de l'hypothèse du Culex Mosquito comme agent de transmission de la fièvre jaune (1881) passe initialement par une réflexion sur une <u>question</u> posée par l'américain Bemiss et sur une <u>réponse</u> proposée par l'anglais Manson.

1. Pour Bemiss (1880), le virus de la fièvre jaune se reproduit dans le corps du malade, mais c'est <u>à l'extérieur</u> qu'il acquiert ses propriétés infectantes. De l'épidémiologie américaine Finlay retient donc cette question : quel est ce <u>milieu</u> propice à l'élaboration du principe morbide ? En quoi consiste cette <u>transformation</u> "extra-corporelle" ?

2. Les travaux de Manson sur la filariose (1877-78), <u>travaux connus de Finlay</u>, indiquaient dans quelle direction chercher la solution. Manson a montré comment les embryons de la filaria adulte, qui loge dans les lymphatiques, passent dans la circulation et, la nuit, sont prélevés par les moustiques -hôtes intermédiaires-, puisque c'est dans leur corps que les embryons achèvent leur croissance. Au moment de la ponte, la femelle vient mourir sur l'eau et libère les filarias adultes. L'homme ingère l'eau contaminée, la filaria retrouve son hôte définitif et le cycle est bouclé.

3. Finlay a <u>changé</u> la question et <u>modifié</u> la réponse sur la base des observations relatives à la fièvre jaune. Pas de parasite dans le sang et intégrité des globules rouges (Sternberg 1879) : <u>la fièvre jaune n'est donc pas une maladie parasitaire</u> et il faut écarter l'hypothèse d'un hôte intermédiaire. Dans la fièvre jaune, pour Finlay une <u>fièvre éruptive</u>, la lésion principale occupe l'<u>endothélium vasculaire</u> (Crevaux 1877). Si le principe morbide est <u>porté</u> (donc aussi <u>prélevé</u>) là où apparaît la lésion, l'éruption, il faut nécessairement envisager, à la différence de la typhoïde (lésion de l'<u>épithélium</u> des voies digestives d'où contamination <u>directe</u>), une contamination <u>indirecte</u> : écarter l'eau et retenir l'<u>inoculation par le moustique</u>, puisque Finlay établit qu'il peut piquer <u>plusieurs fois</u>. Prélever et injecter.

Schématiquement on peut donc dire qu'à la question Bemiss Finlay a appliqué la réponse Manson. Mais c'est parce que cette réponse <u>ne s'accordait pas</u> avec l'anatomo-pathologie de la fièvre jaune que Finlay a été conduit, d'une part, à substituer au problème initial de la transformation extra-corporelle de l'agent pathogène de la fièvre jaune le <u>problème de son mode de transmission</u>, d'autre part, à <u>remanier</u> les éléments de la réponse Manson : abandonner la notion d'hôte intermédiaire, récuser l'eau comme véhicule de propagation et désigner le moustique comme agent de transmission.

Jean THEODORIDES

C.N.R.S., Paris, France

THE HISTORY OF BAT TRANSMITTED RABIES

Bat transmitted rabies is still presently a major medical and veterinary problem in the United States and various countries of Central and South America. As early as the 16th century some authors mentioned that people bitten by bats became "mad" (i.e. rabid).

The modern observations start with the beginning of the present century when it was shown during an epidemic of horses and cattle in Brazil that the disease was identical with rabies, Negri bodies having been found in the brain of the animals. Bats were suspected as vectors and in the 1920's the rabies virus was isolated from hematophagous (vampires) and frugivorous bats from Trinidad and South America and human cases were recorded.

During the 1930's and 1940's it was shown in Mexico that the cattle disease known as derriengue transmitted by bats was identical with rabies.

Between 1950 and 1966 the rabies virus was isolated from insectivorous bats in the United States and its transmission either by their bites or by the air route in caves was reported with human cases. It was also shown that these animals can harbor the virus without being affected.

These recent findings have completely modified our views on the epidemiology of rabies in America.

Shigeo Sugiyama

Lecturer; University of Tsukuba, Japan.

SOCIAL IDENTIFICATION OF 'SCIENTIFIC' MEDICINE;
The case of confrontation of tradional and Western medicinein Japan.

Traditional system of medicine in Japan (Kampo, or Han technique of medicine) was prohibited in the early Meiji era, and Western medicine took the place of it. It was not because Kampo was emprically inferior in its efficacy; Kampo was denunciated through social process of labelling it false and pseudo-medicine, according to the specific (Westernized, or not universal) conception of medicine (and science in general), and Western medicine was in turn designated as authentic, scientific, and effective, by elite doctors and high-ranking officials of the new government after Meiji Restoration.

In terms of their competence to prevent and cure various epidemics, there was nothing to choose between Kampo and Western medicine. Kampo had not a few medicines with real efficacy and were publicly trusted as such. Its lack of surgical technique were not seen as a significant defect. Accordingly, to prohibit Kampo practice altogether, in spite of its emprical or objective effectiveness, the elite doctors and officials resorted to the strategy of denouncing its scientific nature. They claimed that whereas genuine medicine should be not only systematic (anatomy at its foundation, physiology, - - -, and therapeutics at the top of it), but also positive, Kampo did not meet these requisites; Not being founded systematically on the most basic discipline of anatomy, Kampo was a mere aggregation of separate experiences of medical treatments; The reliabilities of most treatments were fully questionable because they were not proved effective with a positive method, and a few anatomical charts described by Kampo doctors were far from positive.

To what extent were these judgements grounded and impartial? First of all, it should be pointed out that Kampo had its own system of medical theory and treatment, where lied at the base peculiar philosophy or principles originated in China and more or less modified in Japan, and therefore anatomical knowledge did not play an essential role. Thus it is unfair to say that Kampo was not systematic; it simply was systematized in not the same way as Western medicine. Secondly, the efficacy of Kampo were confirmed through long time experiences to the same degree of positivity as that of Western medicine. Furthermore, anatomical charts of Kampo tradition express symbolically its theoretical system. However different they might be from those of Western medicine, it does not follow at once that they entirely lack positive contents.

In conclusion, the criticism of Kampo does make sense only when that no other ways of systematization with their own positive nature than those in Western medicine could ensure that a system of medicine is genuine or scientific. This point could have been the matter of controversy and were in fact so to some extent. Nevertheless, those who endeavoured to introduce Western medicine into Japan and make it prevalent looked down on Kampo categorically.

Hu Naichang

Lecturer, The China Institute for the History of Medicine and Medicine Literature

DEVELOPMENT OF TRADITIONAL MEDICINE IN CONTEMPORARY CHINA (FORM 1949-PRESENT)

In the thirty-five years since 1949, traditional medicine has made remarkable progress in China. This paper deals with a study of the development and achievement of traditional medicine in contemporary China as manifested in health policy and principles, medical organization, training of medical personnel, results of studies on theories of traditional medicine, material medical secret and proved remedies, medical history and medical literature and achievment of studies of medicine of the minority nationalities such as the Tibetan, Mongolian, Uigur, Dai etc. The author holds that the reasons for the rapid development of traditional medicine in contemporary China are: the awareness of the traditional medicine as an important component in the national culture heritoge, which contains rich experiences and valuable theoretical knowledge, and the effective policies which have provided dependable foundation and financial support for the development of traditional medicine. Hence, positive factors from among the braod masses of medical workers and allied personnel have been brought into full play. Besides, the adoption of modern scientific methods and means in the research of traditional medicine has opened up a new path for the progress of traditional medicine in China.

Evan M. Melhado

Associate Professor, University of Illinois, Urbana, Illinois, USA

COMPETITION VS. REGULATION IN AMERICAN HEALTH POLICY

In the late 70s, makers of American health policy shifted dramatically in favor of competition (as opposed to regulation) in the delivery of health services. Like the antecedent emphasis on regulation, competition aimed at the twin goals of attaining efficiency in the delivery of health services and improving their accessibility; but as the embodiment of differing values, competition entailed redefinition of these ends. This paper analyzes and explaines the shift to competition among policy makers and explores the underlying change in values.

The paper begins with the regulatory approach, notes its eventual alignment with advocacy of comprehensive benefits under a universal system of health insurance, and then explores the discrediting of both regulation and universal entitlement to comprehensive benefits. Finding conventional political distinctions largely irrelevant to the demise of the traditional policy agenda, the paper identifies a variety of other forces making for a shift among policy makers. These forces include perceived failures of past policy (particularly the persistence of cost escalation under regulation and declining accessibility even for the middle class), diminishing faith in the efficacy of many health services, growing disillusionment with direct governmental intervention in both economy and society, and fear that regulation posed a threat to fundamental values.

As early as a decade before its widespread acceptance, the competitive alternative emerged as a minority view confined to two kinds of policy analysts: Professional economists, who responded to outcries against cost escalation by bringing their professional expertise to bear on the health sector and launching the subdiscipline of health economics; and a small number of observers, located outside the ranks of economists and operating at the margins of public institutions, who perceived regulation as falling short of its ends while embodying an unacceptable authoritarianism in its means. In pressing their case, competitive theorists gave priority to different values from those prominent in the regulatory tradition and accordingly redefined the goals of efficiency and accessibility. As regulatory policy lost legitimacy in the late 70s, the competitive approach was available as a well articulated alternative to be embraced by policy makers.

Gad Freudenthal

Centre national de la recherche scientifique, Paris, France

THE COHESION OF MATTER FROM ARISTOTLE TO AVICENNA

The cohesion of sublunary substances confronted Aristotelian physics with a crucial problem: What is it that binds together the four elements, hindering their motions toward their natural places? The standard medieval account was in terms of <u>unctuous moisture</u> - inseparable and cohesive. The paper traces the origins of this notion back to Aristotle himself: to him, the formal cause of (the cohesion of) plants and animals was their "soul," an idea whose <u>physical</u> foundation he sought to establish through the concept of <u>connate pneuma</u>. Accordingly, fat substances resist decay because they are highly pneumatized (viz. through concoction by vital heat). Forms thus are perdurable because <u>pneuma</u> inheres (living, esp. fat) matter, an idea shared by medical theory too.

A radically different tradition is that of Greek alchemy: it accounted for the cohesion of metals in terms of mercury, identified with (non-Stoic) <u>pneuma</u>, the inert and permanent component of matter (sulphur, or "soul," accounting for colors etc.).

Arab natural philosophers fused together the two incompatible theories. Their synthesis was necessarily incoherent: unctuous moisture was construed as mercury since it was the principle of cohesion, but also as sulphur because (often) inflammable. This syncretic theory, later embraced by Latin natural philosophy too, thus at least had the merit of allowing one to account for <u>all</u> relevant phenomena: for instance, fused metals ("water") do not evaporate because their moisture is unctuous, but can be calcined because of their sulphurous component.

Ca

Newman, William R.

Graduate Student, Harvard University

The Origins of the Summa perfectionis

The Summa perfectionis has long been considered one of the most influential texts of alchemy and mineral technology written in the Middle Ages. Although the medievals ascribed this work to "Geber" (Jābir ibn Ḥayyān), historians since the nineteenth century have suspected that the Summa's author was really an occidental. Unfortunately, the author of the Summa makes this very difficult to prove, for he mentions no previous writers, texts, or even place-names. In addition, he appears to have rewritten his sources, almost completely avoiding literal transcription. Therefore we have in the Summa a text of considerable sophistication which names no authors and whose own author's precise dates, identity, and geographical origin have remained up to now unknown.

Despite the mystery surrounding "Geber" it has recently been possible for me to determine his probable identity. This discovery is based on the coadunation of two facts. In 1935, the historian of science Julius Ruska wrote an article on the Latin translations of Muḥammad ibn Zakarīyā ar-Rāzī's Kitāb al-Asrār (Liber secretorum in Latin). In this essay, Ruska revealed that a reworking of the translation contained a number of chapters on atraments, salts, and other minerals, not found in the original translation. He further noted that these sections were strikingly similar to the Summa perfectionis, and that the reworker even claimed he was planning to write a Summa in which "all his works would be explained." Ruska was led by this to hypothesize that the reworker of the Liber secretorum was also the author of the Summa.

Recently, I analyzed a little noticed text by one Paulus de Tarento, "lecturer of the Franciscans in Assisi," according to the colophon. I have found manuscript evidence placing this author around the end of the thirteenth century. Most importantly, the Theorica et practica (or TP) of this Paul contains the very chapters that Ruska theorized to have been written by the author of the Summa. The TP appears, for several reasons which will be detailed in our talk, to have been the original source of the "Geberian" chapters. Furthermore, the TP shows an undeniable ideological and technical rapport with the Summa, and for various reasons which will also be described, it appears to be of earlier date.

We have therefore discovered a text which helps to confirm Ruska's hypothesis, while also proceeding further, in that it contains a seemingly genuine ascription. We hope, by means of this text and others, to help place the Latin alchemy of the thirteenth century within its proper social and cultural milieu.

Kwang-Ting Liu

Professor of Chemistry, National Taiwan University, Taipei

ON THE QUESTION ABOUT THE KNOWLEDGE OF OXYGEN IN MEDIEVAL CHINA

A French translation of two chapters of a missing eighth century Chinese book "Ping Lung Jen" accompanied with interpretations was published in 1810 by Klaproth.[1] He considered the author, Maŏ Hhóa, had quite clear ideas about oxygen gas which was named "Ký-ȳne." There has been a number of discussion on the question of the authenticity of this book.[2-4] Needham suggested that it could be an eighteenth century work rather than an eighth century one.[5] In the present study a re-examination of Klaproth's paper comes to a different conclusion. The concepts on the change of matter in "Ping Lung Jen" are similar to those described in some other Taoist's alchemy books of medieval China. The Ký-ȳne mentioned by Klaproth has no relation with oxygen gas, and is likely the "Yin" of the Chinese "Yin-Yang" theory.

1. H. J. Klaproth, *Memoires de l'Acad. de St. Petersbourg*, 2, 476 (1810).
2. M. Muccioli, *Archiv. di Storia Della Scienza*, 7, 382 (1926), cited in M. E. Weeks, "Discovery of the Elements," 5th Ed., 1945, Journal of Chemical Education p. 90.
3. H.-C. Yuan, "Chun-Kuo Hua-Hsueh Shih Lun-Wen Chi," 1956 San-Lien, pp 221-231, and literatures cited therein.
4. N.-C. Meng, *Hua-Hsueh Tung-Pao*, No. 3, p. 54 (1978).
5. J. Needham, "Science and Civilization in China," Vol.5 Pt. 3, Cambridge University, 1976, pp 242-249.

Christoph Meinel

Universität Hamburg, Institut für Geschichte der Naturwissenschaften, FRG

MATTER THEORY AND REFORM OF KNOWLEDGE IN JOACHIM JUNGIUS' NATURAL PHILOSOPHY

The early seventeenth-century revival of ancient atomism and its result, the overthrow of the Aristotelian theory of matter, is commonly considered to be a turning point in the history of chemistry. However, the historical impact of these mechanical philosophers has often been exaggerated since their concepts of atoms, elements and chemical changes, sometimes surprisingly 'modern', had so little influence upon the real development of chemistry, that even Herbert Butterfield had to admit a "postponed scientific revolution in chemistry". To some extent this contradiction results from our habit of looking at these early concepts "nostricentrically" (W. Pagel) as if they were immediate precursors of modern chemical notions and not embedded in quite different historical contexts. From a closer and "idiocentrical" perspective however, the century reveals a great variety of matter theories which are often difficult to incorporate into the main current of the scientific revolution.

This is especially true for the reform program of the scientist and philosopher Joachim Jungius (1587-1657) whose role in the history of science is still a subject for debate. Leibniz compared him to Descartes and Galileo; later on Jungius was regarded as the "Bacon of Germany" or as one of the founders of modern chemistry. But his work is little known, and his vast manuscript papers are virtually unexploited.

Shortly before 1630 Jungius outlined a reformed scientific methodology and a sophisticated theory of matter, based on experience, chemical analysis and inductive reasoning. By these means he expected to create an empirical, axiomatically structured natural science, more geometrico, which in turn should be the starting point for a general reformation of philosophy. In the end however, his ambitious program failed because of its strict empiricism, inductivism, and the very nature of chemistry. Nevertheless, it deserves historical attention since it reveals that early-modern natural philosophy, as it was taught at schools and universities throughout Europe, was not as sterile and simple-minded as it might appear if the scientific revolution is regarded as an unequal dialogue between Salviati and poor Simplicio. Within its historical context Jungius' 'chemical reform' of natural philosophy was conceived as a real and feasible alternative to the Galileian way, since, at the time, it was by no means clear which of the two approaches would eventually provide the key to the puzzles of nature and thus open the door out of the labyrinthine natural philosophy of the late scholasticism and Aristotelianism.

Betty Jo Teeter Dobbs

Associate Professor of History, Northwestern University

The Fire at the Heart of the World:
The Dilemma of Science and Religion in Newton's Alchemy

In an unpublished treatise entitled The philosophical origins of gentile theology Isaac Newton argued that the most primitive religious structures were characterized by "a fire for offering sacrifices [that] burned perpetually in the middle of a sacred place." Newton, attributing to the ancients an understanding of Copernican heliocentrism, assumed that these religious structures symbolized the cosmos for the ancients, with the fire representing the sun at the center. At a deeper symbolic level the fire may also represent the fiery Stoic deity, the creative fire immanent in the universe, hidden but active at the heart of the world and at the heart of everything in the world, a concept that may certainly be traced back to Zeno and perhaps even to Heraclitus. The Stoic doctrine of an occult creative fire that was the secret agent in the shaping of matter had been adopted by many alchemists, and it seems quite likely that Newton also identified the creative alchemical fire at the heart of matter with the perpetual sacred fire at the heart of the symbolic prytaneum. That being the case, Newton faced a dilemma. To study the sacred fire in matter might be a religious act, but to bring the sacred fire under the rule of law, as he sought to do in his alchemical work, was to place it within the natural world and so destroy its sacramental function.

Maurice Crosland

Unit for History of Science, University of Kent, Canterbury, U.K.

A HISTORY OF THE GASEOUS STATE - AS REFLECTED IN THE TERMINOLOGY USED IN THE SEVENTEENTH AND EIGHTEENTH CENTURIES

One of the greatest (and perhaps least appreciated) discoveries of the eighteenth century was the discovery of the gaseous state. The concept of gas as a third state of matter depended on the realisation that air was not one substance but [one of] a whole class of substances. Unlike 'spirits', which had an ambiguous status, gases were as fully entitled to be considered as matter as liquids and solids.

There are many ways of approaching the history of gases; some scholars would take a strictly practical approach and concentrate on the difficulties in the preparation and collection of gases. But I consider that <u>conceptual</u> difficulties in understanding the gaseous state were even more fundamental than the practical problems.

A useful (though necessarily superficial) approach to understanding conceptual difficulties is to study the terminology used to describe different instances of gases. Early mentions of gases are recorded under a wide terminological range, from 'vapours' to 'smokes'. Sometimes the physical location determined the name, so that a gas detected in a mine might be known as a 'damp'. Red 'fumes' evolved in a laboratory might become 'smoke' in the open air. It is worth asking why VAN HELMONT's 'gas' was not more generally adopted before the late eighteenth century. Another problem is the continuing use of the term 'air' to denote gases even after gases were seen to be different from atmospheric air.

In his early research, the young LAVOISIER referred to oxygen as "plus air que l'air commun". Later, of course, the French chemist was to give a lead in the reform of chemical nomenclature, insisting that the names of substances should reflect their <u>composition</u>. This marks a change in the nomenclature of gases, which had previously been generally described by their <u>properties</u>. As a final example of how changes in terminology illustrate conceptual changes, we might take the case of carbon dioxide, described successively as 'gas sylvestre' by VAN HELMONT, 'fixed air' by BLACK, 'acidum aereum' by BERGMAN and 'acide carbonique' by LAVOISIER.

A proper history of the gaseous state is a desideratum for the future. Meanwhile a beginning may be made with a study of terminology and some of the related conceptual problems. The actual presentation of the full text of this paper will depend on economic rather than academic considerations, but the above summary is presented as a starting point for some future discussion.

Anders Lundgren

Dr., Office for History of Science, Uppsala university

THE QUANTIFICATION OF CHEMISTRY IN THE 18TH CENTURY

The quantification of chemistry in the time of Lavoisier has always been considered a crucial step in the history of chemistry. However, little attention has been given to this important issue in the history of 18th-century chemistry.

Chemistry differs from "physics" (meaning experimental physics) in its dealing with the unique, not the universal. Still one might ask if in 18th-century chemistry can be found an attitude towards scientific research similar to the attitude found among physicists. E.g. an experimental study of processes, a certain status and use of theories and not the least an extensive use of numbers and mathematics. This would then consitute a new trait in chemistry, which earlier belonged mainly to natural history.

In relation to defining the meaning of quantification the following areas can be proposed for a closer study: First, the balance and the role of weighing - what questions were supposed to be answered with the use of the balance, and what questions were raised by the results. Second, the law of conservation of matter, which made it possible to study chemical reactions as processes by comparing an initial and final state. Third, the mineralogical classification systems, which changed from being purely descriptive to being based on chemical composition. To these might later be added the theory of chemical proportions, which depended on analysing chemical composition needed for classification, and the atomic theory, understood as a practical-oriented instrumental theory, able to explain and summarize experimentally produced results.

From the study of these areas emerge the concepts of "synthetic" and "analytic" quantification that can help explain the characteristics of quantification in 18th-century chemistry. Synthetic quantification, which pre-dates analytic, has the character of a recipe. It attempts to determine how many parts of different substancess are needed in order to create a new substance with desired properties. In general such quantification does not rely on any theory, being based on a long empirical tradition. On the other hand, analytic quantification entails finding the composition, in numbers, of a specific substance. Such analysis requires a theory of chemical composition as well as precise analytical techniques.

Quantification of 18th-century chemistry can be charcterized by a steady growth of analytic quantification, thereby facilitating a quantitative study of chemical processes.

Lissa Roberts

Student Recommended Visiting Faculty, University of California, at Irvine

CHEMISTRY AND MEDICINE: THE LANGUAGE AND PRACTICE OF SCIENTIFIC REVOLUTION IN EIGHTEENTH CENTURY FRANCE

For most of us, the Encyclopédie and revolution are an inseparable pair. Diderot himself stated that the enterprise was intended as a tool with which to change the popular way of thinking. His article "Encyclopédie" repeatedly refers to the immanence of revolution in both thought and politics. While this is well known, little attention has been given to the encyclopedic contributors who particularized the call to revolution by aggresively prophesying radical change in their own disciplines. Two salient cases are Gabriel-François Venel in his article "Chymie" and Jean-Jacques Ménuret in his article on medical observation. Both stated that a scientific revolution was at hand. Further, both claimed that the result would be the overthrow of Mechanism's rational tyranny.

The purpose of this paper is to follow the course of ideas and events in chemistry and medicine from the 1750's when the articles were written, through the revolutionary reorganization of France's scientific societies and educational system which began in the 1790's. These are the questions I will investigate. First, what elements did the approaches of Venel and Ménuret share? Both men were influenced by the Stahlian tradition of natural philosophy which emphasized empirical observation and experimentation and posited the principles of activity (phlogiston and anima) to reside in Nature itself. Secondly, what is the relation between Venel and Ménuret's calls to revolution and the reforms later advocated by Lavoisier in Methode de nomenclature chimique and Traité élémentaire de chymie and by Cabanis in Observations sur les hôpitaux and Du degré de certitude de la médecine? The latter two refered to Condillac's epistemology as a basis for their proposed methodologies. Both called for a stance of epistemological modesty, limiting conclusions to the sensible realm and eschewing a priori theories of causation. Both wanted theory grounded in observation and experimentation. Thirdly, how were the disciplines of chemistry and medicine actually reformed in revolutionary France? Finally, what does this course of ideas and events tell us about the larger course of revoluiona and reform in French thought and society?

Nicholas Fisher, University of Aberdeen, Scotland, UK

THE *ENCYCLOPÉDIE* AS A MODEL FOR CHEMISTRY TEXTBOOKS

The encyclopedias of the eighteenth century, and in particular the *Encyclopédie* itself, were published at a critical period in the history of education when the content and method of university teaching were being re-examined in the light of the Scientific Revolution. The encyclopedists played a crucial role in establishing the boundaries of the specialized science which was to be one foundation of the revival of European university teaching in the early 19th century. In chemistry, this influence extended also to the scope of the discipline as set out in early 19th-century texts: the attempted coverage was complete (or encyclopedic), with monographs a great rarity before 1850. And above all the style of exposition was the 'historical, plain method' of Bacon and Locke, as advocated by Diderot; in contrast to the deductive, 'systematic' texts of the 17th and 18th centuries, the new texts were often historical in approach, and almost invariably inductively structured. In the new pedagogy (again partly based on Lockean epistemology), the facts were supposed to 'speak for themselves', rather than being tailored and arranged according to any particular system. Two particularly influential encyclopedic texts were Fourcroy's *Système des connaissances chimiques* (1800) and Thomas Thomson's *System of chemistry* (1802), both of which evolved from their authors' earlier encyclopedia articles, and which served widely as models in the first half of the 19th century.

Yukitoshi Matsuo

Associate Professor, Doshisha University, Kyoto, Japan

SCIENTIFIC CONTROVERSIES IN THE MID-NINETEENTH CENTURY AND THE PROBLEM OF COGNITIVE MECHANISM

In the history of modern science controversies had a "positive" value, a promotion and consolidation of scientific knowledge, while they often had a "negative" one, a mudsliding into fruitless controversies with regard to priority. Some "negative" controversies, however, sometimes played important roles, if considered in relation to cognitive mechanism in science. In this paper some mid-nineteenth-century controversies which have been often regarded as fruitless mudslidings will be proved to have sometimes played "positive" roles in a cognitive mechanism.

The mid-nineteenth century saw an active development and prompt controversial communication in many fields of natural science. Here three most typical controversies will be treated: Second Water Controversy, Glacier Controversy and Energy Controversy. In these controversies were more or less seen internationality and nationalism, the spirit of "justice," such lessons from mudslidings as co-operation in research and proper citation and reference in publication, and the active role of journals and reviews as field of controversies.

The Second Water Controversy originated from a long Éloge of James Watt by French physicist F. Arago and its English translations published during 1830s. The Controversy was once more threshed out after half a century as for the priority over the discovery of the composition of water. And it helped to get over the wall of nationalism into the spirit of justice: internationalism in the evaluation of results of scientific researches, and the problem of historic facts and historical evaluation came to the front. As to the Second Water Controversy a lot of European scientists opined in reviews and journals, and G. Wilson's treatise attached to his book <u>Life of the Honourable Henry Cavendish</u> was one of the best fruits.

In the early 1840s J. D. Forbes, a Scottish physicist and L. Agassiz, a Swiss naturalist had a warm controversy over the discovery of the structure and motion of glaciers. Their controversy developed into an infamous complicated one, while promoting scientific researches. The Controversy sporadically continued over thirty years and had three main stages, always showing mudsliding aspects. It urged scientists' deep reflection on problems of co-operation in research and of proper reference to preceding works.

The Energy Controversy bitterly conducted chiefly by Tyndall and Tait between the Second and Third Glacier Controversies was closely related to the Glacier Controversy. Many scholars also entered into the Energy Controversy, and it was in the course of the Controversy that Colding claimed his priority as to his independent discovery. While the Controversy was going on, these problems were discussed chiefly in the <u>Philosophical Magazine</u>: patriotism and internationality, objective evaluation in the light of the historic facts, recognition of a proper "field" of controversies (i.e. proper journals), further mudslidings and reflection on them.

Robert J. Morris, Jr.

Associate Professor of the History of Science, Oregon State University

British Concepts and Lavoisier's Caloric Theory, a Study in the Transmission of Ideas

The famous <u>Memoir on Heat</u> by Lavoisier and Laplace, rightly called a classic of modern science, is also a significant step in the development of Lavoisier's conception of heat which in mature form was called the caloric theory. The <u>Memoir</u> was a response to the ideas of the Irish physician, Adair Crawford, and more specifically to Crawford's views as explained in a lengthy summary of them by his London associate and friend, John H. Magellan.

Crawford's and Magellan's work contain the first effective publication of the concepts of specific and latent heat. Crawford fused these concepts into a quantitative theory of capacity change which explained not only heat phenomena accompanying changes of state but all chemical reactions as well. One result of his theory was to identify dephlogisticated air as the source of the heat produced in respiration and combustion.

Crawford's small book was published in London in July 1779. Its significance was almost immediately appreciated by some of his British friends and associates who sent news of it, summaries, evaluations, and sometimes copies of the book itself to their continental correspondents. Demand for the book was such that the edition was sold out before the end of the year leaving many people interested by the news and praise but knowing few of the details which so impressed the books readers. It was then that Magellan set out to prepare his summary to save himself, he said, "the trouble of writing one over and over again" for his many correspondents. This essay appeared in April 1780.

Lavoisier and Laplace heard an account of the theory in a letter from Magellan read to the French Academy of Sciences in February 1780, read the details in mid-1781 when Magellan's essay was reprinted in Rozier's journal, and again somewhat later in Tiberius Cavallo's <u>Treatise on Air</u>. According to contemporaries, Crawford's ideas quickly became the subject of considerable discussion both in Britain and abroad. Indeed by the time Lavoisier and Laplace published their analysis of it in 1783, Crawford's book, Magellan's essay and a variety of extensive summaries of one or the other had appeared in nearly forty separate publications in English, French, Italian, German, Dutch, and Latin. The theory was even made the subject of a prize essay by one scientific group. Although there were criticisms of aspects of Crawford's theory, discussions of it continued beyond the end of the century with Dalton being its most famous nineteenth-century advocate.

Alistair Duncan

Reader in the History of Science, Loughborough University of Technology
Loughborough, Leicestershire, LE11 3TU, UK.

AFFINITY AND LATE EIGHTEENTH CENTURY CONCEPTS OF CHEMICAL COMBINATION

In trying to understand the thinking of chemists it is necessary to know the mental pictures which they hold of how chemical changes take place rather than the philosophical or mathematical theories which physicists and mathematicians tell them to hold. The two views are often rather different. In this paper descriptions of chemical combination by various writers of the late eighteenth century are considered, and a conventional model emerges of particles of constant mass having constant short range properties of affinity or attraction. It is suggested that chemists of the period used such a model, which may be implicit rather than explicit, both to explain and to predict experimental results.

Seymour H. Mauskopf

Professor, Duke University

The <u>Anales del Real Laboratorio de Quimica de Segovia</u> of J. L. Proust

       Joseph Louis Proust (1754-1826) spent the major portion of his career in Spain. After a brief stay in 1779-1780, he was in continuous residence for the period 1785-1806. It was during this sojourn that Proust produced the analytical chemistry for which he is most famous, notably the formulation of the Law of Definite Proportion. Until 1799, Proust was Professor of Chemistry at the Royal artillery school in Segovia.

       During Proust's stay in Segovia, he published two volumes of a journal which he called the <u>Anales del Real Laboratorio de Quimica de Segovia</u> (1791, 1795). These volumes are extremely rare; Dr. Ramón Gago Bohorquez, of the Universidad de Granada, and I are preparing a re-publication of them. Their contents are almost entirely Proust's work and they provide excellent material for examining the devleopment of his thought and techniques in the period before he began his famous series of papers in the <u>Journal de Physique</u> and the <u>Annales de Chimie</u>. The journal also gives insight into the utilitarian components of Proust's chemical activities in Spain. I shall explore both of these aspects of the journal.

Kiyohisa Fujii

Research Associate, Tokyo Institute of Technology

The Berthollet-Proust Controversy and Dalton's Chemical Atomic Theory

The Berthollet-Proust controversy and Dalton's atomic theory appeared almost concurrently at the very beginning of the 19th century. Therefore it is likely that between the theory of definite propotions, one of the main subjects of the controversy, and Dalton's atomic theory there was an important interrelation. In the latter-1803 Dalton could propose a list of relative weights of ultimate particles. In the same year Berthollet's <u>Essai de Statique Chimique</u> was published, then the English translation in 1804. It is not known whether Dalton read the <u>Essai</u> in French or in English between 1803 and 1808. At any rate, Dalton replied to Berthollet's attack on his theory of mixed gases, in Part I of his <u>New System of Chemical Philosophy</u> published in 1808. Although Dalton opposed Berthollet's opinion, he never referred to Proust's doctrine of definite proportions in his <u>New System</u> or in any other paper. Did Dalton know nothing about the debate or Proust's theory? It may be somewhat difficult to suppose that Dalton did not. In fact, not a few English abstracts of Proust's papers appeared in <u>Nicholson's Journal</u> in those years. Even before 1803, a crucial year when Dalton recognized the importance of the inquiry into the relative weights of the ultimate particles of bodies, he must have encountered Proust's ideas and results through Thomson's textbook. In 1802 Thomas Thomson published the first edition of <u>A System of Chemistry</u>. In volume one of this book Thomson reviewed Proust's researches on metals. In the volume three of <u>A System of Chemistry</u>, Thomson discussed the laws of affinity in great detail. First he suggested one to one combination in neutralization reactions. Then he proceeded to examine whether bodies were capable of combining with each other indefinitely or in certain determinate proportions, on the basis of the models he devised. Had Dalton read Thomson's comments on Proust's theory or Thomson's theoretical treatment on chemical proportions? Certainly he had. Because, Dalton explained again his new theory of the constitution of mixed gases of the atmosphere in the <u>Philosophical Magazine</u> and <u>Nicholson's Journal</u> of 1802, citing Thomson's remarks from the volume three of <u>A System of Chemistry</u> as one example of those who had misunderstood his hypothesis of mixed gases. In my opinion, Dalton's paper on the combination of oxygen and nitrous gas, read 12th November 1802 to the Manchester Lit. and Phil. Society and published in 1805, might support the possibility that Dalton could have encountered Proust's ideas during the formative period of his atomic theory. Judging from Dalton's statement in this important paper, he had at least some interest in whether the elements of oxygen might combine with a intermediate quantity of nitrous gas between the two extreme terms, which was one of the conflicting issues in the debate.

Cc

E. T. Florance

Creative Research Analysis, PO Box 1621, Monterey, CA 93942, USA

THE FOURTH STATE OF MATTER IN THE COSMOLOGY OF HUMPHRY DAVY

Throughout the chemical philosophy of Sir Humphry Davy, from his youthful essay of 1799 to his metaphysical posthumous work Consolations in Travel, ran a persistent emphasis on the importance and nature of light. Davy believed that light was an imponderable body with a special relationship to ponderable matter: it existed in a fourth state, beyond the conventional three states of solid, liquid, and gas. Although often acknowledging his debt to Newton for the corpuscular theory of light and for conjectures on the interconvertibility of light and ordinary matter, Davy himself apparently conceived the idea of a fourth state of existence beyond the gaseous form.

In his more mature works Davy included radiant heat with light in a category he called etherial substance or radiant matter. The deep chemical significance of radiant matter in Davy's view was based on its relation to the hypothetical forces or powers which, he felt, held together the ultimate particles of matter in different arrangements to form the chemical elements.

Davy's arguments used to explain and justify the role of radiant matter in his chemical system will be analysed to clarify the underlying cosmological assumptions which permeated that system. It will be demonstrated that neoplatonism, which is known to have influenced many Romantic philosophers, poets, and other literary figures, can be detected implicitly and explicitly in Davy's writings and his lectures. A hierarchical, neoplatonic cosmology may have served to inspire the notion of a fourth state of matter.

Sally Newcomb

Asst.Prof.,Phys.Sci.Dept.,Prince George's Comm.Col., Largo, Md. 20772

LABORATORY EVIDENCE OF SILICA SOLUTION SUPPORTING WERNERIAN THEORY

There was strong disagreement among geologists at the end of the 18th and the beginning of the 19th centuries about the topic of rock origin and, ultimately, a theory for the origin of the earth. Both James Hutton (1726-1797) and Abraham Gottlobb Werner (1750-1817) reasoned from empirical evidence in order to provide an explanation for the origin of rocks and the processes that form them. Both theories postulated methods of formation of granite and basalt. Huttonians considered them to have been solidified from a molten state. The Wernerian view was that they crystallized from solution in the waters of a primordial ocean. As geologists of the early 19th century distanced themselves from fanciful, theological accounts of earth origin, laboratory evidence became more important. Some of this evidence had been available for a number of years, and must have contributed to the plausibility of the Wernerian position.

This paper will examine what was known about the properties of silica in water solution. A number of Wernerians were laboratory chemists and mineralogists, and had experience in analysis as well as in the investigation of properties of mineral substances. Since the time of T. Bergman laboratory procedures for the determination of silica in mineral waters were well known. Quantitative water, rock, and mineral analyses as practised by R. Kennedy, J. Murray, R. Jameson, R., Kirwan and others not only detected the silica in solution, but also demonstrated conditions for its solubility. D. Brewster and others quantitatively analyzed the fluids found in mineral cavities. Analogies to rock formation were made from the condition of silicified wood, the form of silica deposits observed in the analytical train, and procedures necessary to redissolve precipitated silica.

Whether supporting Werner or Hutton, some geologists believed that in the past earth processes were similar to those seen in the present, but they occurred at different rates. Therefore, the Wernerian position of postulating a primordial ocean both warmer and more basic than at present was not unreasonable. The laboratory results of R. Kirwan, R. Kennedy, J. Murray, D. Brewster, R. Jameson, J. Black, and H.M. Klaproth bear this out, and help explain the long and strong influence of Wernerian thought.

Hornix, W.J.

Faculty of Science, University of Nijmegen, The Netherlands

DYESTUFFS AND THE DEVELOPMENT OF ORGANIC CHEMISTRY 1850-1890

The growing family trees which connect the marketed dyestuffs via intermediary products with raw materials gives not only a good starting point for the description of the industrial development around dyestuffs. Also the study of the impact of organic chemistry on the production of dyestuffs can be fruitfully structured through these trees.

The development of organic chemistry can be described in an analogous way through the catalogue of known chemical reactions resulting in a network of substances. The constitutional theories can be considered as a classification of reactions and of substances that brings system in the catalogue, and structure in the network. A consitutional theory originates partly in the failures of earlier systematisations in describing the steadily growing network of substances and reactions, partly in presuppositions imported from inorganic chemistry, physics or philosophy.

The analysis of the growth of the network of known substances and reactions around fuchsin, alizarin and indigo and the discovery of the class reactions diazotation and coupling, essential for the production of azodyes, will illustrate the characteristics of different constitutional theories and their relative advantages and weaknesses. This analysis results also in a characterisation of dyestuffs chemistry in different periods and in an evaluation of the contributions of Hofmann, Graebe, Baeyer and Griess.

Kekulé's structural theory, that appears as superior to its historical alternatives will be explained as a scheme open to corrections and important extensions: new class reactions, new chemical functions, new structural elements. I will stress the important role that Baeyer plays in the extension of Kekulé's theory and his central position in the development of dyestuffs chemistry and dyestuffs production. Finally, I will try to evaluate the importance of dyestuffs chemistry for the development of organic chemistry in general.

Yasu Furukawa

Associate Professor of the History of Science, Yokohama College of Commerce, Yokohama, Japan

CAROTHERS, MACROMOLECULES, AND THE EMERGENCE OF AMERICAN POLYMER CHEMISTRY

This paper examines the scientific work of the American chemist, Wallace Hume Carothers (1896-1937), and his place in the history of American polymer chemistry. Between 1928 and 1937, Carothers carried out with his collaborators a series of investigations on polymers and polymerization within the fundamental research program at the Du Pont Company. His study in this field was based on the macromolecular concept proposed in the early 1920s by the German chemist, Hermann Staudinger. But from the outset of his research, he departed from Staudinger's approach.

A close examination of his background shows that there were no direct ties between Carothers and Staudinger or his German school. Trained as an organic chemist without the background of German chemical education from which earlier generations of American chemists often had benefitted, Carothers was a product of American pragmatic education immediately after World War I. A top student of Roger Adams' Illinois school of organic synthesis, he fully used the synthetic approach to the study of polymers, in contrast to Staudinger's rather analytic approach. By the use of established organic reactions (such as condensation), Carothers aimed to build up artificial macromolecules, the existence of which was then still controversial in scientific circles. In the Company's framework, his theoretical work was soon led to practical applications such as the production of the first synthetic fiber, nylon, although this was an unforeseen consequence of his initial basic research.

Carothers' macromolecular synthesis convinced his contemporaries of the macromolecularity of polymers as well as the great possibility of applications in this area. Under his influence, there emerged the first generation of American polymer chemists in the university, who included Carl S. Marvel at the University of Illinois and Paul J. Flory at Cornell University. Through close contact with Carothers, they introduced the study of macromolecules into their teaching and research, and trained many Ph.D. students. Thus, stemming from Carothers' industrial laboratory, this field was established as a new academic discipline in this country by the late 1940s.

Jeffrey L. Sturchio

Associate Director, Center for History of Chemistry

Civic Chemistry in Metropolitan New York: The Chandler Circle, 1870-1910

Recent research by urban historians offers new perspectives on roles for technical experts in the cities of the Gilded Age. Studies of the city-building process, including the provision of urban services such as gas, water, and sewage disposal and the improvement of public health, illustrate the expanding markets for expertise during the period. Other historians have investigated the social and institutional contexts of intellectual life in late nineteenth-century cities, with an emphasis on the emergence of professionalism in various pursuits. But few scholars, with the notable exception of Charles Rosenberg's 1967 essay on "The Practice of Medicine in New York A Century Ago," have combined these perspectives by examining the independent practice of an applied scientific profession in major American cities of the period. By focusing on the central figure of chemist Charles Frederick Chandler--Dean of the Columbia School of Mines from 1864 to 1897-- and the circle of his students, former assistants, and other colleagues, this paper illuminates the cultural milieu of civic chemists in metropolitan New York from 1870 to 1910.

My main objective is to focus on the configuration of career choices available to ambitious young chemists during these decades. What was the structure of the chemical community? How did aspiring professionals use the network of clubs, scientific societies, colleges, and other institutions of urban culture to establish themselves as gentlemen of probity and civic virtue, and thereby achieve upward mobility? And--especially for consulting chemists-- how did they use their expertise to legitimate claims to professional authority? Even leading academic chemists had diverse ties to the metropolitan world of commerce, while industrial consulting, public health regulation, and water, food, and drug analysis were more characteristic activities of municipal chemists in the urban economy. Chandler and his circle, like their Mancunian counterparts discussed in Robert Kargon's <u>Science in Victorian Manchester</u> (1977), combined "enterprise and expertise" in fashioning careers at a time when professional roles for chemists had not yet been securely established. An examination of the credentials, consulting practices, and social networks of Chandler and other metropolitan chemists illuminates the complex relations of chemistry and culture in Gilded Age New York.

David J. Rhees

Smithsonian Institution, Washington, D.C.

The Chemists' Crusade: The Chemical Foundation and the
Popularization of Chemistry, 1919-1937

    The Chemical Foundation was established in February 1919 by executive order of Woodrow Wilson. Conceived by Francis P. Garvan, who served as its president from 1919 until his death in 1937, the Foundation was set up to license to American manufacturers about 5,000 German chemical patents which had been confiscated by the U.S. Alien Property Custodian during World War I; all income above expenses was to be used in advancing American chemistry and chemical industry.
    The Chemical Foundation gave away large sums for chemical research, yet the bulk of its considerable earnings was spent on an ambitious public relations campaign. This campaign was devoted to three main causes: raising tariffs on synthetic dyes and other chemicals, establishing a "chemo-medical institute" (which took the form of the National Institute of Health in 1930), and boosting the "farm chemurgic" movement (which advocated the development of industrial uses for surplus farm products). Some of the activities utilized in promoting these causes included the funding of a nationwide chemistry essay contest for high school and college students, and the distribution of millions of popular books and pamphlets (including 143,000 copies of Edwin E. Slosson's book, Creative Chemistry, published in 1919).
    This paper documents the Chemical Foundation's often controversial use of such popular science texts and techniques in their crusade to establish an independent American chemical profession and chemical industry. The paper will also highlight links between the Chemical Foundation's activities and similar crusading efforts of the American Chemical Society and the DuPont Company.

Harm Beukers

Assistant Professor, Division of History of Medicine, University of Leiden, Leiden, The Netherlands

DUTCH INFLUENCES ON CHEMISTRY IN LATE TOKUGAWA JAPAN

In 1720 the strict isolation policy of the Tokugawa shogunate was somewhat relaxed, creating a possibility for Japanese scholars to renew acquaintance with Western sciences. To avoid trouble with the government, the scientific interests tended at first to be confined to jitsugaku - practical learning - such as medicine.
Through Von Siebold's teachings (during his stay from 1823-1829) the scope and depth of understanding of the Japanese students expanded. In the beginning chemistry was noticed as a part of medicine, especially for a better knowledge of the properties and preparation of drugs. However, gradually chemistry was recognized as an independent science.
The isolation policy limited the possibility to select freely chemical literature. Therefore, chemistry was not introduced into Japan through original classics but through annotated translations in Dutch, such as A. Ypey's translation (1804-1807) of Henry's Elements of Experimental Chemistry or G.J. Mulder's translation (1834-1842) of Berzelius' textbook. To provide a broader base the shogunate founded laboratories for chemistry employing Dutch specialists such as W.K. Gratama, A.J.C. Geerts, P.H. Eijkman and P.C. Plugge.

Walter Kaiser

Hochschulassistent, Johannes Gutenberg-Universität Mainz, Fed.Rep. of Germany

CRITICISM OF KINETIC THEORY AND EARLY STEREOCHEMISTRY

It is well known that the kinetic theory of gases met with criticism from various sources. One should mention here Rudolf Clausius' distinction between a mechanical theory of heat in general and the special hypothesis of the kinetic theory. There was also Ernst Mach's rejection of atomism. In addition Wilhelm Ostwald favoured energetics and phenomenological thermodynamics. One should consider also internal criticisms like Josef Loschmidt's reversibility paradox and Ernst Zermelo's recurrence theorem. In the case of Max Planck and Zermelo it is especially obvious that the seemingly pure theoretical arguments are connected with the philosophical arguments. Thus the questionable mechanical explanation of the second law of thermodynamics was tied together with the problems of the reality of atoms and molecules.

Sometimes the historical development of early stereo-chemistry is described in the same way. Especially Hermann Kolbe's furious attack against Jacobus Henricus van't Hoff's stereo-chemistry is considered to represent the influence of positivism. In fact, Kolbe explicitly dealt with the question of whether van't Hoff's hypothesis of an asymmetric carbon atom is an observable structure or not. The influence, however, of Kolbe's extremely polemic writings on van't Hoff's stereo-chemistry was remarkably low. Whereas Mach's arguments are philosophical and rather objective Kolbe's arguments seemed to be mainly of a personal nature.

On the other hand there was severe criticism of van't Hoff's theory of a close relation between an asymmetric carbon atom and the observable optical activity of an organic compound. One should mention here Pierre Eugène Berthelot. To this story also belong inversion reactions first observed by Paul Walden. It is interesting that van't Hoff discussed these awkward reactions in a letter to Wilhelm Ostwald. Very characteristic are also Adolf von Baeyer's attempts to determine the structure of the terpenes, especially the structure of limonene. Obviously von Baeyer though he had failed to reconcile his analytical findings with van't Hoff's theory of asymmetric carbon atom and optical activity. Thus this type of criticism was quite different from criticism in the field of the kinetic theory of gases. A close connection with the positivistic philosophy of Ernst Mach is barely recognizable within these discussions.

H.A.M. Snelders

Institute for the History of Science, University of Utrecht, Utrecht, The Netherlands

THE DUTCH PHYSICAL CHEMIST J.J. VAN LAAR (1860-1938) VERSUS J.H. VAN 'T HOFF'S 'OSMOTIC SCHOOL'.

The 'forgotten' Dutch physical chemist Johannes Jacobus van Laar (1860-1938) studied physics and chemistry at the University of Amsterdam (1881-1884). His teachers were a.o. J.D. van der Waals, J.H. van 't Hoff, and the mathematician J.D. Korteweg. After his studies Van Laar became a secondary school teacher, at first in Middelburg (1884-1895), later in Utrecht (1895-1897). Because of a severe illness, however, he then had to go into retirement. He lived on private lessons and on the products of his pen. In 1898 he became a Privat-docent in mathematical chemistry; in 1908 he was appointed lecturer in propaedeutic mathematics in Amsterdam. But already in 1912 he had to retire for health-reasons. He left The Netherlands for Switzerland where he spent the rest of his life.
   Van Laar was a fervent propagandist of the microscopic interpretation of thermodynamic equations of state of systems on the basis of the interaction of the constituent particles. He opposed to what he called 'the osmotic school' of physical chemistry. All his life he fought the abuses of osmotic pressure, viz. the extension of that concept considered as a physical reality, to isolated homogeneous solutions, and its practical application to non-diluted solutions and even to solutions of strong electrolytes. When dealing with a general theory of solutions Gibbs' thermodynamic potential is the only obvious and certain way to solve the problems.
   Already in 1895 Van Laar gave as his opinion that the (apparent) deviations of Oswald's dilution law are caused by the extremely strong electrostatic field of the ions. In 1900 he assumed a practically complete dissociation of strong electrolytes.
   Van Laar immediately accepted P. Debye and E. Hückel's theory of strong electrolytes (1923); ten years later, however, he rejected Debye's concept of an ionic atmosphere and accepted - for theoretical and practical reasons - I.C. Gosh's space lattice model of 1918.
   Although his influence on the development of physical chemistry was relatively small, Van Laar's great merits were that he repeatedly confronted the Dutch chemical community with the possibility and necessity of the use of the thermodynamic potential on physical-chemical problems.

Akira YOSHIDA

Université Meiji, Tokyo, Japon

W. Ostwald et la théorie des acides d'Arrhenius

Dans des livres de chimie contemporaine, on parle de la définition des acides et des bases selon Arrhenius. Certains historiens des sciences attribuent l'origine de cette définition au mémoire d'Arrhenius de 1887:"Ueber die Dissociation der in Wasser gelösten Stoffe." Cependant, il n'en donne la définition ni dans ce mémoire ni dans celui publié en 1884:"Recherches sur la conductibilité galvanique des électrolytes", ni dans aucune de ses publications. Il est certain que la définition est faussement attribuée à Arrhenius. L'introduction de cette erreur est, nous semble-t-il, relativement récente, puisqu'encore en 1940, W.F. Luder a appelé "théorie d'Arrhenius-Ostwald" ce que nous appelons aujourd'hui la définition d'Arrhenius. Ce dernier établit en 1883 sa théorie de la dissociation spontanée des électrolytes et proposa une explication de la chaleur constante de neutralisation entre un acide et une base forts. Ostwald démontra expérimentalement que la chaleur dégagée lors de la neutralisation correspond effectivement à la chaleur de formation de l'eau, qui est constante. C'est Ostwald qui donna en 1894, par suite de ces recherches, la définition classique des acides et des bases appelée actuellement et à tort la théorie d'Arrhenius.

BLONDEL-MEGRELIS Marika

Centre National de la Recherche Scientifique, Paris, France

TYPES ET STRUCTURE CHIMIQUES: HISTOIRE D'UNE ABSTRACTION.

La volonté de se "libérer de la notion de type comme fondement des considérations de constitution chimique" pour lui substituer le "principe de la structure chimique" marque une importante coupure dans la chimie du XIXe siècle.
　　La distance qui sépare la notion de type chimique, telle que DUMAS s'est efforcé de la définir, telle que WILLIAMSON et HOFMANN l'ont utilisée et même telle que GERHARDT a tenu à la préciser, de celle de structure telle qu'elle semble en voie d'élaboration chez COUPER et chez KEKULE (avec bien des nuances), puis telle qu'elle est définie par BOUTLEROV, est celle-là même qui sépare deux objectifs: d'une part la volonté de représenter les métamorphoses chimiques, de traduire le comportement des corps, d'autre part la volonté de donner la raison de ces métamorphoses en cherchant cette raison dans l' "arrangement des atomes", c'est-à-dire dans la disposition des atomes dans l'espace et/ou dans la "relation chimique" des atomes entre eux: ce que Auguste LAURENT tentait d'exprimer par sa formule "prendre la place et jouer le rôle".
　　L'idée de type reste encore, et paradoxalement, très liée à la nécessité d'une permanence matérielle, même si cette matière est susceptible de varier, dans certaines limites. L'idée de structure met au contraire l'accent sur l'aspect fonctionnel des éléments en jeu dans les métamorphoses. Par structure, BOUTLEROV voudra "désigner l'enchaînement chimique ou les façons dont les atomes sont liés entre eux dans un corps composé".
　　Pour cela, il faudra passer de l'observation des "équivalents" chimiques à l' "idée" des atomes; de la considération des substances chimiques à celle de leurs liaisons. Il faudra donc passer de l'objectif de description et représentation des phénomènes à la tentative d'une compréhension profonde des phénomènes par l'énonciation d'une idée à leur sujet, la description devant alors être produite comme résultante des considérations de structure.
　　Cela marque très exactement l'inflexion du concept de théorie en chimie: la fin de la théorie n'étant plus de lier les faits et de les unifier, donc de les réduire, mais de les rationaliser, c'est-à-dire les faire dépendre de quelques principes généraux dont ils devraient absolument résulter, multiples et divers.

Kenneth L. Taylor

University of Oklahoma

TURGOT, DESMAREST, AND THE "OBSERVATIONS GÉOLOGIQUES" OF 1760

Some geological notes made in 1760 during a journey from Paris to Lyon, originally believed to be by Anne-Robert-Jacques Turgot (Oeuvres de Mr. Turgot, ed. P. S. Dupont de Nemours, III [1808], 376-447), were subsequently claimed by Nicolas Desmarest as notes he had prepared and given to Turgot. This was conceded by Turgot's editor Dupont de Nemours (Oeuvres, I [1811], 52-53), but has remained open to question.

Assembly of evidence on the whereabouts of Turgot and of Desmarest during the period indicated in the notes makes it appear highly probable that they were written by Turgot. This is supported by certain details in the notes themselves.

Confident attribution of these notes to Turgot augments out knowledge of his scientific activities and ideas. It also affords an expanded view of a small group of French geological thinkers around 1760 carrying out actualist-minded field observations in a manner reflecting the ideas of G. F. Rouelle.

James R. Fleming

Ph.D. Candidate, Princeton University, Princeton, NJ, USA

The Clouds that Gather 'Round the Setting Sun:
    Antebellum Meteorological Networks and the "Problem of American Storms"

    In 1849, Joseph Henry and James P. Espy, under Smithsonian sponsorship, organized a network of volunteer observers extending across the United States and into Canada in order to explore the problem of "American storms." Using standardized instruments and guidlines for observers, the far-flung network of Smithsonian observers, numbering over 600 by 1860, reported by telegraph and by mail to Washington.
    The Smithsonian Meteorological Project was self-consciously based on the successful model of an earlier, smaller project directed by Espy in the 1830's: the Joint Committee on Meteorology of the American Philosophical Society and the Franklin Institute. At the focus of both networks, theorists such as Espy and Elias Loomis observed the continent's weather and used the data in their studies of winter storms.
    This paper describes and compares the westward-looking networks of the Joint Committee and the Smithsonian. Of central interest and importance is the relationship between the physical theories of the meteorologists and the data collected by the observational networks. It is argued that theoreticians began with the hope that a few case studies of prototypical weather phenomena would allow them to "solve the problem" of storm dynamics by discovering the laws governing the continent's weather. As massive amounts of data were accumulated and compiled, and weather patterns showed no apparent regularities, meteorologists moved toward theories of weather emphasizing stochastic processes. The Civil War severely crippled the Smithsonian's meteorological work and marked the sunset for the era of volunteer observational networks.

Walter E. Pittman, Jr.

Professor of History, Mississippi University For Women

Soil and Geology - Contributions of
Eugene Woldermar Hilgard 1833-1916

In 19th Century America, Eugene W. Hilgard pioneered in the application of Science (geology and chemistry) to agriculture. Renowned as the American "father of soil Science" Hilgard originated much of the discipline and pioneered in the transfer of the new technology to farmers. Hilgard's greatest success was his creation of the California Experiment System and associated extension services which permitted the successful transfer of technology from experimental science to the farming public. Hilgard first demonstrated the connection between soil type and climate, first delineated soil horizons and devised most of the standard soil testing methods, instruments and standards. His studies led to an understanding of how alkali soils originated and became the basis of Federal policy.

Hilgard was also a noted geologist, known as the "father of Gulf Coast" geology. He originally outlined the stratigraphy of the Gulf Coast region in the 1850's as well as being a key figure in outlining the American Tertiary. His work is still basically accepted.

During the Civil War, Hilgard, who was head of the Mississippi Geological Survey and Professor of the University of Mississippi worked as a scientific advisor to the Confederate state government.

Michele L. Aldrich, American Association for the Advancement of Science, Washington, DC

Alan E. Leviton, California Academy of Sciences, San Francisco, CA

YOSEMITE VALLEY: Controversy in Grandeur

Yosemite Valley and Yosemite National Park lie nestled in the central Sierra Nevada, the "Serrate" or "Snowy Mountains," named by the early Spanish explorers who entered the region in the mid-18th century. Yosemite Valley was long used by native American Indians, but its existence was not known to the western world until 1851 when a vigilante group, "the Mariposa Battalion," entered the valley in pursuit of Chief Tenaya and a band of Uzumati Indians who had been raiding settlements in the vicinity of Mariposa.

In 1864 Yosemite Valley was deeded to the State of California by the Federal Government to become the first State park. But the valley was surrounded by lands to which the Federal Government retained title and in 1905 it was returned to the government and a national park, the size of the State of Rhode Island, was set up to preserve some of the most magnificent glacial and alpine scenery in North America.

Visitors to Yosemite are awed by the scenic whole of the valley, its botany, zoology (especially the bears), and even meteorology, not just by the geology. Geologists, on the other hand, have studied the site preoccupied by the theories and specialities of their day, starting with Josiah Dwight Whitney in the 1860s, who argued that Yosemite was a giant block-faulted chasm. John Muir challenged Whitney's assertions, contending that glaciation was a major factor in shaping Yosemite, and by 1900 some professional geologists came to support Muir's ideas. For decades Yosemite became a favorite case study for the relatively new science of geomorphology. Current geologic scholarship, however, has turned to investigating the granite and tectonic basis of the valley's history, as well as clarifying the sequence of glacial episodes responsible for its present configuration. The recent synthetic work of Huber and Wahrhaftig emphasizes the central portion of the Park as representative of most of the major elements of Sierran geology.

At the turn of the century, Yosemite and its sister valley to the north, Tuolumne, became the center of a controversy between those who wanted to use these valleys as vast water reservoirs to serve the growing population of San Francisco, and those, like John Muir, who wanted to preserve the scenic and scientific value. Today, a similar controversy rages, between those who want to develop the valley as a recreational center and those who want to conserve and preserve its scenic wonders.

Saleh Billo, King Saud Univ., Geology Dept., ad-Dir'iyah, Riyadh, Saudi Arabia

UNIFORMITARIANISM IN GEOLOGY

Hutton's concept of limitless time and the acceptance of his uniformitarian principle as the alternative to catastrophism theory in interpreting the earth's geologic history are part of the basic foundation of modern geology. To determine the conditions under which ancient animals and plants lived requires understanding their modern counterparts. And just like the Danish Steno recognized the fossil tongue-stones from Malta and other fossils in Italy to resemble modern shellfish, his Scottish follower formulated his more comprehensive uniformitarian doctrine-the present is the key to the past.

Since historical geology is based on physical geology and biology, the history of the earth is constructed from data on fossils, rocks, and their origins.

In a renaissance of science, Hutton was a lawyer, a physician, a farmer, and finally a geologist whose observations about the earth matched the discovery of the Polish astronomer N. Copernicus that the earth revolves around the sun and not vice versa to account for the prevailing days and nights.

Although the status of the term uniformitarianism is uncertain, methodological or inductive uniformitarianism remains unchallenged, while uniformity of rate is untrue.

The concept of unlimited time is of course relinquished now through the progress achieved in geochemistry, radiometric dating, and astrophysics.

Examples of present-day depositional facies at the continental margins of North America, the Arabian Peninsula, and other areas have paved the way in the interpretation of older sedimentary facies.

William B. N. Berry

Professor, University of California, Berkeley

ON CHARLES DARWIN'S BIOSTRATIGRAPHIC STUDIES OF THE SOUTH AMERICAN TERTIARY FORMATIONS

By the late 1820s, some earth historians realized that it was fossils in rocks and not some aspect of the rocks themselves that must be used to develop a time scale by which the earth's history might be read. As a part of his natural history observations while on the voyage of the BEAGLE, Charles Darwin followed in the footsteps of William Smith, as did many other geologists of the day, in collecting fossils stratigraphic layer by stratigraphic layer to obtain the data basic to development of a geologic time scale based on fossils. In 1833, Charles Lyell proposed the principle of using the percentage of still-living species (primarily molluscan) in rocks of the Tertiary interval to divide that interval into shorter-duration time intervals. Lyell pointed out that relatively few molluscan species were present in the stratigraphically lower part of the Tertiary sequence of rocks and that relatively many species occur in the upper part of the Tertiary. Lyell considered primarily marine molluscan species and noted that the fossil taxa should be compared with extant taxa from relatively nearby shores. The principle, when first enunciated, was applicable to western Europe, although Lyell suggested that it might be applicable worldwide. Darwin's studies of the marine Tertiary rocks and their contained fossils in Patagonia and from the western side of South America provided tests of Lyell's principle and of its applicability in areas outside of Europe. Darwin made carefully stratigraphically located collections of marine mollusks from the South American Tertiary rocks he encountered. The fossils were identified by the paleontologist G. B. Sowerby. These taxa were compared with identifications of extant molluscan species from relatively nearby shores made by Mr. Cuming. Darwin noted that of 59 fossil species found at a number of localities along the coasts of Peru and Chile, only 3 were identical with living species. Darwin concluded that the rocks bearing these fossils belonged to the older part of the Tertiary (an interval to which Lyell had given the name Eocene). Darwin reached the same conclusion from his analyses of fossil and modern molluscan species from the Tertiary formations in Patagonia. Darwin then noted that the fossil species from the Chile-Peru Tertiary rocks differed markedly from those found in the Patagonian Tertiary. Nevertheless, he considered the rocks in both areas Eocene, using Lyell's principle. Darwin went on to conclude that two major marine molluscan provinces existed about South American shores in the Eocene and stated: "we must descend to the apex of the continent, to Tierra del Fuego, to find these two great conchological provinces united into one." Darwin's South American Tertiary studies demonstrated the applicability of Lyell's principle in an area outside of Europe as well as enhancing its validity. As well, Darwin recognized the existence of faunal provinces in the geologic record when few such provinces were known.

Tore Frängsmyr

Professor, Office for History of Science, Uppsala University, Sweden

THE DISCOVERY OF THE ICE AGE: A SCANDINAVIAN VIEW

It is well documented that the break-through for modern geology took place in the middle of the nineteenth century. Yet historians of geology continue to engage in lively discussion concerning where to place the turning point: in the theoretical debates in England about Lyell's uniformitarianism or in the continental empirical insights? For geology in Scandinavia and Sweden it is quite simple when this step was taken, the discovery of the ice age. Here the landscape provided empirical findings that could not easily be explained: the land elevation phenomenon, the long ridges of deposited material ('åsar', eskers), the scratches on primary rocks, and the erratic boulders. Many attempts were proposed to explain the origin of these phenomena, the most famous one being Nils Gustaf Sefström's 'petridelaunic flood'. which had many similarities with Lyell's drift theory. The key to all these problems was Louis Agassiz' glacial theory, published in 1837 and 1840. As we know this theory caused strong opposition among geologists, even in Sweden. Soon, however, Swedish geologists began to see the advantage of this kind of explanation. And, gradually it became clear to them that this theory could explain the problematic phenomena mentioned above.

As a leading figure in this research works, Otto Torell (1828-1900) not only accepted but also developed and completed Agassiz' theory. In addition to studying the Swedish landscape forms he made field trips in Switzerland, Norway, Iceland, Greenland, Spitzbergen and North America. Many valuable contributions also were made by Hampus von Post, a chemist and director for a glass industry, and Axel Erdmann, first chief of the Swedish Geological Survey established in 1858, but Torell was the first with a total view of all the processes responsible for shaping the surface of Scandinavia. He took also the initiative to the Swedish polar expeditions, which reached a peak with A.E. Nordenskiöld's successful journey through the north-east passage in 1880.

Without being too patriotic it can be claimed that Swedish geological research contributed greatly to the success of the glacial theory, yet these contributions are almost totally overlooked in modern histories of nineteenth-century geology.

H.E. Le Grand

University of Melbourne

RECEPTION OF CONTINENTAL DRIFT THEORY IN AUSTRALIA 1920-1940

Few models of theory in science take into account the importance of disciplines and specialities as both formal and cognitive divisions of the scientific community. Emphasis is usually placed on the comparative merits of rival theories as judged by a relatively undifferentiated disciplinary community. The history of the modern revolution in geology, focussing upon "continental drift" and plate tectonics, indicates that a more fine-grained analysis -- one taking into account differentiation, specialization, and concern with particular theoretical or empirical problems -- is required.

Geologists, particularly in the English-speaking world, were more concerned with the detailed study and elucidation of specific geological regions and structures than with over-arching synthetic theories. Henry Frankel has recently labelled this as geological "provincialism". In Australia, specialization and provincialism contributed to the scepticism with which most scientists greeted A. Wegener's integrative, speculative theory according to which the present continents were the fragments of a former supercontinent which had broken up in relatively recent geological times and which were now drifting through the ocean floors. The theory to them seemed to be of scant aid in solving the problems on which they were working. Some few scientists interested in _other_ problems of Australian geology and biology reacted far more positively. For them the theory promised solutions in _their_ problem-fields.

Opponents and proponents of Wegener's theory rarely engaged in comprehensive, systematic evaluations of it with respect to its rivals. The critical points of comparison for each scientist were the specific research topics which he or she was pursuing. From this perspective, differentiation, specialization, "provincialism", and the study of discrete problem fields within a given scientific community are cognitive and social strategies which play an important role in the reception of novel theories. This phenomenon, if it be a general feature of scientific change, is readily assimilable to some recent models of the growth of scientific knowledge.

Zuraya Monroy Nasr

Full Time Prof., Univ. Nal. Aut. de México, México,D.F.

RATIONALISM AND EMPIRICISM IN THE CONSTITUTION OF
SCIENTIFIC PSYCHOLOGY: R. DESCARTES AND J. LOCKE

Being interested in the constitution of Psychology as an independent science, this work examines the place of psychological aspects in the cartesian rationalism and Locke's empiricism. Through their examination it is possible to understand the trends that psychological aspects took in the XVIIth century. The assertion of Psychology as an specific region of knowledge began in this period.

R. Descartes states the priority of reason as a philosophical foundation and a principle of science. As it is in the reason where the foundations of true and certain knowledge are, it becomes a <u>sui generis</u> substance. Reason and innate ideas also depend on the existence of God. And, in the cartesian epistemology, God is not a thesis that can be ignored. God is the source and absolute guarantee of indubitable knowledge. Conceived like this, the human mind could not be studied by the procedures of modern science. It could only be investigated in a metaphysical manner.

The empiricist theory of knowledge, that began with J. Locke, develops epistemological and psychological matters. Starting from the conviction that all knowledge has its origin in the experience, he denies any innate or trascendental principle. In order to found his epistemological project, he begins his study of the human understanding with a psychological research. If experience is the source of knowledge of the physical world and, if it is the main constituent of the understanding, then experience and its origin must be investigated.

We may conclude from the cartesian epistemology claims that the psychological aspects do not reach any specificity as a domain of knowledge. However, in the terms of Locke's empiricism demands, the psychological aspects are recognized in their specificity.

Being so, it's possible attain a better comprehension of the reason why experimental Psychology, that emerged during the XIXth century, grows up within the context of the empiricist Philosophy.

Karin Johannisson

Assoc. prof. Depart. of History of Science and Ideas, Uppsala Univ.

POLITICAL ARITHMETIC, STATISTICS, AND 18th CENTURY SOCIAL SCIENCE

The quantifying ideals penetrating physical science from Galileo onwards, soon spread to the study of social phenomena. In England, the latter part of the 17th century brought forward the new science of political arithmetic, meaning a numerical description of "state facts" (land, population, industries, trade etc.), analyzed and generalized into alleged indices for national welfare. This quantitative way of evaluating the material and human resources of a country, therefore was closely connected to the state's interest. It illustrates an important stage in the history of political economy, here hesitating at the cross-roads of practical and theoretical goals. Not all European countries, however, were convinced that numerical description was the best way of analyzing state resources. In Germany the science of <u>Statistik</u> or <u>Staatenkunde</u> disputed quantification, considering it an instrument all too crude to grasp the innermost truths of state. In Sweden, on the other hand, quantitative analysis enthusiastically was linked to the national welfare programme. The dream of decoding all social phenomena by numbers, was soon, however, domesticated: political arithmetic was reduced and transformed into strictly empirical vital statistics. These statistics, famous all over Europe, were the first step of institutionalization of statistics as a science.

ERNESTINA JIMENEZ-OLIVARES

Universidad Nacional Autonoma de Mexico   Dr.

THE ARCHIVES OF THE INQUISITION OF MEXICO AND THEIR IMPORTANCE IN THE HISTORY OF PSYCHIATRY.

The Inquisition of Mexico as in many other countries involved a large group of the mentally ill. Some were possessed by the Devil; others called "Alumbrados" belonged to a magic religious sect. Several were burned alive.

By means of a careful study of the Inquisition documents, it is easy to discover mental pathology in such persons. The vast majority were never given a medical exam. This paper deals with several cases of patent psychopathology deserving treatment in an asylum, rather than a cruel death sentence for heresay.

Lee Mintz

Department of the History of Science, Harvard University,

TOWARD A CRITICAL HISTORIOGRAPHY OF THE HUMAN SCIENCES:
Some Programmatic Comments

In the United States, the historiography of the human sciences remains the pariah of the history of science community. Why? The history (and prehistory) of the physical and the biological sciences has always provided the central subject matter for the history of science in this country. (Related ventures have encompassed the history of mathematics and logic, the history of technology, the history of medicine and public health, and associated fields.) In general, and in sharp contrast, the human sciences -- as another great class of sciences -- have been excluded, largely because they have not been regarded as sciences.

Is this exclusion warranted? I think not. But if the historiography of the human sciences is to be accepted as a legitimate domain of substantive inquiry within the history of science, a number of crucial questions must be addressed. How, if at all, can the various disciplines and sectors of the human sciences be understood as sciences? Does the great range of types of human sciences (often differing vastly in character, in myriad ways, from the physical and biological sciences) necessitate the development and use of new investigative programs and new methods of inquiry? If so, what are the most striking philosophical difficulties, logical and methodological quandaries, operational impediments, and practical puzzles that the envisaged innovative efforts must confront? Overall, what are the prospects of success, if any, in a quest for an effective and systematic program of research into the history of the human sciences?

The present paper, drawing on other work in progress, attempts to suggest possible preliminary answers to these and related questions. It sketches central concerns and principles for a critical, self-reflexive, social-theoretically informed historiography of the human sciences. To a degree, these proposals are introduced through critical encounters with earlier programs and positions. The essay identifies a few of the chief problems and a number of the most exciting prospects for the proposed program.

Joan Leopold

University of California, Los Angeles

THE PRIX VOLNEY AND THE HISTORY OF GENERAL AND COMPARATIVE LINGUISTICS IN THE NINETEENTH CENTURY

This paper will report on an international research project to study and publish selected Prix Volney essays with commentaries upon them. The Prix Volney, awarded from 1822 onwards by the Académie des Inscriptions et Belles-Lettres in Paris, was the first major international prize to recognize work in general and comparative linguistics. Entries were submitted anonymously from all over the world. It was the major linguistic prize of the nineteenth century and the first of its kind. Some of the essays submitted were by well known authors (including those in other fields such as politics, military science, law and anthropology); some were by authors not so well known or very obscure. Essays will be accompanied by commentaries which will cover 1) the life of the author, 2) the author's contribution to linguistic theory and to other fields of study, 3) the importance of the essay in demonstrating the development of the author's thought in the course of his career, 4) the importance of the essay in showing the development of general and comparative linguistics in the period 1822-80, and 5) the role of the Volney Prize Commission in fostering and channeling linguistic research and professionalization in the nineteenth century.

These successful and unsuccessful prize essays are significant indicators of the rapid changes and reorientations of paradigm which were taking place within the greater and lesser linguistic minds of the time. The "Concours Volney" was a forum within which paradigms were sought and were sent to battle one another. These paradigms likewise influenced the changing conception of the goals of the prize competition itself. Thus while the competition channeled linguistic scholarship, it was itself reflective of the pressure for change and professionalization within the linguistic community on an international scale.

Heikki J Hakkarainen

Prof., Helsinki School of Economics and Business Administration

NATURWISSENSCHAFTLICHE MODELLE IN DER DEUTSCHEN SPRACHWISSENSCHAFT AM BEISPIEL DER DIFFUSIONSTHEORIEN

Seitdem Johannes Schmidt 1872 (Die Verwandtschaftsverhältnisse der indogermanischen Sprachen) seine Auffassung von einer wellenförmigen Ausbreitung sprachlicher Innovationen (sog. Wellentheorie) vortrug, hat sie die Vorstellungen der Sprachforscher von dem räumlichen Aspekt der sprachlichen Veränderungen beherrscht.

Die Idee Schmidts, von ihm bescheiden als "Bild" verstanden, denn er vertrat die Meinung, dass "bilder in der wissenschaft nur sehr geringen wert haben", war gemeint als Erwiderung auf bzw. Gegenpol gegen die darwinistisch beeinflusste Stammbaumtheorie Schleichers.

Aus der gleichzeitigen Opposition Schmidts gegen die ebenfalls naturwissenschaftlich bzw. -gesetzlich orientierte junggrammatische Schule, die ihn mit Schuchardt verband, könnte analytisch geschlossen werden, dass seine eigenen Ideen nicht aus naturwissenschaftlichen Modellen stammten.

Die Wellentheorie Schmidts dürfte auch keine direkten naturwissenschaftlichen bzw. überhaupt wissenschaftlichen Vorbilder bzw. Parallelen in ihrer Zeit besitzen, obwohl ein allgemeiner Bezug auf Physik verfolgt werden kann.

Die Wellentheorie hat in der Sprachwissenschaft ihr Eigenleben bis heute fortgeführt, ohne Anschluss an die Theorien von der Diffusion von Innovationen, die - ursprünglich im Bereich der amerikanischen Agrarsoziologie (Tarde & Parsons, The Laws of Imitation, 1903) entwickelt - immer weitere Kreise der Wissenschaft als eine Dachtheorie mit starker Erklärungskraft einbezogen hat.

LENNART OLAUSSON   Dep. of History of Ideas and Science
University of Göteborg, Göteborg, Sweden

AUSTROMARXISM AND SOCIOLOGY IN FIN-DE-SIÈCLE VIENNA

In the first decade of 1900 a new tradition within Marxism was developed in Austria by Max Adler, Otto Bauer, Rudolf Hilferding and Karl Renner, the so-called Austro-Marxism. In 1907 some of the Austro-Marxists were co-founders of the Viennese "Soziologische Gesellschaft" together with R. Goldschied, W. Jerusalem, L.M. Hartmann, among others. In my paper I will discuss the strain between on the one hand Marxism as Sociology and on the other hand as the ideology of the working class.

The Austro-Marxists all studied at the juridical faculty at the university of Vienna. Through these studies their idea of what constitutes the social sciences was founded. Max Adler tried to formulate this in philosophical terms in his <u>Kausalität und Teleologie im Streite um Die Wissenschaft</u> (1904). He developed a new kind of ideology of science, which would make it possible to see Marxism both as a social science and as an ideology of the working class.

To understand Adler´s ideology of science we must relate his formulations to the intellectual milieu of fin-de-siècle Vienna, and to the Neo-Kantian and positivist philosophy. We also have to consider that the social sciences in Austria developed out of jurisprudentia.

It was necessary for Adler and his fellow Austro-Marxists to convince the academics of their time that Marxism was <u>the</u> Social science, and that Marx was the "Newton" or "Darwin" of society. But they also had to show that Marxism was a necessary element in the ideology of the Social Democrats. Here the discussion of the relationship between ethics and science was very important. Adler took the position that a scientific theory can tell people how to act and at the same time give them moral guidance in doing what they are doing.

The key concept in Adler´s ideology of science was the "Vergesellschaftete Mensch", which made Marxism to the basic social science.

John E. Chappell, Jr.

1212 Drake Circle, San Luis Obispo, California 93401, U.S.A.

EXTRA-SCIENTIFIC INFLUENCES ON PARADIGM FORMATION IN SOCIAL AND HUMAN SCIENCES: CULTURAL CONCEIT VERSUS ECOLOGICAL AWARENESS

Creation and acceptance of paradigms are often influenced by extra-scientific factors, such as conditions of life in society. This is especially true in social and human sciences. Varying relationships with environment have induced varying paradigms, in three stages. Social sciences are thus interpreted in terms of attitudes to environment.

(1) STAGE OF PRE-MODERN HUMILITY. Acceptance of natural limits prevailed, and was reflected in political theory by a tendency to accept regimes in power. Yet even in antiquity a thread of human determinism can be recognized (C. Glacken, Traces on the Rhodian Shore, 1967).

(2) STAGE OF EXPANDING DOMINANCE. As chances to expand exploitation of nature followed upon discovery of the New World (extensive expansion) and growth of modern science and technology (intensive expansion), static and limited paradigms gave way to those emphasizing change and human creativity. "Human exemptionalism" from natural limits characterizes such thought (R. Dunlap in Amer. Behav. Scientist, 1980 no. 1). Social Contract theories (Rousseau, etc.) saw society as an artifact made from a state of nature, subject to revolutionary change. Expanding environmental wealth led to less emphasis on moral virtue and more on wealth and property (J. Rodman, ibid.); thus economics emerged.

The substage of cultural conceit began to develop as an antithetical feature within dominance theories as the Industrial Revolution spread, and new lands abroad were first fully exploited; it has reached its fullest expression only recently. Cultural conceit tends to overestimate human powers in relation to (a) exploitation of nature; (b) ability to shape society; and (c) ability to create truth and value without being limited by real experience or logic (This latter tendency is found also in art, and especially in physical science; see my paper in Physics at this Congress). Perhaps its most dangerous manifestation is in growth-oriented capitalist economics and its accompanying political ideology, which claim capitalism creates wealth; it would be closer to the truth to say wealth creates capitalism. Even the ecologically-oriented physician-humanist R. DuBos practices cultural conceit when, in Celebrations of Life, 1981, he explains human interaction with the natural environment too exclusively in terms of culture and ingenuity, ignoring or misinterpreting applicable ideas on climatic causation in the writings of E. Huntington (Civilization and Climate, 1924).

(3) STAGE OF ECOLOGICAL AWARENESS. New ecological paradigms now vying for attention throughout the social and human sciences are not limited to mere acknowledgement of human-environment interactions, but go further in focusing on harmful side-effects from technology, and limits on resources. Beyond this, serious re-evaluation of once-popular ideas on climatic influences is needed, yet resisted even in geography (ch. by J. Chappell, Jr. in M. Harvey and B. Holly, eds., Themes in Geographic Thought, 1981). But steps in this direction are being made by some archaeologists; and by "climate and history" scholars, as exemplified by interdisciplinary conferences held in 1979 (arts. in J. of Interdisc. Hist., spring 1980; T. Wigley et al, eds., Climate and History, 1981).

Dong-Hyun KIM, Ph.D.
Professor & Chairman, Department of Public Administration
SUNG KYUN KWAN UNIVERSITY

COMPARATIVE ANALYSIS OF PERCEPTIONS OF THE ETHOS OF SCIENCE

The issues addressed in the research effort to be reported derive from the Mertonian tradition of the sociology of science. However, the present research is designed to question and to test the very basic premise of the Mertonian tradition of an undifferentiated and unchanging ethos (normative system) of science. It attempts to answer these questions: "Is the scientific ethos perceived as 'real' by scientists related to the conditions of the practice of science experienced by these scientists?" Or, "if individual scientists are exposed to different conditions of the practice of science, will they perceive adherence to the ethos of science differently?"

The focus of the present research is solely upon the ethos of science, or the moral norms of assumptions of science, as opposed to the technical norms of the scientific method. These moral norms and assumptions define appropriate patterns of interaction both among scientists themselves and between scientists and those who are representatives of the institutions of society.

Finally, the current research effort may be justified as "filling a gap in the literature" in that it attempts to explore a question which no other single piece of research has explored. It will examine scientists' perception of the entire package of the ethos of science - i.e., both the assumptions of purity, autonomy, and impartiality and the norms of universalism, organized skepticism, communism, and disinterestedness.

José María Infante
Professeur, Faculté de Philosophie
Universidad Autónoma de Nuevo León, México

LÉVI-STRAUSS ET LE STRUCTURALISME

À cause de la polémique très intensive qui le structuralisme a détaché dans le sein des sciences sociales, cette travail a de l'intention de placer exactement cette courant, et aussitôt le rôle de Claude Lévi-Strauss dans son développement. Il s'agit d'une discussion des idées par eux-mêmes et pas d'etiquettes ou bavardages sur les différences. Le travail présente les idées structuralistes dans le contexte d'histoire des idées, et il étude dans un moment posterieure le structuralisme linguistique et ses dérivations et influences postérieures, et ainsi que son extension au reste des sciences sociales.

Étant donné que, comme le travail l'essayera, la figure de Claude Lévi-Strauss occupe un lieu central dans tout ce développement, le travail contraste ses idées avec celles de ceux-là qui aussi ont reçu le surnom de structuralistes, d'une manière spécial ceux des Althusser, Barthes, Foucault et Lacan.

Le travail est centré sur le structuralisme comme notion et conception méthodologique et ne conteste ni analyse les concepts et théories structuralistes dans son extension aux sciences sociales.

Hc

Dr.Guy Ahonen, Swedish School of Economics and Business

Administration, Helsinki, Finland.

THE PARADIGM OF ECONOMICS: EXPLAINING RESPONSES TO CRITICISM

The paper aims at presenting some empirical evidence for the "realism-to-instrumentalism" thesis by employing the case of economics. The analysis assumes that sciences are structured like paradigms, which means - among else - that the validity of criticism to a certain degree is internally determined (i.e.relative). Accordingly, it supports the idea that criticism of sciences cannot be assessed by referring to analytical and empirical arguments only; also the paradigmatic state of development of the science criticized has to be considered.

For more than two centuries economics has been criticized for not having realistic implicit assumptions (recently e.g. Kaldor 1978, Balogh 1982), for not having an object of knowledge which is relavant for policy-making (Kristol 1981, Karmarck 1983), for not producing methodologically satisfactory theories and hypotheses (Blaug 1982), and so on. And the criticism claiming that economics is in crisis still goes on (Bell and Kristol 1981).

Strangely enough, the criticism does not seem to hit its target. The same arguments are frequently repeated without having practically any effect on the research made by the economists. One explanation to this is provided by Friedman's famous essay from 1953, which manifests the instrumentalist methodological position of economics. Though this position has been severely criticized (c.f. Frazer and Boland 1983), we argue that it contains the essence of how economics 'immunizes' itself from criticism. We do not agree with Friedman's conviction that instrumentalism is a quality generally possessed by sciences. Actually we do not accept at all the unitary science idea, which Friedman's essay seems to assume. But we do argue that sciences in certain phases of their development seem to tend to promote an instrumental mode of reasoning, having originally been based on a realistic type of argumentation. Making this point we give some support to the paradigm-theory of sciences, generated by Kuhn (1962). We employ especially the notion of normal scientific and internalistic determination of research within sciences. The idea of sciences developing from realistic to instrumental modes of reasoning resembles Brante's(1981) interpretation of Kuhn's paradigm-theory, according to which sciences are originally externally determined and subsequently become more and more internally determined.

Hc

Michael M. Sokal, Worcester Polytechnic Institute

Worcester, Massachusetts 01609 USA

LIFE-SPAN DEVELOPMENTAL PSYCHOLOGY AND THE HISTORY OF SCIENCE

This paper argues that recent scholarship in life-span developmental psychology can, when focused precisely, help historians of science interested in biography better understand many aspects of their subjects' lives and careers, including the origins of their scientific ideas and the institutions within which they worked. In particular, in using some of the concepts developed by Daniel J. Levinson in The Seasons of a Man's Life (Knopf, 1978) to illuminate the science and organizational work of James McKeen Cattell -- distinguished experimental psychologist, creator of psychological tests, and long-time (1894-1944) owner and editor of Science -- this paper offers a case study of the development of science as the interplay of an individual's personal life, his ideas, and the institutions with which he was associated. It concludes that historians of science can find in this recent psychological research a valuable analytical tool.

In analyzing these developments, this paper demonstrates the direct influence of the circumstances of an individual's life on the evolution of his scientific ideas. Cattell, for example, was raised in a particular family setting and social milieu, both of which helped determine his personal interests and his view of the world around him. Levinson's ideas shed light on how this background helped focus Cattell's attention on a specific set of philosophical concerns and helped steer him towards particular influential scientists. These ideas and individuals did much to shape Cattell's scientific ideology, and thus his scientific work. In fact, Cattell's major scientific activity -- his development of a psychology that could be applied through his tests (in the 1890s) and his Psychological Corporation (in the 1920s) -- sharply reflected the philosophical ideas that his family concerns had led him to in the 1870s.

Furthermore, this paper argues that, in providing a framework of understanding for the development of an individual's approach to those around him, Levinson's ideas do much to illuminate the ways in which his personal life can influence the course of scientific institutions. That is, as Cattell's style of dealing with others evolved through time in response to the changing circumstances of his life, so too did the journals he controlled. As a result, the emergence of Science as the most influential early-20th-century general American scientific periodical can not be grasped fully unless it is seen (in part) as the articulation of Cattell's own personal development. Consequently, this paper concludes that the ideas of life-span developmental psychology have much untapped potential importance for historians of science.

Jan Barmark

Ass. Prof, Theory of Science, University of Gothenburgh

Eastern Influences into Western Psychotherapy: From Poul Bjerre to Arthur Deikman

What kind of practice and science is psychotherapy? What should a paradigm for psychotherapy look like. Ther has been a lot of answers to questions such as these in the history of psychotherapy. Some psychotherapists have gone outside the domain of science in order to get ideas from philosophy and religion, much as a reaction to a dominating positivist paradigm in science. Some of these psychotherapists have found a rich soil of insights about the human mind in the eastern philosophies. We will start our historical investigation with the Swedish innovator Poul Bjerre (1876-1964) who in many respects was an early forerunner of humanistic psychology, and continue with C.G. Jung, Erich Fromm, Abraham Maslow up to the contemporary discussions between psychotherpists such as A. Deikman and buddhists such as Tarthand Tulku and Chogyam Trungpa. Poul Bjerre takes his stand in the ideal of man to be found in the Chinese philosophy, C.G. Jung sees the many simmilarities between his practical experiences as a psychotherapist and the Chinese image of man. Erich Fromm writes about zen-buddhism and psychoanalysis as two ways leading to mental health. A. Maslow finds in eastern philosophy inspiration for his criticism against positivistic ideals of science and calls his own methodological alternative for a "taoistic" science. We will follow up this thematic discussion (using Gerald Holtons term) using empirical material from the history of psychotherapy and cathegories from the theory of science. Our topic has not only a historical interest but is also relevant for todays discussions about psychotherapyresearch. A central topic here is ethnopsychiatry. That means comparative studies of knowledge production about the human mind in different cultures and times.

David E. Leary

Assoc Prof of Psychology & the Humanities, Univ. of New Hampshire

EXCHANGING THE COINS OF THE REALM: THE METAPHORICAL AND RHETORICAL CONSTRUCTION OF THE NEW PSYCHOLOGY IN AMERICA

Over the past several decades, there has been a notable rise of interest in the role of metaphor in the construction of scientific knowledge. More recently, the role of rhetoric in the scientific enterprise has begun to attract serious attention.

This paper will discuss the fundamental role played by metaphor and rhetoric in the construction and establishment of scientific psychology in America between approximately 1880 and 1920. The central character in this story will be William James, who "figured out" his influential psychology--as well as his intimately related philosophy--on the basis of explicitly metaphorical conceptual categories. In addition, he played a fundamental role in arguing the so-called New Psychology into existence, on both the intellectual and institutional levels. That is, he not only convinced himself and others of the relevance and potential of the new conceptual framework that he represented, he also persuaded administrators and trustees that the new scientific psychology deserved a place in the academic curriculum. Without such a place, the new science would have had little ground for growth.

What makes James a particularly salient figure in this story is that he was so explicitly self-conscious about what he was doing. His metaphors were not the product of mere verbal facility or unintended conceptual sloppiness. Metaphors, he claimed, were the only means of advancing thought in any scientific (or non-scientific) realm. In addition, he won the endorsement of the Harvard administration by virtue of strategically devised arguments about the need to stem the flood of materialist science, not on the basis of the scientific merit of the New Psychology.

(To explicate the metaphorical title of the talk: Metaphors are coins, used to gain a purchase on reality (Nietzsche). Rhetoric is the exchanging of such coins. The New Psychology proposed a new economic system, trading in old coins for new. James argued for the new economy in terms of its superior cash-value, particularly as regards the conservation of spiritual and moral reality.)

Jack D. Pressman, University of Pennsylvania, Philadelphia, USA

INNER TENSIONS: CONFLICTING VIEWS OF SCIENTIFIC
PSYCHIATRY IN AMERICA, 1935 TO 1960

In the mid-1930s, psychiatry in America was ridiculed as the most backward of medical specialties, a clinical art devoid of experimentally derived knowledge. Leaders of the profession, foundation officials, and prominent researchers across a wide variety of fields argued that the one path to progress in the theory and treatment of mental disorders lay in building a truly scientific foundation. Although the psychiatric profession in America at this time was split into widely differing schools of thought, ranging from somaticists to Freudians, a shared conception of the individual as a psychobiological organism helped unite these disparate groups into a united reform campaign.

By the early 1950s, scientific psychiatry appeared to be on a firm footing: many states had built neuropsychiatric research institutes; massive federal funding for research was available; mental hospitals were brought into closer contact with university medical schools; and the profession as a whole had been greatly expanded through training programs. The introduction of the major tranquilizers in the mid-1950s, which dramatically changed the institutional care of the mentally ill, was viewed as the capstone of what the new experimentally based psychiatry could accomplish.

However, at the same time as this "infrastructure" was being erected, the tenuous alliance between the somaticists and the analysts was disintegrating, with the two camps now promoting incomensurable views of what scientific psychiatry constituted. Underlying these heated discussions, which led to upheavals within the American Psychiatric Association, were varying conceptions of institutional and professional service roles as well as purely intellectual considerations. The dimensions of this debate has had lasting consequences for a profession which has since faced even greater scrutiny concerning its "scientific" credentials.

Fernando Vidal

University of Geneva, Switzerland

JEAN PIAGET. A BIOGRAPHICAL CASE-STUDY OF THE IMPACT OF
PROTESTANTISM ON PSYCHOLOGY.

Whenever he referred to his own development, Swiss psychologist and epistemologist Jean Piaget (1896-1980) minimized the importance of his religious concerns and experiences. It is possible, however, to show that religion played a key role in his development, that Piaget can be justly considered as a religious thinker, and that he even became a spokesman for liberal Protestantism against the dogmatic revival that followed World War I.

After having been active as a naturalist, Piaget turned into a "metaphysician and theologian" (as he described himself in 1917). In his 1915 philosophical poem *The Mission of the Idea*, he integrated the themes of Christian socialism into a metaphysical system of Plotinian overtones. Recently discovered documents reveal that *The Mission* was created within a Christian students movement animated by the liberal Protestant values elaborated in the second half of the nineteenth century. They also show that Piaget's system corresponded so much to the questions and ideals of this movement, that he was celebrated as a model of the Christian youth supposed to build a new world on the ruins left by the War. Imbued with the theological, social, and philosophical problems of a movement traditionally open to rational analysis, Piaget elaborated a project for a scientific psychology of values.

This was among Piaget's first projects in psychology, and it is manifest in many of his writings up to the 1930s, including his first books on child development (language and thought, 1923; judgment and reasoning, 1924; conception of the world, 1926; physical causality, 1927; moral judgment, 1932). Piaget proposed to study the logic of value judgments. Although considering all values equal from a subjective point of view, he claimed that his science of values could establish the "psychological or biological superiority" of some of them, by determining a developmental hierarchy. In 1928, for example, Piaget publicly attributed this type of superiority to the liberal Protestant preference for an immanentist conception of God, *versus* the transcendentalism of the orthodox revival headed by Swiss theologian Karl Barth.

Piaget's biography illustrates how closely connected with developmental psychology a religious *problématique* could be. Piaget, however, was (with C.G. Jung, for example) one of the last psychologists whose work directly bears the stamp of liberal Protestantism. For a deeper understanding of the impact of this movement on the history of psychology, one must also turn to such figures as W. James and G.S. Hall in the US, and to Piaget's predecessors and colleagues T. Flournoy, P. Bovet, and E. Claparède in Switzerland.

Per Sörbom

Assoc. Professor, Uppsala University, Uppsala, Sweden

The Diffusion of Theories of Diffusion:
Exemplified by the Mariner´s Compass in Western Literature.

The earliest description known to us today of how to make a mariner´s compass, as well as of how it is used for navigation purposes, is to be found in Chinese literature. These facts have led such eminent scholars as Joseph Needham and Lynn White, Jr. to draw the conclusion that the mariner´s compass was invented in China and that it spread from there to the rest of the world. An investigation of the sources and arguments used by advocates of such a theory of diffusion does show, though, that there are no positive proofs of a spread from China to the West.

The purpose of this paper, which is a report from a work in progress, is not to take up arms in that old and nowadays only sporadically fought battle between diffusionists, isolationists, and other adversaries in the anthropological, ethnographical and historical fields. The question of when, and where, the mariner´s compass was invented, and whence and how it spread, is of no real concern here. The compass is taken as an example, as gunpowder and printing will be used in the same way further on, of the attitudes Western authors and scholars have taken towards different claims of priority regarding important inventions and the possibility of a diffusion from their place of origin to the rest of the world, as well as what is thought of diffusion as such.

The survey of Western literature from the 17th into the 20th where the origin and the spread of the compass is discussed sheds light on how European attitudes towards other cultural spheres have changed over the centuries: from incredulous fascination, to enthusiastic appreciation of the Enlightenment, to a depreciation during the age of imperialism, which still can be traced in our time. Joseph Needham´s long, and at certain points debatable list of inventions that he maintains have spread from China to the West, must be seen in the light of his desire to counter such lingering 19th century attitudes. At one time much cherished theories of once great peoples such as the Celts and the Goths, as well as badly disguised chauvinism does naturally influence many authors.

No real body of theories of diffusion in a modern scientific sense can be discerned until anthropology, ethnography, and archaeology came of age in the late 19th century. Diffusionism has appeared in a great many varieties, but all diffusionists start from the assumption that man, basically, is non-creative and that civilization must be seen as a result of the spread of a limited number of inventions.

Most anthropologists today show no interest in diffusionism, but in other fields it is still much in vogue and is often used uncritically. The idea of diffusion is taken over wholesale by historians from different fields, but they seldom seem to be aware of the complexity of the various theories of diffusion they use.

Anne S. Fernald

Master's Candidate, University of Colorado, Boulder, Colorado, U.S.A.

THE CHANGING MOSAIC: A LAND USE HISTORY OF THE BABACOMARI CIENAGA

European explorers and invaders discovered an inhabited land (Jennings, 1975; Cronon, 1983). During the last 1000 years the Babacomari Cienaga, Santa Cruz County, Arizona, has been the site of recurrent human activity. The analysis of archaeological records and archival materials containing references to the Babacomari Cienaga use in the 18th century indicate prior settlement by the Sobaipuris Indians in the 13th and 14th centuries. Comparison of photographs taken in the late 1800's with photographs of many of the same scenes taken in the 1980's indicates the magnitude of disturbance and the relative vegetational change. The analysis of current vegetational composition and species distribution reflect varying degrees of disturbance relative to the type of land use. The population structure of the cienaga reflects ongoing changes in species composition triggered by disturbance initiated nearly 1000 years ago. The major pre-Spanish disturbances include agriculture, fire, and flood. Post-Spanish disturbances include agriculture, automobiles, cattle grazing, erosion, fire, and flood.

Solveig Widén

Åbo Akademi library, Åbo, Finland

Hilma Granqvist - a female sociologist with pioneering ideas

Hilma Granqvist, who lived between 1890 and 1972, was a Swedish sociologist in Finland. She defended her doctoral thesis on 13th January 1932, being a pupil of the world-renowned sociologist, professor Edvard Westermarck of Åbo Akademi. At Åbo Akademi, which is the university of the Swedish-speaking population of Finland, founded in 1918, she was the second female Doctor of Philosophy.
The doctoral thesis of Hilma Granqvist is entitled: Marriage Conditions in a Palestinian Village, I. It is based on notes made during many years of research in the village of Artas not far from Bethlehem.
During her 82 years of life Hilma Granqvist often faced opposition at home in Finland and had to contend with economic difficulties. However, as early as in the 1920's her energy and drive took her as far as to Palestine of that time, where for many years she collected extensive material, which in addition to the doctoral thesis resulted in four other works written in English on the conditions of the Palestine woman. They are entitled: Marriage Conditions in a Palestinian Village, II, 1935; Birth and Childhood among the Arabs, 1950; Child problems among the Arabs, 1950; Muslim Death and Burial, 1965.
International experts judge Hilma Granqvist as an outstanding sociologist with pioneering ideas.
At the death of Hilma Granqvist a large portion of her literary remains came into the possession of the Library of Åbo Akademi, while part of them, e.g. her large collection of photographs, were taken over by the Palestine Exploration Fund in London on the initiative of the English anthropologist Shelag Weir. These photographs have been published by Karen Seeger in the book Portrait of a Palestinian Village: the Photographs of Hilma Granqvist.
Hilma Granqvist´s collection in the charge of the Åbo Akademi library comprises letters, diaries, newspaper cuttings, drafts for and the manuscripts of both published and unpublished works, essays, articles, and lectures. The collection of letters is especially valuable, as Hilma Granqvist often preserved the copies of the letters she wrote. They as well as the diary notes - often extremely detailed - give a valuable picture of the difficulties and problems of e female researcher only some decades ago not least in economic respects. The manuscripts and the drafts reveal in an interesting way with what exactness Hilma Granqvist worked on her publications and the notes bear witness to the close contact she had with the Arabs.

Stephen J. Cross

Doctoral Candidate, The Johns Hopkins University, Baltimore, USA

THE NEW WORLD OF SCIENCE: SOCIAL SCIENCE AND THE RESEARCH IDEAL IN 1920s AMERICA

This paper discusses the institutional remaking of social science in 1920s America. It adopts the framework of an "organizational" interpretation of the decade.
The professional social sciences arose in the late nineteenth century under the joint impulse of social reform and academic professionalization. WWI marks a crucial point of transformation for both of these complexly related contexts. For the organizers of progressive reform the legacy of the war was predominantly one of frustration and decline. By contrast, national mobilization for war provided pre-war modernizers of the institutions of American science with substantial opportunity to pursue their goals, and valuable experience of bureaucratic innovation. The development of social science in the 1920s manifests both of these contrasting legacies. The social sciences offered a terrain for the realization of earlier dreams of social reconstruction through novel institutional strategies.
This paper attempts to characterize the main features of the institutional growth of social science after WWI. Of principal importance was the participation of activist social science influentials and their foundation patrons in voluntary associational activities of the type that characterize the decade. In the social sciences professional association was seen above all as an institutional mechanism to promote interdisciplinary cooperation. Interdisciplinary research promised a break with the received formalisms of the traditional academic disciplines: it was "realistic" research, offering both genuinely scientific knowledge and a focus on concrete social problems.
Underlying this interdisciplinary strategy was the ideal of basic research. With their colleagues in the natural sciences, social scientists linked their professional identity to the research ideal by valuing production of fundamental scientific knowledge above the immediate application of science to practical problems. The goal of providing a rational, guiding social intelligence was by no means abandoned; indeed the rhetoric of professional programmatics in 1920s social science is rife with scientistic hubris. But the production of technologies of social control (to use the language of the day) now took the mediated form of a long march through the institutions of scientific research.
Though social scientists after WWI did not make the world anew as the reconstructionists hoped, they remade the institutional framework of their field, and laid lasting foundations for the modern world of science. And a subsequent question is raised as to how the anti-formalist theories of the new social science served to recast the world of experience itself within the categories of modernity.

Mei Rongzhao
Vice-professor. Institute of History of Natural Science, Academia Sinica, Beijing, China.

THE LOGIC IN THE MO JING AND ITS INFLUENCE UPON THE ANCIENT CHINESE MATHEMATICS

Mo Jing, the classic book of Mohist School which appeared during A Hundred Schools of Thought Contend of Warring States (457-221 B.C.), deals systematically with logic and gives a convincing analysis of the problem of definition and the topics of the same and the difference, the infinite and the finite, the motion and the static, and so on. This logic exerted great in--fluence upon the ancient Chinese mathematics. This paper in--troduces briefly the logic of Mo Jing and enumerates fifteen mathematical definitions and mathematical topics in it and also explains that Liu Hui's theoty of mathematics (3 A.D.) inherit--ed and developped the logic of Mo Jing in the words, the method of demonstration and the mathematical content.

(Joseph) Cheng-Yih Chen

Professor of Physics, University of California, San Diego

ON THE DATING OF THE OLDEST EXTANT ILLUSTRATION OF A CHINESE ABACUS

An illustration of an abacus is found in an undated Chinese illustrated primer, "Hsin-Pien Tui-Hsiung Saû-Yen" 新編对相四言 Newly Compiled Illustrated Four-Word Primer, acquired in 1941 by the East Asian Library of Columbia University. A one-time owner, Chu Po-Tzǔ 祝伯子, in his handwritten note at the end, dated 1922, considered the primer to be an exact copy of the work of the Sung period (960-1279) for in one place the character 匡 in k'uang 筐 (basket) is written with one stroke less as 筐 in respect to the taboo for the character of the First Sung Emperor Chao K'uang-Yin 趙匡胤 (927-976). In describing a copy of this book owned by the Library of Congress, A.W. Hummel wrote in the Library's Quarterly Journal (1946) that: "This Chinese reader is an early sixteenth century reprint of an edition that appeared in 1436. Far older Chinese primers exist, but none with illustrations." There is no explanation how these dates were arrived. In favoring the argument that abacus was only just becoming popular in China by the time of Ming, L.C. Goodrich considered the date 1436 to be reasonable (ISIS 39, 239, 1948). This dating is accepted for the earliest illustration of a Chinese abacus by J. Needham (Science and Civilization in China, vol. IV, 1959) and by Wang T'ing-Hsien 王庭顯 (Chu-Suan Hseŭn 珠算学, 1962). Not until 1977, Chang Chih-Kung 張志公 argued (Wen Wu 文物 11, 1977), based on an identical copy of the primer that he found in 1965 in China, that the primer is an early Yuan edition of Southern Sung work.

Since then, a Ming copy of the primer, "K'uei-Pen Tui-Hsiang Ssŭ-Yen" 魁本对相四言 Best Edition of the Illustrated Four-Word Primer dated 1371 has surfaced in Japan, permitting a direct comparison. The primer of 1371 with two additional entries in the illustrations is apparently a later edition of the undated primer. The new illustration for the abacus has ten parallel columns, one more than that of the undated primer. Based on the comparative analysis of the two primers, it is probable that the illustration for the abacus first appeared in the original version of the primer of the Southern Sung (1127 to 1279). This dating is also consistent with recent studies on the evolution of mathematic rhymes devised to facilitate the recall of arithmatic rules and to speed up calculations as the art of arithmetics was transferred from the counting rods to the abacus from late T'ang to the Sung times.

Dr. Sachidanand Das

* Visiting Fellow, Department of Physics, the University of Birmingham, P.O. Box 363, Birmingham B15 2TT, UK.

ANCIENT INDIAN VEDIC MATHEMATICS AND ALGEBRAIC COMPUTATIONS

Solutions of linear algebraic equations (in one, two and three variables) and sums of certain types of fraction-additions have been obtained using the system of Vedic Mathematics (Bhāratī Kṛiṣna Tīrthajī Mahārāja, Vedic Mathematics or Sixteen Simple Mathematical Formulae from the Vedas, 1965). The unique features of this novel axiomatic approach to mathematical computations of which Vedic sutras form the basis are described elsewhere, and the possibility of its computer applications is shown therein (Das, Computational Algorithms and Vedic Mathematics, 1980).

In the present discourse, we show how solutions of certain types of equations can be arrived at more simply and faster using certain formulae from Vedic Mathematics. For this, five algorithms in the form of four sutras (aphorisms or theorems or formulae) and one sub-sutra (corollary) have been used. The four formulae are: Paravartya Yojayet ("Transpose and adjust"), Anurupye Shunyamanyat ("If one is in ratio, the other one is zero"), Sankalana-vyavakalanābhyām ("By addition and by subtraction"), and Sopantyadv-ayamantyam ("The ultimate and twice the penultimate"). The corollary is Antyayoreva ("Only the last terms").

The cryptic meanings of all the above sutras and the sub-sutra have been explained, and their modus operandi when applied to the solution of algebraic equations illustrated by means of examples. A comparison has been made of the calculational time between Vedic methods and the theoretical methods that are in use at present in school and college algebra. It has been shown that for the solution of the same set of problems, the time taken by the former can be a half, a third or even a tenth of the time taken by the latter. Other drawbacks of the conventional methods versus Vedic methods while performing numerical computations are also pointed out. The many applications of the sutra Paravartya Yojayet in the present context has shown how the cryptic nature of Vedic sutras can lend them to many interpretations.

While usefulness of Vedic Mathematics as a simple, compact and fast computational tool is stressed, attention is also drawn to its intellectual side. It is suggested that Vedic Mathematics can form part of a course on numerical analysis.

* Permanent address: Scientific Officer (SE), Bhabha Atomic Research Centre, Neutron Physics Division, Trombay, Bombay-400085, India.

S. D. SHARMA

Reader in Physics, Dept. of Physics, Punjabi University, Patiala, India

Old Indian Methods for Solving Indeterminate Equations of Second Degree

The indeterminate equations of first degree (the Kuttaka equations) were solved for integer variables by Aryabhata (499 AD) using Valli method. (Valli being an array of coefficients A_js in $S_{n+1} = A_n S_n + S_{n-1}$. In the Kuttaka equation $(ax+b)/c = Y$, Valli is obtained from continued fraction of a/c. The Pell-equation (Vargaprakrti equation) of the form $Nx^2 + A = y^2$ (N and A being integers) was solved first by Brahmagupta (7 AD) for integers x and y. Later, for A=1, cyclic method (Cakravala) was developed by Jaideva (10th AD) and Bhaskara II (11th AD). The Cakravala method amkes use of the techniques of Brahmagupta and Aryabhata simultaneously in a cyclic way. Later on this equation was soved by Bapudeva Shastri (19th AD) using a generalized version of Valli in that very specific Indian style.

It may be added that for the equation $Nx^2 - 1 = y^2$, Bhaskara arrived at the result that N has to be expressible as sum of two squares for integer solutions to exist. This result was later arrived at by Deron Pale in the 18th Century AD. How Bhaskara might have arrived at the result, is discussed in this paper.

Besides this, the Bhavita equation of the type $ax+by+c = xy$, (a,b,c being integers) was solved by Bhaskara II for integer x and y using geometrical and indeterminate algebraic techniques. This equation was later solved by Lagrange. This paper exposed how Valli method can be used to solve this equation. In general consistent advancements in Valli methods and their applications are reviewed.

Mogens Esrom Larsen

lektor, Københavns universitets matematiske institut

The theory of proportions by Archimedes and Euclid.

In his formulation of Eudoxos' theory of proportions in Book 5 of "The Elements", Euclid missed a point: Until theorem 16 (included) the theory does not require the existence of a difference btween two magnitudes of the same kind. Archimedes was the first to recognize and exploit this fact. In his treatise "The Sphere and the Cylinder", Archimedes reformulates the Euclidean axiom as follows: "One of two different magnitudes of the same kind exceeds the other with (at least) so much, which added to itself can exceed any of the two." In our language:

If $A < B$, then there exists C, such that

1) $A + C < B$

2) $B < n \cdot C$ for some integer n.

This is the very basis for comparing curves with straight lines and a sphere with a plane disc.

So, when we talk about the Archimedean axiom of order, it is (as usual) slightly misleading. We think of an axiom due to Eudoxos found in Euclid's Elements, while Archimedes based his work on a much more general assumption.

Wilbur Knorr, Program in History of Science

Associate Professor, Stanford University, Stanford, Calif. 94305

THE METHOD OF INDIVISIBLES IN ANCIENT GEOMETRY

Archimedes' "mechanical method", as illustrated in his Method, makes prominent use of indivisibles. Less well known are examples of the same conception in discussions by Hero and Theon of Alexandria. Further, Dionysodorus' rule for solids of revolution is likely to have been derived through an analysis based on indivisibles comparable to Archimedes', while Pappus' measurements of the Archimedean and spherical spirals turn on an ingenious adaptation of the usual manner.

It is clear that the method of indivisibles was not an isolated aspect of Archimedes' work, even if the other examples cited can all be associated with him. In alternative proofs in Quadrature of the Parabola and Conoids and Spheroids, based on proofs in the Method, Archimedes removes the indivisibles. But the Eudoxean "exhaustion" method, with its elaboration by Archimedes, renders virtually trivial the elimination of such infinitist assumptions. Thus, even if indivisibles would not be admitted into formally rigorous proofs, ancient geometers hardly viewed them with "horror." We may well suppose that within a less formal stratum of the ancient geometry, now only sparsely represented in our sources, indivisibles appeared as a valuable heuristic method.

The notion of constituting figures as aggregates of indivisibles formed by parallel sectioning appears not to be a novel feature with Archimedes. His account in Meth. (pref.) suggests a precedent in the earlier Eudoxean geometry, and this can be supported from the "cone fragment" of Democritus. One can argue that an indivisibilist treatment of the cone, rigorized by Eudoxus, could provide Archimedes his model, while the alternative treatment in Euclid's Elem. XII 5 would represent an effort to improve the earlier treatment.

The Legacy of the Ancient Method

Certain Chinese applications of indivisibles in the 3rd cent. A.D. can be separated from any known Greek treatment, despite the close conceptual ties with our passages from Hero and Theon.

Cavalieri's use of indivisibles may be viewed as an elaboration based on the forma efforts of Archimedes. Aspects of terminology indicated an explicit carry-over, while the rigorous finitist method in Archimedes naturally suggests, as a simplification, a preliminary form using indivisibles. In effect, the 17th-cent. analysis of indivisibles can be seen as retrieving the heuristic base on which Eudoxus and Archimedes had built their formal "exhaustion" versions.

DR A.K. BAG

Head, History of Science Division, Indian National Science Academy, New Delhi, India.

The Knowledge of Calculus in India in the Pre-Newtonian Period.

The tradition of observing fast and various religious rites by the Hindus in India on the occasion of eclipses till today indicates that the accurate prediction of time and duration of eclipses beforehand was considered to be of great national importance in ancient and medieval times. In problems of this nature, it was essential to determine the instantaneous motion (<u>tat-kālika-gati</u>) of planets or stars at any particular instant. Āryabhaṭa (476), Brahmagupta (628), Āryabhaṭa II (950), Bhāskara II (1150) and others in their attempt to find instantaneous motion of moving objects have given many results which, in modern differential notations, can be written in the form like:

$$\delta(\sin x) = \cos x \, \delta x$$
$$\delta(\cos x) = -\sin x \, \delta x$$

Some glimpses of the Rolle's Theorem and the Mean Value Theorem of Differential Calculus are also noticeable.

The application of an idea of infinitesimal (integral) calculus has also been successfully applied in finding the surface, volume of sphere and the value of 'π'. The result, though it does not give limit of a sum as it is at present, is undoubtedly the nearest approach to the method of integral calculus. The results as dealt by Indian scholars will be discussed in detail in the original paper to emphasize its importance in history of mathematics.

Tetsuo Tsuji

Professor, Tokai University, Hiratsuka, Japan

CARACTERISTICS OF JAPANESE MATHEMATICAL METHOD IN THE 17TH CENTURY

In Japan the lord Tokugawa subjugated the whole country and established the powerful feudal government in the beginning of the 17th century. Afterward, the domestic industry and commerce were getting more and more flourished in a peace time and the learned activities of citizen became energetic. In particular, medicine, natural history and mathematical learning made great progress. Japanese people studied, as fast as they could, the learned knowledge accumulated in China by that time and endeavored to develop some new method for themselves.

One of the most prominent scholar in a domain of mathematical learning was Takakazu Seki (1642? -1708), who raised rapidly the level of Japanese mathematics. His great achivements covered a wide field of mathematics: 1) the invention of characteristic expressions of equations similar to the Western algebra, 2) how to solve algebraic equations with numerical coefficients like one called the Horner's method in the West and its completion, 3) the discovery of the determinant, concerned with simultaneous equations, 4) how to get the total sum of various series, and so on.

For the first time in the Orient, Seki changed the method of calculation from a use of computing rods <u>Sangi</u> on the board <u>Samban</u> to a operation of written figures on paper. He did not know any method of expression by means of mathematical symbols in the Western algebra. Seki used only chinese numerals and characters. Accordingly, he did not express an algebraic equation with an unknown quantity x but always dealt with just series of numerical coefficients, whose arrangement served as an equation. Moreover, he devoted his mental power to the solution of every difficult equation since he had no interest in the mathematical theory of number and the logical framework like Euclidian geometry, without knowledge of the Western mathematics. After all Seki's method of calculation became very complicated.

I would like to discuss in detail how Seki operated logically the calculating process with his unique method. (Feb. 22, 1985)

Ryousuke Nagaoka

Associate professor, Tsuda College, Tokyo, Japan

Another Look at the Rigorization of Analysis

    Synthetic geometry in Greek style dominated as a model of rigor in mathematics until late in 17th century, but algebraic symbolism gradually showed surprising power of discovery and infinitesimal calculus, armed with algebraic symbolism, therefore qualified as analytic, made a great advance during less than 150 years without adequate foundations. The calculus of variations by Lagrange marked the climax of this fertile combination of calculus with algebra.
    To give calculus logical foundation was achieved in the forth quarter of the previous century. From then on, calculus has rapidly been a most well founded branch of mathematics and sophisticated treatments on infinities and limits, which, based on inequality arguments, dispenses with logically problematic concepts, has now become even a model of mathematical rigor. And Cauchy is often said as a first mathematician who took a step forward of this kind of foundation. So he is considered sometimes as a hero who eliminated ambiguity in calculus and sometimes as a ringleader who made calculus so complicated and so far apart from natural intuition.
    A little examination of his text shows, however, that his original motivation for rigorization should no more be identified with the one of later so-called arithmetization movement than d'Alembert's or Lagrange's. Let me quote an example. It is well known that Cauchy presented his famous criterion of a series to converge, but that he proved only its necessity. Logicall speaking, he could not prove the sufficiency, because this is nothing but the completeness of the set of reals, so some definition of reals is indispensable, which Cauchy lacked. But the definition later established by Cantor and others is reduced finally to mere a phrase: "a rational series satisfying the Cauchy condition defines a real." And Cauchy himself composed this sort of arguments in case of his definition of a definite integral. So it is likely that Cauchy did not prove the sufficiency because he did not feel obliged to prove. And the aim of the criterion was to exclude divergent series and to put calculus on a sound base for further development of the theory: "a divergent series does not have a sum." (Cours d'analyse) Cauchy's rigorization is thus a criticism against too much free appeal to algebraic formalism, and is a little different from the one affected by philosophical inclination to put mathematics formally on purely logical grounds.

MAIERU' LUIGI

Professor of Mathematics - Liceo Scientifico 'G.SCORZA' di COSENZA - ITALY

LE 'NARBONENSIS' et LES AUTEURS ITALIENS DU XVIe SIECLE.

Dans le contexte des études sur les miroirs ardents se pose - comme suite - le merveilleux problème de la convergence d'une ligne et d'une droite sur un plan,même dans le cas où les deux lignes sont prolongées à l'infini. Vers 1550 paraîssent deux petits traités: l'un d'Antoine GOGAVA (1548), l'autre d'Oronce FINE'(1551); entre ces deux publications se place la traduction italienne d'un petit traité hébreu ayant comme sujet seulement le merveilleux problème dont sa construction diffère de tout ce qu'on avait fait jusqu'à ce moment-là.L'ouvrage est publié à Mantou en 1550: je l'appelle 'Narbonensis' parce qu'il a été attribué à Mosé de Narbonne(vécu au XIVe siècle).Le petit traité ne passe pas inaperçu. Parmi les auteurs italiens, c'est Jérome CARDANO qui,le premier, lui prête attention dans son ouvrage DE SUBTILITATE(de la troisième édition de 1554 jusqu'aux suivantes);puis c'est François BAROZZI qui en fait una argumentation spécifique(en 1586).CLAVIUS le rappelle dans son commentaire aux Eléments d'Euclide (à partir de la deuxième édition de 1589) et Pierre Antoine CATALDI s'en inspire pour son traité 'des lignes équidistantes et non équidistantes'(de 1603-4).
Les auteurs cités jusqu'ici sont d'importants points de repère pou le contrôle d'une bonne connaissance du Narbonensis, mais on peut le lire aussi à la lumière des Eléments Coniques d'Apollonius (où sa construction est présentée dans l'édition juive et non pas dans l'édition italienne). Dans ce cas le Narbonensis ne présente rien à remarquer,et même il entre,en explication,dans la présentation de la XIVe Proposition du deuxième livre des Eléments Coniques d'Apollonius (soit selon la traduction de Commandino de 1566, soit dans la traduction faite par Maurolico).La lecture que J.CARDANO fait du Narbonensis,tout en entrant dans la présentation des propriétés des sections coniques, est telle qu'elle mérite une attention différente:en effet,elle représente - à ses yeux -quelque chose de nouveau dans le domaine des Eléments Coniques. On trouve une confrontation de ces données même dans l'étude qu'en fait Jacques PELETIER en 1556 et en 1563,dans le sillage de Cardano. Pour BAROZZI,c'est une attention 'essentielle' que il faut dédier à notre petit traité:celui-ci devient le point de départ d'une adéquate manière de traiter le problème dans le contexte des Eléments Coniques.En même temps, parmi les mathématiciens,on lui donne une place de premier plan. Apparemment on saisit seulement un aperçu dans l'ouvrage de CLAVIUS, mais si l'on suit sa démonstration du cinquième postulat d'Euclide,on s'aperçoit que ce ne pas secondaire l'argumentation sur la convergence distincte de l'incidence entre lignes droites. Ce n'est pas négliger l'attention que Cataldi prête à ce problème dans l'étude spécifique sur le lignes équidistantes et non équidistantes (voir notamment l'appendice à son petit ouvrage dans lequel on saisit l'importance qu'il donne à notre problème).
Il faut remarquer;enfin,que pour une étude scrupuleuse et pour l'élaboration du problème,c'est d'une certaine consistence soit la mise au point de ce qui signifie 'faire de la géométrie', soit la recherche d'une terminologie philologiquement exacte.

J.V. Field

Science Museum, London, U.K.

Linear Perspective and the Geometry of Girard Desargues

Desargues has earned his place in histories of mathematics chiefly for his invention of what is now called projective geometry (<u>Brouillon project d'une atteinte aux evenemens des rencontres du cone avec un plan</u>, Paris 1639). However, he earned his living not as a pure mathematician but as a practitioner of the (by then) mathematised crafts of military engineering and architecture. Apart from the treatise on conics, all Desargues' mathematical works belong to the practical tradition. The two published before the treatise on conics are a page on music (in Mersenne's <u>Harmonie Universelle</u>, Paris 1636) and a short work on linear perspective (Paris 1636). Constructing a perspective image is a form of conical projection, and although Desargues' work is mainly concerned with a worked example, its final paragraphs do state some general results (without proof). These give strictly analogous treatments of sets of parallel lines and lines that converge to a point. They thus foreshadow the projective treatment of pencils in the treatise of 1639. Desargues' geometry must, of course, be seen as a development of ideas found in ancient geometrical works, such as those of Euclid, Apollonius and Pappus, but his distinctive innovation, in relating figures one to another through the apex of the cone, seems to be derived from the perspective (optical) tradition. It is Desargues' relation to this practical tradition that will be discussed in my paper.

Albert Dou
Seminario de Historia de las Ciencias. Universidad Autónoma de Barcelona. Bellaterra (Barcelona). Spain.

LOCAL THEOREMS IN SACCHERI'S GEOMETRY (1733)

How much rigorous is Saccheri in his <u>Euclides ab omni naevo vindicatus</u> (1733)?. P. Staeckel remarks that in the case of the obtuse angle the proofs are insufficient, because Saccheri uses I.16 of Euclid's <u>Elements</u>; but C. Segre remarks that the proofs are fully rigorous.

1. The first thirteen propositions of <u>Euclides</u> are correctly proved, provided that some of them be interpreted locally. Because I.16 and I.17 are locally true; while I.27 and I.28 are global and therefore necessarily false.

2. Saccheri was somehow aware of this situation. He says: "I shall never use I.27 of I.28 of the <u>Elements</u>, but not even I.16 or I.17, except where clearly it is question of a triangle every way restricted".

3. Saccheri proves that in the case of the obtuse angle there are no parallels.

4. Theorems that are locally true, even if they are globally false, are not sufficient to deduce the inconsistency of the hypothesis of the obtuse angle. Saccheri does refute this hypothesis in Prop. XIV, but he tacitly uses propositions (I.27 or 1.16 in the large) that are not locally true.

Eduard Glas

Delft University of Technology, Delft, the Netherlands

MONGE'S INNOVATION, AND ITS METHODOLOGICAL SIGNIFICANCE

Lagrange, Laplace and Monge were commissioned by the revolutionary government of 1794 to teach at the newly founded Ecoles the proper methodologies of their respective fields of eminence. It was believed that all mathematics, however advanced, technical or specialistic, must be liable to exposition in such a systematic and logical way as to be generally understandable and learnable. To meet these requirements, they undertook to 'rectify' the conceptual framework of the discipline. Generally, important sources of scientific concepts include the perception of the object and problem field of the science, linguistic possibilities of characterising these objects and problems, and the inherent (logical) structures of the language as such. In spite of considerable differences in emphasis, both Lagrange and Laplace founded their rectificatory programme on the logical structure of mathematics, conceived as a self-contained, analytical language. More in particular they wanted to reconstruct mathematics as a formal calculus on symbols whose meaning had to be defined implicitly by their interconnections in the formal system, without appealing to any intuitive interpretations. Especially appeals to geometric imagery were considered foreign to the proper language of mathematics, and had to be excluded.
In sharp contrast with this linguistic (formalistic) programme, Monge insisted on the objective meaning of mathematics, and expressly denied that its contents and methods could formally be expounded and accounted for. We construe his alternative programme as a conceptual innovation since it hinged crucially on a highly consequential new perception of the object and problem field of mathematics as a whole. Monge conceived of mathematics as effective conceptualization of the 'moving geometrical spectacle' that is reality, with analysis merely as a general 'script' to denote, ultimately, geometric transformations, the latter being the most appropriate and universal tools for the exploration, the exposition and teaching, and the application of mathematics, including even the most abstract parts of analysis.

We intend to present an adequate characterization of Monge's fundamental new conceptions, to clarify historical constraints accounting for their viability and propagation, and to refer specially to the ways in which in the historical process internal and external factors bear upon, and affect, one another. There is, we feel, a disproportion between the elaboration of the latter sort of interrelations and interactions, and the vast amount of historical detail that has been assembled already on the internal and external issues of mathematics separately.

Christian GILAIN

Université Paris VI, France

INTEGRATION EN TERMES FINIS ET CLASSIFICATION DES FONCTIONS CHEZ CONDORCET

Il faut le reconnaître, le mathématicien français Condorcet (1743-1794) est beaucoup plus célèbre pour son action politique (notamment durant la Révolution), voire pour ses idées philosophiques, que pour ses travaux mathématiques proprement dits. A l'encontre de l'opinion de contemporains comme d'Alembert, Lagrange ou Laplace, condidérés comme complaisants, la plupart des historiens des sciences ont estimé que les travaux de Condorcet n'avaient rien apporté d'original dans le champ des mathématiques pures.

Il est vrai que les écrits mathématiques de l'encyclopédiste sont difficiles à lire et à comprendre; son style qui allie souvent au goût des généralisations, l'imprécision du langage et l'instabilité de la terminologie en est largement responsable. Cependant, nous pensons que, sur le fond, le jugement des historiens doit être révisé, et réévaluée la place de Condorcet dans l'histoire des mathématiques.

En particulier, en utilisant plusieurs sources dont son Traité, inédit, du Calcul intégral (1778-1782), nous montrerons que Condorcet a créé la théorie de l'intégration en termes finis des expressions et des équations différentielles. C'est surtout Liouville, qui, publiant abondamment sur ce sujet à partir de 1833, a attaché son nom à cette importante théorie qui reste d'une grande actualité, tant en liaison avec la géométrie algébrique qu'avec le calcul formel sur ordinateur. Nous serons donc amenés à comparer les travaux de Condorcet avec ceux de Liouville, ainsi d'ailleurs qu'avec certains écrits de Laplace et d'Abel.

C'est autour de la distinction fondamentale entre intégration "possible" et intégration "en termes finis", qui apparaît en 1772 dans ses travaux, que Condorcet organise son Traité. Ainsi sa classification des fonctions donne-t-elle un rôle central à celles qui sont formées de manière finie à l'aide d'opérations algébriques ou transcendantes élémentaires, fonctions qu'il appelle "analytiques", mais en les distinguant des autres fonctions utilisées en Analyse.

Certes, on ne trouve pas chez Condorcet la rigueur et la clarté dans la définition des concepts et dans les démonstrations qui caractérisent les mémoires de Liouville. Cependant, il pose des problèmes, dégage des concepts, donne des méthodes et des résultats, qui dessinent déjà les principaux contours de la théorie. Pour nous en tenir à l'intégration des fonctions d'une variable, indiquons par exemple que Condorcet possède les énoncés (mais pas les démonstrations correctes) de trois théorèmes essentiels qui, pour z fonction algébrique de x, donnent la forme de l'intégrale $\int z\,dx$ lorsqu'elle est algébrique et plus généralement celles de $\int z\,dx$ et $\int e^x z\,dx$ lorsqu'elles s'expriment en termes finis.

Nous essaierons d'établir, à la lumière de ces travaux, le rôle réel de Condorcet dans les mathématiques de son temps.

GRAHAM FLEGG

Reader & Chairman, History of Mathematics, the Open University,

                                                    Gt. Britain

NICOLAS CHUQUET: A NEW ASSESSMENT

For the most part, Nicolas Chuquet is known only for the work entitled <u>the Triparty</u>, which comprises roughly one third of his 1484 manuscript.   Until the last few years, other parts of the manuscript have remained unpublished, and until 1984 no English translation existed.   Now that most of the manuscript is available in French and a substantial portion in English, it is possible to reassess Chuquet's role in the history of mathematics, and to correct some of the exaggerated claims that have been made on his behalf.   He emerges well from such a reassessment, though in some areas he is not the innovator which some parts of his manuscript suggest.   De la Roche, for a long time accused as a plagiarist of less than full competence, also emerges rather better from reassessment than might have been expected.   The suggestion that Chuquet's influence was small because of the innovatory nature of his work is challenged in the light of the actual history of the manuscript indicating that it was virtually unknown.   It is, however, a work of more general interest to historians than has been suggested in the past on account of the light which it throws on social conditions of the day, particularly through the problems solved and the section which comprises a commercial arithmetic.

Warren Van Egmond

Institut für Geschichte der Naturwissenschaften, Universität München

THE STUDY OF HIGHER ORDER EQUATIONS IN EUROPE BEFORE PACIOLI

Luca Pacioli's Summa de arithmetica of 1494 was the first printed book to contain a treatment of algebra. Because of its ready availability as a printed book, its size, and the seemingly comprehensive treatment of mathematics it offered, historians of mathematics have generally taken it to be representative of the state of algebra in Europe at the end of the 15th century. The picture thus presented is clearly unimpressive. Pacioli's discussion hardly included anything not already contained in the algebra of the 9th-century Islamic mathematician Muhammad al-Khwarizmi; only the six basic linear and quadratic equations were treated in detail. Although Pacioli did include six biquadratic or quartic equations, these were all reducible to quadratics. Cubic equations were said by Pacioli to be impossible to solve, because "it has so far not been possible to formulate general rules because of the disproportionality between [the terms]."

Since the only other work of European algebra before Pacioli that was available in print until recently was Leonardo Pisano's Liber abbaci, and Leonardo did not discuss higher order equations at all (a single cubic equation that arose in the course of a problem was not treated as a general type), historians of mathematics formed the view that the study of algebra in Europe had lain fallow for 300 years and never advanced beyond the solution of the simple quadratic equations.

A study of the sizable manuscript material remaining from this period, however, reveals that this picture is grossly inaccurate. The science of algebra was in fact diligently studied in Europe during the period between Leonardo and Pacioli, and one of its most distinctive features was the considerable attention devoted to the study of higher order equations. Cubic equations are already found in the oldest surviving manuscript from 1328 and later manuscripts regularly contain lists of equations up to the sixth degree. Although most of these are easily reduced to lower degrees and the proposed solutions of the few irreducible equations are generally incorrect, it is clear that European algebraists of the 14th and 15th centuries went well beyond the limits of al-Khwarizmi's elementary treatment. A careful study and comparison of the manuscripts further reveals several distinct traditions which indicate that the solution of higher order equations was a topic of continuing interest among European mathematicians well before the publication of Pacioli's book. Contrary to the way many historians have presented it, the Summa is not the beginning of the revival of algebra in Europe, but a very inadequate summary of the stubstantial work that had preceded it.

Owen Gingerich and Robert S. Westman

Professor of Astronomy and of the History of Science
  Harvard-Smithsonian Center for Astrophysics, Cambridge, MA, USA
Professor of History, University of California, Los Angeles, USA

PAUL WITTICH AND THE PREHISTORY OF LOGARITHMS

A traditional story relates that Lord Napier got his initial inspiration for the invention of logarithms from a fellow resident of Edingburgh, John Craig, an astronomer and royal physician [1]. Craig had reported a method used on the continent to replace multiplication and division by addition and subtraction.

We shall confirm this story and fill in the background relating to the invention of the prosthaphaeresis method by Paul Wittich sometime before 1577. Around 1577-8 Wittich conferred with Craig, then Dean at Frankfurt an der Oder, where Craig copied some of Wittich's mathematical examples into his own copy of Copernicus' <u>De revolutionibus</u>, which he eventually carried back home to Edinburgh. Wittich subsequently visited Tycho Brahe on Hveen, where the Danish astronomer promptly acquired the prosthaphaeresis technique. Already in 1916 J.L.E. Dreyer had suspected Tycho's claim that he had played an important role in this invention [2], and our present researches establish that the method was indeed Wittich's own discovery.

---

[1] Florian Cajori, p. 100 in <u>Napier Tercentenary Memorial Volume</u> edited by C.G. Knott (London, 1915), quoting Anthony à Wood (1632-1695).

[2] J.L.E. Dreyer, "On Tycho Brahe's Manuel of Trigonometry," <u>Observatory</u> 39 (1916), 127-131: "Is not the conclusion irresistible, that similarly the invention of the method in 1580 was due to Wittich alone?"

Md

Maria Sol de Mora-Charles

Professeur titulaire d'H^e de la Science, Université du Pays Basque

Leibniz et le problème des partis. Quelques manuscrits inédits

Depuis son voyage à Paris, en 1672, Leibniz se montre intéressé à l'aspect mathématique de la théorie de la probabilité. En automne 1672, il fait la connaisance de Huygens, qui va l'initier en Mathématiques, rencontre Malebranche et Arnauld et peut examiner, en 1673, quelques uns des manuscrits de Pascal.

De retour à Hanovre, il essaye de faire ses propres calculs. Dans un autre travail, j'ai commenté déjà le manuscrit du 5 janvier 1676 "Sur le calcul des partis", et l'édition de K. Biermann et Margot Faak:"G.W.Leibniz' De incerti aestimatione" est bien connue, un texte de septembre 1678.

J'essayerais dans mon travail de commenter deux autres manuscrits de Leibniz; le premier contient deux problèmes non datés, en latin, et le deuxième, de 1686, en français, sur un problème proposé par Mr. Bernoulli et apparu dans le Journal des Savants de 1685.

E. J. Aiton

Manchester Polytechnic, Great Britain

LEIBNIZ'S JOURNEY TO SOUTHERN GERMANY AND ITALY

Leibniz set out on the journey that eventually took him to
Italy in October 1687, soon after his project of draining the
Harz mines using vertical and horizontal windmills had ended
in failure. While the primary aim of the journey was the search
for historical documents relating to the origin of the House of
Brunswick-Lüneburg, he lost no opportunity for meetings with
scholars and visits to museums and other places of interest,
for the purposes of which he made many detours. Some of the
details are described in a diary he kept during the journey.
On leaving Hanover he visited the Landgrave Ernst von Hessen-
Rheinfels, with whom he discussed the reunion of the Churches.
Leibniz saw this as a political problem, whose solution was
necessary if Europe was to regain unity. During a short stay in
Frankfurt he viewed the natural history collection and books on
caterpillars of Maria Sibylle Merian. In April 1688, he
made the most significant historical discovery of his tour.
For in the Benedictine monastery in Augsburg he found the
vital evidence concerning the common origin of the Guelfs and
the Counts of Este. Early in May he arrived in Vienna, where
he visited the Bishop of Wiener-Neustadt to discuss problems
of Church reunion and through him secured introductions to the
Imperial Court. Leibniz remained in Vienna until February 1689,
awaiting permission from the Duke of Modena to use his
archives for historical and genealogical researches. While in
Vienna, Leibniz was given a copy of the Acta Eruditorum in
which he found a review of Newton's Principia. Three topics
especially caught his interest, namely optic lines, resisting
media and planetary motions, subjects on which he had worked
himself. Without having access to his notes, he quickly
drafted papers on these topics, which he sent for publication
in the Acta Eruditorum. They include some of his best work on
the applications of the infinitesimal calculus. In Rome he
met the Jesuit China missionary Grimaldi and became a member
of the Accademia fisico-matematica. Here he saw Newton's
Principia for the first time and started writing his Dynamica,
intended as a treatise to correct and complete Newton's work.
At the end of 1689, he arrived in Modena, where, on behalf of
the Duchess Sophie, he attempted after 600 years of separation
to arrange a new alliance of the two Houses by marriage. This
came to pass in 1695. Leibniz arrived back in Hanover in June
1690 after an absence of over two-and-a-half years of intense
activity in pursuit of his multitudinous interests. During
this time he had sent regular reports to his friends the
Landgrave Ernst and the Duchess Sophie besides undertaking a
number of political commissions for the Duke of Hanover.

Javier Echeverría

Prof. Philosophie de la Science. San Sebastián. Univ. du Pays Basque. Espagne.

Recherches inédites de Leibniz concernant la Topologie et les fondéments de la Géométrie.

    Leibniz a commencé en 1679 (Characteristica Geometrica) à développer son projet d'une analyse des Elements d'Euclide et des notions géométriques fondamentales: situs, distantia, continuum, congruentia, similitudo, etc.. Il pense qu'on peut démontrer tous les axiomes d'Euclide, ce qui permettrait la construction d'une Géométrie nouvelle: la Characteristica Situs, où l'on n'aurait plus recours aux figures grecques ni aux équations cartésiennes.
    Pendant toute sa vie, et notamment en 1679-80, 1682-83, 1685-86, 1691-93, 1695, 1698, 1700, 1710, 1712 et 1715, il a travaillé à ce genre de questions. La plupart des fragments qu'il a laissé écrits et sans publier (à cause du mauvais accueil de Huygens à ces recherches) restent encore inédits.
    La communication fait partie de la transcription des manuscrits inédits de Leibniz concernant l'Analysis Situs et essaie de dégager les fondements de la Géométrie leibnizienne de même que le sens d'une telle quête, qui pointe maintes fois vers la Topologie Générale du XXème siècle. Chez Leibniz on ne trouve plus une Géométrie de la règle et du compas, mais une Géométrie de n'importe quel corps rigide, beaucoup plus générale que celles d'Euclide et de Descartes. Chemin faisant, on y trouve des recherches très intéressantes concernant les notions de connexion, de frontière, d'intérieur et d'extérieur, de voisinage, de continuité, etc., de même que des essais de formalisation de la Géométrie par le moyen des characteres. En essayant de démontrer chacun des axiomes d'Euclide et de simplifier les définitions et les constructions des Elements, Leibniz arrive à se poser des questions qui n'auraient une réponse précise qu'au XXème siècle.
    On conclût qu'il faudrait ouvrir un chapître nouveau de l'Histoire de la Géométrie et de la Topologie, en tenant compte des contributions de Leibniz, tout à fait inouïes à l'époque.

* * * * *

Lorraine J. Daston

Assistant Professor   Princeton University

The Domestication of Risk: Mathematical Probability and Insurance, 1650-1830.

Despite the efforts of mathematicians to apply probability theory and mortality statistics to problems in insurance and annuities in the late seventeenth and early eighteenth centuries, the influence of this mathematical literature on the voluminous trade in annuities and insurance was negligible until the end of the eighteenth century. I argue that the combination of profitable pre-probabilistic practices and a legal notion of risk as "genuine" uncertainty (as opposed to the quantified uncertainty of probabilities) were largely responsible for this neglect, and that even in the first applications of probability theory to insurance practice fiscal considerations all but overwhelmed the mathematical methods. The emphasis upon uncertainty as the defining element of an aleatory contract made for an identification of insurance, particularly life insurance, with gambling, and many insurance and annuity schemes of this period exploited the association to attract customers. Only with the advent of new middle-class attitudes that placed provisions for one's family above provision for oneself; private self-sufficiency above public charity; the fear of downward above the hope of upward social mobility; security above surprise could life insurance allegedly based upon the certainty of mathematics and the regularity of mortality statistics compete with life insurance conceived as a wager.

Lorenz Krüger, Institut für Philosophie, Freie Univ. Berlin

Berlin (West), Germany

THE SLOW RISE OF PROBABILISM

A remarkable feature of 19. century science is the coexistence of an expanding statistical practice with unshaken deterministic theoretical beliefs. This paper attempts to contribute to an explanation of this apparent paradox.

(1) Traditional methodological ideals (mos geometricus) and successful scientific explanation (Newton, Laplace) merge in Kant's epistemology. Kantian ideas prove influential in science all through the century (e.g. Helmholtz).

(2) The distinction and combination of constant and accidental causes created the impression that determinism and probability were compatible (Laplace, Quetelet, Cournot et al.). But this reconciliation became possible only because the ontic and the epistemic levels were not clearly distinguished. If the constant cause is a cause at all, then in a sense which is entirely foreign to the usual deterministic picture. Hence a priori assumptions about the development of probabilities in terms of frequencies or Bernoulli's mathematical theorem replaced the missing dynamics.

(3) The critics of classical probability (Fries, Ellis, Venn) undermined the unwarranted identification of statistical correlation and causal determination. But no more than the Laplacians could they explain the objective relevance of "constant causes" or the mutual compensation of "accidental causes". They speculated instead about the internal coherence of statistical aggregates , in terms of genus and species (Ellis), or of individual variations within natural classes (Venn in 1866, i.e. after Darwin).

(4) A clarification became accessible only in a field in which both deterministic dynamics and probabilistic statistics could be related to the same objective theory. Maxwell's distinction between dynamical and statistical knowledge no longer coincides with that between the ontic and the epistemic aspects of a theory. But it remained unclear how the "two kinds of knowledge" were to be integrated. Peirce's indeterminism revived the fusion of epistemic and ontic elements in his specific brand of idealism.

(5) Probabilism in science had to await not only quantum theory but also ideas concerning the complexity of classical systems (e.g. liquids, organisms), ideas which were only hinted at by Maxwell and are becoming current today.

Theodore M. Porter

Department of History, University of Virginia

Number and Diversity: The Fruition of Statistical Thinking

W. S. Jevons' observation in his 1874 **Principles of Science** that "Number is but another name for diversity" exemplifies the viewpoint that inspired the development of modern statistics. Before the late nineteenth century, the use of probability in science was limited to the reduction and estimation of error. The same function was assigned to probability in relation to the social science of statistics by such writers as Quetelet, whose aim was precisely to imitate the success of astronomy and physics. The most important developments in statistical thinking during the late nineteenth century grew out of the need to adapt probability to deal with what came to be recognized as very different problems and issues of biology, social science, and gas physics. It was increasingly supposed that variation, especially in the human domain, is not mere error, but a phenomenon of fundamental importance. In the context of these studies, and partly also in reaction against the outlook that would reduce the study of society to problems just like celestial mechanics, there emerged two statistical movements of the first importance. One called for recognition that science in fact is not capable of reducing all variation to law, and suggested, at first tentatively, that the appearance of chance and freedom in the world is not due merely to our ignorance. This rejection of determinism was influential in physics, biology, and social thought. The other intellectual program involved the search for techniques by which variation could be seized and its causes understood—not in detail, but statistically. It was this search, directed to certain particular problems in social science and, especially, biometry, that gave rise to mathematical statistics.

John Beatty

Arizona State University, Tempe, Arizona, USA

The Importance of Chance in Evolutionary Biology

As sure as there was a "probabilistic revolution" in evolutionary biology in the 1930s and 40s, there was a "counterrevolution" in the 50s and 60s. In the earlier period, evolutionary biologists attributed considerable significance to stochastic evolutionary changes. That is, while they attributed many evolutionary changes to natural selection, they attributed many others to chance instead. So-called "random drift" was, however, invoked less and less frequently in the period that followed. During this period, natural selection was the increasingly favored agent of change. Empirical grounds cited in support of the shift consisted primarily of selectionist reinterpretations of changes originally attributed to random drift. Proponents of the importance of selection thus increased their store of exemplary cases at the expense of proponents of the importance of drift.

Throughout the shift in opinion concerning the relative importances of random drift and natural selection, evolutionary theory retained the stochastic character it had assumed in the 30s when random drift was first explored and considered to be so significant. The stochasticity of evolutionary theory was never at issue among proponents of the importance of drift and proponents of the importance of selection. Something very different was at issue. It is historically and philosophically quite important to recognize that agreement over the stochastic formulation of a theory is consistent with wide disagreement over the extent to which phenomena are due to chance, even within the domain to which the theory is agreed to apply. Stochasticity issues in science concern not only whether chance (on whatever interpretation) must be taken into account in theory, but also the importance of chance relative to the deterministic causes described by the theory. The latter issue may not be addressed by the theory itself.

Gerd Gigerenzer

The "Intuitive Statistician": On Egon Brunswik's Metaphor of Man

Few experimental psychologists deviated so far from their contemporaries in their answers to the problematic questions "What to look for?" and "How to proceed?" as Egon Brunswik. This paper is concerned with Brunswik's attempt at probabilistic revolution, the first such attempt in 20th century psychology. The question why this revolution failed so dramatically is analyzed by means of a comparison with L.L. Thurstone's first great success using a probabilistic conception. My answer is based on the division of psychology into two more or less unrelated research programs, the "experimental" and the "correlational" schemata. Each was composed of different ideas about subject matter, methodology, and purpose, and founded on a different metaphor of man. The connection of the particular ideas in the historically evolved schemata was fairly arbitrary (as indeed was the division itself). My thesis is, that Brunswik's probabilism was both misunderstood and finally rejected, since it could not be assimilated into either of the schemata. The reasons suggested are (1) he attempted a consistent integration of elements from both schemata, that is, to integrate what was considered to be unrelated; and (2) his probabilism was based on a different metaphor of man, the "intuitive statistician". Thurstone, interested in application rather than in systematic theoretical view, adapted to the established schemata. His ideas could be assimilated and were celebrated as success. Brunswik's probabilism, however, died in the desert between the two rigid schemata.

Andreas Kamlah

Prof., Universität Osnabrück, FB Kultur- und Geowiss.

The Gestalt Switch in the Theory of Probability at the Beginning of the 20th Century

Frequentists like to portray subjective or logical probability as concepts not to be taken seriously and wonder how they could have been so popular in the 19th century. They may ask why frequentism had to wait so long for becoming the dominating interpretation of probability. In fact very few people in Germany before World War I claimed that probability is relative frequency though most scientists were familiar with the law of large numbers. We have to consider the sudden change in interpretation of probability as a genuine gestalt switch à la Kuhn. There are a lot of reasons why the old paradigm of Laplace's theory was dominant in Germany for more than a whole century, for example:

1. Philosophers used to analyze scientific concepts in psychological terms of "ideas". The rise in psychology in the second half of the century strengthened this psychologism concerning concepts.
2. The Bayes formula which is hardly thinkable without subjective or logical probabilities was considered as indispensable for the empirical determination of aposteriori probabilities. Frequentists did not know any equally effective method for estimating probabilities.
3. There are no infinite sequences of events ever observed in reality. These are but counterfactual idealizations.

There are sophisticated arguments put forward against frequentism by Stumpf, von Kries, and other German philosophers or scientists. But about 1900 there is a change. Hausdorf, Bruns, and some other mathematicians start their expositions of probability by defining it as relative frequency. They consider it as a physical quantity like energy or electric charge. The change in interpretation cannot be understood as a result of newly discovered facts or mathematical insights. The old facts are rather seen in a different light. In my paper I try to describe the forces which conserved the old theory quite successfully for a long time.

MARIANO HORMIGON . UNIVERSIDAD DE ZARAGOZA, ZARAGOZA, SPAIN.

Profesor Titular, Universidad de Zaragoza.

LES PREMIERES REVUES MATHEMATÍQUES EN ESPAGNE (1848-1916)

Dans cet article on expose les données principaux qui expliquent la difficile existence des premières revues mathématiques espagnoles et qui représentent le lent et compliqué procesus d'incorporation des mathématiciens espagnols à la communauté mathématique internationale.

La période parcourue va de 1848, l'année de publication du *Periódico Mensual de Ciencias Matemáticas y Físicas* à Cádiz, jusqu'a 1916, quand la *Revista de la Sociedad Matemática Española* est disparue. Entre ces deux bouts, on considère les élèments fondamentaux de quatre publications périodiques: *El Progreso Matemático*, édité à Zaragoza entre 1891 et 1900 et dirigé por Zoel García de Galdeano; *Archivo de Matemáticas*, publié à Valencia pendant les annés 1898 et 1899 et dirigé par Luis G. Bascó; *Gaceta de Matemáticas Elementales*, apparue à Vitoria entre 1903 et 1906; et *Revista Trimestral de Matemáticas* qui, sous la direction de José Ríus y Casas, a continué à Zaragoza l'effort de *El Progreso Matemático*.

Mf

Santiago RAMIREZ

Professeur, Université Nationale du Mexique.

LE COURS DE JEAN CAVAILLES

Le propos principal de cette communication est de présenter le contenu du cours que Jean Cavaillès a donné à la Sorbonne en 1941.

Dans ce travail nous avons utilisé les notes prises par Mme. Marie Louise Gouhier et par Mme. Geneviève Rodis-Lewis.

Il s'agit, ici, de présenter un ensemble de thèses sur la philosophie mathématique. Ces thèses sont contenues dans la première partie du cours.

1. <u>L'objet de la science: atteindre l'absolu.</u>

Cette proposition peut être pensée de deux façons: a) du point de vue de la transcendentalité kantienne, la quête de l'absolu est conçue comme le resultat d'une force qui aspire à dépasser les limites que la critique fixe à la raison. b) Du point de vue théologique, la quête de l'absolu est liée à la fin des temps modernes telle qu'elle fut annoncée par Romano Guardini.

2. <u>L'histoire de la philosophie mathématique oscille entre l'affirmation cartesienne de l'absolu et l'absolutisme "hyppocrite" de Kant.</u>

Deux solutions historiques ont été proposées: a) celle qui veut reduire la mathématique à la logique et b) celle qui veut la reconstruction simultanée des mathématiques et de la logique.

3. <u>L'expérience mathématique ne se situe pas dans l'histoire.</u>

De la deux problèmes posés par Cavaillès: a) celui de la necessité de prendre conscience du processus par lequel l'objet mathématique s'engendre et b) celui du mouvement de la pensée mathématique conçu comme changement du sens.

4. <u>La mathématique est la science de l'infini.</u>

On peut encore poser le problème de deux façons: D'un côté, la pensée mathématique est la pensée du monde. "Est-ce que celà a un sens?", demande Cavaillès. De l'autre, on peut tracer l'histoire de l'infini: pour la modernité, l'infini n'existe pas (d'où l'hyppocrisie de Kant). C'est à partir de la fin des temps modernes, annoncée par Kierkegaard et Nietzsche, que la recherche de l'infini se pose à nouveau, mais cette fois-ci en sachant que l'on ne pourra jamais le trouver.

5. <u>En mathématique, "je ne pense pas ce monde".</u>

L'affirmation signifie que l'objet de la physique n'est plus absolu. De plus, elle signifie que la pensée physique n'est possible que grâce à la pensée mathématique.

Roger L. Cooke, University of Vermont, Burlington VT, USA

V. Frederick Rickey, Bowling Green St. Univ., Bowling Green OH, USA

MATHEMATICS AT CLARK UNIVERSITY, 1889-1921

Clark University, founded as a purely graduate university in 1889, played a significant role in the emergence of America from the mathematical backwater of the nineteenth-century into the worldwide mathematical community of the twentieth. After an unsuccessful attempt to recruit Felix Klein for the faculty, the university chose William E. Story, who had been on the faculty at Johns Hopkins throughout Sylvester's stay, to lead the department. He was joined by three recent PhD's: Henry Taber from Hopkins, and Henry S. White and Oskar Bolza, both students of Klein. Completing the faculty was the eccentric Joseph de Perott. This blend of British and German influence, which was unique in America, was a source of its strength. Active faculty, solid coursework, a good small library, and an exciting seminar provided an environment congenial for yound American students who previously had few opportunities for advanced studies besides Europe. During these first few years the department at Clark was current with the latest developments in European mathematics. Consequently, Clark's new departure in education quickly gelled to produce the best department for mathematical research in this country.

Disaster, in the form of financial and morale problems, struck Clark University in 1892. The mathematics department suffered the loss of Bolza to the new University of Chicago and White to Northwestern. Story and Taber held the department together, and between them produced two dozen respectable PhD's, most notably Solomon Lefschetz. However the financial problems prevented the infusion of new blood, so a gradual decline set in. Finally, in 1921, the faculty were retired and the department was closed.

Alejandro R. Garciadiego

Departamento de Matemáticas, Facultad de Ciencias, U.N.A.M. (México)

Georg Cantor's influence on Bertrand Russell.

          Russell explicitly enounced the logicist thesis, the fact that all Mathematics is Symbolic Logic, in his book The Principles of Mathematics published in May 1903. Historians of mathematics, philosophy and science have attempted to explain the origin of this thesis by discussing an intellectual trend that contains two major causes. The first source includes the analysis of the works of some forerunners (i.e., Leibniz, Frege and Peano, among others). This is even more evident after reading some of Russell's later recollections where: he praises the work of Leibniz, claims to have been the first one to have read Frege and, describes a trascendental turning point in his intellectual development after meeting Peano during the summer of 1900. The second element incorporates the legendary polemics demanding the reexamination of the foundations of mathematics as an outcome of the newly discovered set theoretic paradoxes and, the outgrowth of three philosophical schools in the mathematical community: Logicism, Formalism and Intuitionism.

          Nevertheless, recent studies have suggested an alternative explanation. Unpublished material, kept at the Bertrand Russell Archives (McMaster University, Hamilton, Ontario, Canada) have unvailed how Russell came upon his new ideas as a reaction towards the previous influence exerted on him by Kantian and Neo-Hegelian philosophers lecturing at Cambridge University. In order to understand Cantor's influence, it is necessary to discuss the writing of The Principles of Mathematics as a complex enterprise evolving from previous attempts to write a book on the principles of arithmetic. The project, contrary to Russell's later reminicences, suffered a long and slow metamorphosis mainly motivated by the gentle and gradual absortion of Cantor's notions, principles and methods. Published and unpublished works allow us to assess how Russell shifted from his first negative reaction towards Cantor--as a consequence of his Neo-Hegelian "indoctrination", to his almost complete acceptance of Cantor's ideas. This change took place between 1895 and 1901.

Carlos Alvarez.

Universidad Nacional Autónoma de México, México.

LA LOGIQUE DES PROPOSITIONS MATHEMATIQUES DE FREGE A WITTGENSTEIN.

Le *Tractatus* de Wittgenstein nous donne les éléments d'une théorie de la représentation. Celle-ci a comme élément central les "tableaux" que nous nous faisons des faits (T. 2.1). Ces tableaux, la proposition en étant un, établissent deux types de relations représentatives: celle des noms aux choses et celle des propositions aux faits.

La théorie des tableaux est le sous-sol de la critique de Wittgenstein à la théorie représentative de Frege et à la sémantique qui l'accompagne. Mais déjà dans les *Carnets* de 1914-1916, Wittgenstein **fait une** première critique a la théorie de Frege. Dans ces *Carnets* et dans les *Notes sur la Logique* (1913), avec une conception distincte des termes sens/référence et non sur l'idée des tableaux, Wittgenstein commence à prendre distance par rapport a son ancien maître.

Ceci nous invite à reexaminer l'influence de Frege sur la pensée de Wittgenstein sous l'image de l'evolution de ses idées à partir des *Carnets* jusqu'au moment où la théorie des "tableaux" fut élaborée. Avec ceci, nous voulons essayer une analyse comparative des conceptions de Frege et Wittgenstein sur la nature des propositions mathématiques.

Eduardo L. Ortiz, Imperial College, London, England

The Idéologues and Mathematics in Argentina
in the early XIX-th Century

The French shcool of the Idéologues had a dominant influence on
the teaching of Philosophy at the University of Buenos Aires
in the period 1820-1840.

The comprehensive view of society and government offered by the
Idéologie had already attracted the attention of Argentina's state
men and is clearly visible in many of the reforms they introduced
after independence from Spain.

In the same period Mathematics held a predominant position in Argentina for reasons which are not entirely clear. In this paper
we discuss some aspects of the impact of Ideological thought in
Mathematics in Europe and try to identify them in the image of
Mathematics which was transmitted to Argentina in the early years
of the XIX-th century.

Hourya BENIS-SINACEUR

C.N.R.S., Paris

SUR LE THEOREME D'ALGEBRE DE Ch. F. STURM

Artin et Schreier ont utilisé le théorème de Sturm sur le nombre de racines réelles d'un polynôme entre deux limites réelles pour démontrer l'unicité de la clôture réelle d'un corps ordonné. On sait aujourd'hui qu'on peut obtenir le même résultat autrement. La démonstration de E. Becker et K.J. Spitzlay (1975) a un intérêt particulier pour l'historien. Car en simplifiant une preuve faite par M. Knebusch dans le cadre de la théorie abstraite des formes quadratiques, elle retrouve un critère de réalité des racines semblable à celui que J.J. Syvester avait construit, en 1853, au terme de ses propres recherches et de celles de Ch. Hermite sur le théorème de Sturm. Un lien est ainsi établi entre théorie abstraite et théorie classique des formes quadratiques.

J'ai ici pour objet
1) d'examiner les Notes ou Mémoires de Sylvester inspirés du théorème de Sturm. On verra que Sylvester cherche avant tout à formuler un énoncé qui fasse totalement l'économie de la notion de fonction, et a fortiori de celle de fonction continue. Cela le conduit à un critère différent de celui de Sturm, dont le ressort est la loi d'inertie des formes quadratiques.

2) d'examiner les principaux manuels qui ont fait état des travaux de Sylvester et de Hermite sur le théorème de Sturm. Parmi les auteurs consultés, seul Oskar Perron (Algebra, II, Satz 3) donne, en 1927, l'énoncé d'un critère qui fait une synthèse des idées de Sylvester et de Hermite, et se trouve avoir la forme exacte, particularisée à $\mathbb{R}$, de l'énoncé, valable pour tout corps réel clos, de Becker et Spitzlay (Satz 1, ii).

GOMEZ PIN Victor

Directeur du Département d'Histoire de la Philosophie et de la Science de l'Université du Pays Basque. Chargé d'Enseignement au Collège International de Philosophie de Paris

"Robinson et Hegel: deux approches ontologiques du Calcul Différentiel".

L'on sait que dans la première moitié du XIX siècle, le calcul différentiel paraît sortir de son ambivalence par rapport au statut catégorial de ce dont il est question (à savoir les pôles mis en rapport dans la formule de la dérivée) de telle manière que l'introduction par Weierstrass de la méthode ε-δ a pu être identifiée par un auteur contemporain à "l'execution and burial of infinitesimals"

Il est curieux de constater que, presque au moment où se prépare sur le plan mathématique le processus qui conduira au Cours d'Analyse du baron de Cauchy (1821), un philosophe développe dans le cadre de sa discipline d des conclusions allant dans le sens du rejet de l'infiniment petit en acte: entre 1812 et 1816 Hegel soutient que la formule de la dérivée exprimerait un lien dans lequel ceux (dy, dx) qui s'y trouvent rattachés n'ont aucune subsistance en dehors du lien lui-même et donc sont irréductibles au régistre strictement quantitatif.

Avec le non-standard-analysis l'essence quantitative des différentielles est rétablie, et ceci dans le cadre d'une charpente théorique irréprochable. Cette possibilité étant exclue par la thèse de Hegel, l'oeuvre d'A. Robinson représente donc, non seulement une révolution sur le plan mathématique, mais aussi un revirement sur le plan ontologico-catégorial.

Confronter Hegel et Robinson par rapport au calcul différentiel révient à mettre à l'épreuve sur un terrain concret les puissances respectives de la logique des modèles et de la denomée logique dialectique.

Li Zhaohua

Tianjin Normal University, Tianjin, China

A STUDY ON LI SHANLAN'S DUO-JI BI-LEI AND FAN-YUAN SHAN-YOU

Li Shan lan (1811--1882) is a celebrated Chinese mathematician. His works "Do-ji Bi-lei" (the summation of piled up heaps, 1859 A.D?) and "Fang-yuan Shangyou" (the explanation of the mystery of square and its inscribed circle, 1845 A.D) are two masterpiece in the history of Chinese mathematics.

In the first work, he lays emphasis on the summation of series. There are mainly:

1) $\sum_{r=1}^{n-k} r^{p+1}$

2) $\sum_{r=1}^{n-k} \{\frac{1}{p!} \Gamma(r+1) \cdots (r+p-1)\}^2$

3) $\sum_{r=1}^{n-k} \frac{1}{(p-1)!} \Gamma(r+1) \cdots (r+p-2) \cdot r^s$

In the second work, he gives forth the important result in integration: $\int_0^h (\frac{a}{h})^p x^p dx = \frac{1}{p+1} a^p h$

The present paper discusses the derivation of these results and their relations. It points out that the theory and method result from the method of piled up heaps and limit idea of the traditional Chinese mathematics. We can conclude that the traditional Chinese mathematics would have developed into the period of variable mathematics had it developed its own way.

Alexandru Giuculescu

Central Institute for Management and Informatics, Bucarest, ROMANIA

On ROUMANIAN GEOMETERS G.TZITZEICA and D.BARBILIAN

Gheorghe Tzitzeica(1873-1939) and his former student Dan Barbilian(1895-1961) have taught at the University of Bucarest. In the creative activity they differed: while Tzitzeica worked on classical differential geometry with analytic tools, Barbilian was a partisan of the Erlangen Program and applied the axiomatic method. Despite of their allegeance to schools of the 19th century, they were able to create new concepts, to discover remarkable properties and relations and to validate tendencies which proved to be extremely fruitful.

Tzitzeica's major contributions are around the concept of tetrahedral surfaces or affine spheres, which he defined in 1907, and the concept of a class of space curves that were defined in 1911 and called by G.Loria "Tzitzeica curves". He found the invariants K/d^4 (K=Gauss curvature in a point M of the surface; d=the distance from a fixed point to the tangent plane in M) and Td^2 (T=torsion radius in a point M of a curve; d=the distance from the origin to the osculator plane in M). Both tetrahedral surfaces and "Tzitzeica curves" were proved to be invariant to centro-affine transformations. In a book from 1923 Tzitzeica dealt with nets and congruences in a projective space and their dual character was used by him to interpret geometrically the Laplace sequences. Besides his disciples and before many mathematicians continued to work on Tzitzeica's ideas: S.Finikov, T.Kozmina, R.Smirnov, S.Chern, R.Calapso, Kentaro Yano, Buchin Su, etc

Barbilian worked first on the problem of a metric for hyperbolic geometries that he dealt with in a paper on "Jordanian" geometries from 1934, reproduced and commented by L.M.Blumenthal in 1938 under the title of "Barbilian spaces". In 1936 Barbilian defined a Riemannian metric which generalizes Cayley's and Klein's metrics. After P.J.Kelly described in 1954 "Barbilian geometries and the Poincaré model", Barbilian resumed the same topic in 1959,1960 and 1962, and established a general principle of metrization consisting in the formalization of the Möbius abstract continuum. The resulting "oscillating geometries", based on a logarithmic valuation, are including his "Jordanian" geometries from 1934 as well as Kelly's geometries from 1954. Another Barbilian's topic was the construction of geometries on rings described in two papers from 1940 and 1941. Barbilian's ideas of geometries on rings were resumed recently in a different way by H.J.Arnold, and by W.Leissner, who developed the concept of affine Barbilian planes.

Tzitzeica and Barbilian were prominent personalities in the cultural life of their country, the former as an outstanding organizer of the academic activity, the latter as a brilliant writer under the pen-name of Ion Barbu, whose lyrical poems are praised for their originality and beauty.

Kazuo Shimodaira

Professor of Kokushikan University, Tōkyō, Japan

Mathematics in the Early Edo Period of Japan

After the Meiji Restoration (1868) Japan very rapidly imported many kinds of Western science. The government established the first university, University of Tōkyō, in 1877. In the same year the first society, Tōkyō Sūgaku Kaisha, or Tōkyō Mathematics Society, was organised. (A compulsory education system began in 1872.)

Most foreigners usually wonder why Japanese have been able to learn Western science, especially mathematics, during a few years. It is not strange. Most Japanese studied arithmetic in the Edo period (1603-1867) when they were young at private elementary schools. They also could easily use Japanese abacus, or <u>soroban</u>. We can find many kinds of mathematical texts which were published in the period. Some mathematicians could solve very difficult problems. (Some problems were solved by them before Europeans did.)

After opening Tōkyō Sūgakū Kaisha, almost all people stopped studying the traditional Japanese mathematical problems and theorems, and began to study European mathematics. At first they translated European mathematical texts. Surprisingly it was very easy for them. However, this can be explained by the fact that many European problems were the same as, or similar to traditional Japanese ones. In addition, most Japanese learned arithmetic with the same enthusiasm as they now do.

When you want to know the reason why Japanese studied mathematics so much in the Edo period, you need to read some texts from the early Edo period. Nowadays, we can find only a few mathematical texts before the publishing of the "Jingōki". It was a very excellent text which was used by most Japanese. There were many revised editions of it (more than 500 versions). Most scholars who were not only mathematicians said that they studied many kinds of subjects when they were young, which included the "Jingōki". (The fitst edition was in 1627.)

In the Edo period there were a large number of novels, essays, dramas, many kinds of poetical works, etc. When you read these works, you should be acquainted with arithmetic and mathematical recreations. Because these works usually used arithmetical or mathematical problems. I'd like to explain what kinds of problems were used in those works. I'll also show you what kinds of mathematical problems most Japanese knew purely trough their common sense.

Eiji Chigira
Professor of Mathematics
Yonezawa Women's Junior College, Yonezawa, Japan

HOW WASAN (THE TRADITIONAL JAPANESE MATHEMATICS) WAS LEARNED BY
LOCAL FARMERS IN THE 19TH CENTURY

This report is mainly based on an investigation that was conducted in Fukushima Prefecture, Japan. According to the book "Outline of the history of Wasan in Fukushima" by Akira Hirayama, it may safely be assumed that in the 19th century approximately 10,000 farmers learned Wasan in this district.

Kazu Watanabe (1767-1839) is regarded as an eminent teacher who stimulated some farmers to learn Wasan. He himself was a farmer and learned Wasan in Edo (Tokyo) from Yasuaki Aita (1747-1817), who was the founder of the Saijo school of Wasan. His view of Wasan can be summarized as follows: Man, by learning Wasan, is able to build up his character and perceive the laws which are found in all things in the universe and gain understanding in three doctrines—Confucianism, Buddhism and Shintoism.

In succession to Watanabe, three teachers made a great contribution to the learning of Wasan by the farmers. They were Tsuzuki Sakuma (1819-1896), Shigeharu Tanji (1836-1909) and Zenzaemon Ueno (1846-1919). Sakuma compiled good textbooks, Tanji was interested in Enri (theory of circles), especially Tenkyo Kiseki (problems related to locus), and Ueno was deeply concerned about the calendar. They each had a number of disciples. To cite one example, in a notebook which was left by Sakuma, over 2,000 members of Sakuma's group were registered.

The quality and degree of Wasan that the farmers learned was shown by some books which were written and compiled by several teachers, and by the Sangakus (mathematical tablets dedicated to shrines or temples).

The reason why so many farmers learned Wasan is, of course, that the eminent teachers were located in this area and taught it to them. However, we should not forget the fact that the farmers had the corporate spirit of mutual assistance.

Mh

Shen Kangshen

Professor, Hangzhou University, Hangzhou, Zhejiang, P.R. China

TITLE OF TALK   A FURTHER STUDY ON PARALLELISMS
BETWEEN CHINESE AND INDIAN MATHEMATICS

The international learning circles have deeply concerned
for years about the parallelisms in the development of
mathematics in China and in India. The paper gives thorough and objective comparisons depending upon sound
foundation of original mathematical literatures of the
both nations. The paper deals with three branches of Mathematics: arithmetic, geometry and algebra.
China and India have generalised their own science and
civilisation in different circumsferences themselves. But
it is marvellous that there are enough examples to explain
rules, alogrithms, problems and their solutions appearing
in historical mathematics work are similar at all.
Sino-Indian envoys of friendship contacted again and again
from the beginning of Christian Era. It would be inevitable to have cultural exchange and learn from each other.
To discuss the relationship of transmission is complex,
hard but interesting.

Ko-Wei Lih, Institute of Mathematics, Academia Sinica,
Taipei, Taiwan, Republic of China

SOME CHINESE CONTRIBUTIONS TO COMBINATORIAL MATHEMATICS
IN THE 19TH CENTURY

Graph labelings have lately aroused considerable atten-
in combinatorial mathematics. Usually labelings are
done on vertices of a graph so that conditions on edges
are met. However, when a graph is plane, i.e., it has
been laid out on the Euclidean plane with crossings of
edges only at vertices, it is natural to impose requirements on regions. It came as a surprise that such constructions already appeared in a treatise published in
A.D. 1275 by Yang Hui. This theme was further developed
by Chang Chhao circa A.D. 1670. It finally reached the
amazing achievements by an almost unknown amateur mathematician Pao Chhi-Shou, circa A.D. 1880, of constructing
magic labelings for the platonic polyhedra and icosido-
decahedron. Except labelings for the cube, Pao's constructions were illustrated largely by plane net representations of polyhedra. In view of the appearance of
repeated labels, their true significance was commonly
overlooked. In this talk, I am going to clarify Pao's
concepts and exhibit their modern generalizations.

Nguyen Dinh Ngoc

Professor of Mathematics, Hue University, Hue, VIETNAM

Mathematics in Vietnam

   The study of mathematics has a long and honorable history in Vietnamese culture. Mathematics was cultivated even during some of the darkest periods of the Vietnamese past; mathematicians contributed to the struggle for national liberation from the French and later the Americans. The encouragement of mathematical endeavor, and the recognition of the Vietnamese government that mathematicians can aid in the economic and intellectual reconstruction effort, has led to a flowering of Vietnamese mathematics since 1945. Vietnamese specialists have received training at some of the finest mathematical centers in the world-- in the USSR and Eastern Europe as well as France, Japan, and elsewhere. World famous mathematicians have lectured in Vietnam, and number Vietnamese among their students.
   This paper will outline the history of mathematics in Vietnam, especially concentrating on the period after 1945. The subdivisions of the Hanoi Mathematical Institute and other research institutions will be discussed, as will the most developed and productive areas of contemporary Vietnamese mathematics (algebraic topology, classical and functional analysis, optimization theory, etc.).

Masa-yoshi Tagawa

Professor, College of Agriculture, Nihon Nuiversity, JAPAN

Naturalization of Mathematical Languages

   In early years, mathematics had been introduced from West via China to Japan. The word Geometry becomes <u>Chi-Ho</u> by transliteration in China. Japanese read these Chinese characters as <u>Ki-Ka</u> which merely means How-What. This example of intermission gap shows that orient people are not easy to understand mathematics under their own trditional languages. One must get use to realize concept by strange mathematical language like foreign language in studying mathematics.
   Recently, we have a better arrangement. Topology being Toh-Ba-Shue in China, while Japanese use Ih-So-Ki-Ka(Geometry of Phase Phenomena).
   Moreover, there are no article and plural sign of names in most orient grammer. If a student in a test faces a problem : "To get equation(s) of tangent line from a point to a curve", he must think how many answers are there and then solve the rest.
   Though we faced such difficulities for many years, we naturalized many terminology to our language. For example ,negulu(to neglect),Dai-Ki(Algebra and Geometry). These effected or been effected with daily using imported vocabulary,such as Baito(Arbeit),Departo(department store),aparto(apartment).
   Today, we are no longer to be bothered by intermission gap. Most uneasy terminology are well naturalized. There are many new mathematics made in Japan,e.g. Sasakian manifold,Almost Tachibana space.
   Now, we are in computer age. Many new terminology are born day after day. There must be an arrangement of these new languages and new conception. Well naturalization yields progressive intermission.

J. Bruce Brackenridge

Professor of Physics, Lawrence University, Appleton, WI, U.S.A.

A FLUCTUATING FLAW IN THE FIGURE FOR THE "KEPLERIAN" PROPOSITION IN NEWTON'S PRINCIPIA.

Newton has structured the opening sections of the Principia as a pedagogical device. In Section I he presents the basic definitions and mathematical relationships and in Section II he provides a series of problems as examples. These introductory problems at the close of Section II thus prepare the reader for the "dignified" Keplerian Problem with which Section III opens. These problems all seek to find the mathematical nature of a force necessary to maintain a particle in a given orbit with the force directed toward a given point. The Keplerian problem has an elliptical orbit with a force directed toward the focus of the ellipse. The answer that Newton finds is that the force must vary as the inverse square of the distance of the particle from the focus of the ellipse. The problem that immediately precedes the Keplerian problem also has an elliptical orbit, but the force is directed toward the center of the ellipse instead of toward the focus. The answer that Newton finds for this preliminary problem is that the force must vary directly as the distance of the particle from the center of the ellipse.

In all of these examples Newton begins with the geometry of the path, manipulates that geometry, and in the final step obtains the force from a unique critical ratio. The important role that this critical ratio plays in Newton's solutions has long been clear within the context of the work and has been a subject of interest to commentators on the work. It has been demonstrated, in fact, that this ratio reduces to the orbital equation employed by contemporary physicists. (Whiteside, Mathematical Papers, VI:42-43.)

It is of interest, therefore, to follow the diagram from which this critical ratio is derived into the figure for the Keplerian proposition and to note how related constructions change with successive editions. For example, there is a particular portion of the diagram that undergoes successives changes: in the third Latin edition of 1726 an important line begins to fade; in the Motte edition of 1729 it appears correctly and incorrectly in two variant sets of figures; and in the Cajori edition of 1934, with all of its reprints, it is incorrect. In this flawed diagram it is impossible to follow the proof of the Keplerian proposition as presented in the Principia because a triangle employed in that proof is not defined. The paper concludes with an inspection of this diagram as it appears in the two most recent versions of the English Pound note issued by the Bank of England.

Klaas van Berkel

Open University Heerlen / University of Utrecht, Holland

ISAAC BEECKMAN AND THE RAMIST BACKGROUND OF HIS PRINCIPLE OF INERTIA

The Dutch natural philosopher Isaac Beeckman (1588-1637) was one of the first adherents of the so-called mechanical philosophy. In fact, he may well have been the first to develop such a new way of treating physical problems and he quite sure introduced Descartes to it.
   One of the corner-stones of this new philosophy was Beeckman's principle of inertia, which resembled Galileo's principle in that he thought it to be applicable to circular movements as well as to rectilinear movements. It is not sure at what date Beeckman discovered his principle. The first time he mentions it in his scientific diary (Journal) is in May or June 1612. There he uses it to give an explanation of the perpetual movement of the celestial spheres. However, in a letter to Mersenne, written in June, 1629, Beeckman says:

> "What moves in a vacuum, will always continue to move in the same way. Nothing has ever come to my mind which is more certain than this, and in twenty years I have not read, heard or thought up anything which could have given me the smallest suspicion of error on this point".

If we are allowed to take this literally, this statement would take us back to 1609 as the year of Beeckman's discovery of his principle of inertia.
   This dating could give us a clue to the intellectual background of Beeckman's principle of inertia. In 1609, he was studying theology at the University of Leiden, but he was also reading widely in the field of mathematics. In that field, his tutor was Rudolf Snellius, professor of mathematics at Leiden (not to be confused with his son Willebrord, well-known for his discovery of the law of refraction). Snellius was a professed adherent of the philosophy of Peter Ramus and through the influence of Snellius, Beeckman became a Ramist too. It is our contention that the Ramist background of Beeckman's study with Snellius was decisive for his conception of a principal of inertia. The arguments Beeckman put forward in defense of his principle always come down to the arguments from simplicity and intelligibility favored by Ramists throughout the sixteenth and seventeenth centuries. The Ramist background to Beeckman's principle of inertia is a clear refutation of the supposed scientific sterility of the Ramist philosophy.

Lit.: K. van Berkel, <u>Isaac Beeckman (1588-1637) en de mechanisering van het wereldbeeld</u> (Amsterdam, 1983)(in Dutch, with a summary in English)

Christopher B. Burch

Visiting Assistant Professor, University of Pittsburgh

THE PROPER DEFINITION OF 'PLACE'; OR WHY DESCARTES' EARTH STOOD STILL

Setting aside the question of motivation, Descartes, knowing of the astronomy of Copernicus and the fate of Galileo, created a cosmology in which the Earth stood at rest. This position flows serenely from Descartes' fundamental definitions.
For a number of concepts in natural philosophy Descartes distinguishes between common and proper definitions. The proper definitions represent Descartes' coherent schematization, and are for his purposes precise and correct. The common definitions represent a sometimes convenient façon de parler, containing erroneous reference or contradiction with some properly understood concept.
The concepts of motion and rest are defined in terms of Descartes' unusual definition of place. The common definitions of rest and motion encompass positions that will later be shown by Descartes to be erroneous. These concepts are compatible with Galilean relativity, but are not the conceptions utilized in Principia Philosophiae, specifically in the laws of nature, the rules of impact, and the arguments for a motionless Earth. We have there instead Descartes' proper definitions of rest and motion.
Descartes justifies his proper definitions by way of an Aristotelian logic of opposites, in which motion is taken as the opposite of rest. One significant consequence of this stance is the privleged status given to the state of rest. For Descartes rest is not a different sort of motion; it is a different kind altogether. Another, tacit, consequence is that rest is, properly speaking, impossible; and the definitions break down for bodies at the point of impact. These consequences lead to conceptual difficulties within the rules of impact, and later in Principia to difficulties with maintaining the rest status of Earth.
This paper identifies the sources and consequences of Descartes' proper definitions of motion and rest.

Dr. James S. Perlman

Prof. of History of Science--San Francisco State University

The Rainbow: Studies in Development of Scientific Inquiry.

OVERVIEW

	SENECA (3 B.C.-65 A.D.)	ROGER BACON (c.1214-1294)	NEWTON (1642-1727)
I. OPTICAL KNOWLEDGE PREMISES OF THE TIME	1. Reflection: Geometrics of image formation. 2. No knowledge of refraction or dispersion as concepts or principles. 3. Experimentation not yet a socially developed or socially recognized knowledge tool. 4. Aristotle's "Meteorologica" and its treatment of the rainbow as a distorted reflection of the sun.	1. Geometrical knowledge of reflection. 2. Qualitative geometric knowledge of refraction. 3. Knowledge of eye optics.	1. Geometrical knowledge of reflection and refraction. 2. Snell's laws of refraction as expounded by Descartes and Archbishop Dominis.
II. THESIS OR HYPOTHESIS OF RAINBOW MAINTAINED OR DEVELOPED	1. Rainbow as image of the sun distorted by hollow clouds acting as imperfect mirrors.	1. Rainbow as reflected image from clouds of sun's rays, forming the base of a cone of which the apex is at the eye.	1. Raindrops and moisture as refracting, reflecting and dispersing agents of component colors of sun's rays for rainbow formation.
III. MAIN BASIS FOR OBTAINING DATA OR INFORMATION.	1. Memory of own past observations; writings of others.	1. Direct inductive investigation of phenomena.	1. Controlled laboratory set-ups of considerable inferential character.
IV. MAIN BASIS FOR FORMING CONCLUSIONS	1. Reasoning from observations, from analogous situations and from ideas of contemporaries.	1. Inductively generalizing from empirical data of investigation.	1. Verifying original hypotheses by controlled experimentation.
V. MAIN METHOD OF PRESENTING THE MATERIAL OF THE RAINBOW	1. Poses, discusses and analyses various aspects of the problem.	1. Outlines a series of inductive investigations of various aspects of the rainbow.	1. Systematically tests hypotheses through a hypothetico-deductive system built up from definitions to axioms to the propositions tested. The rainbow emerges as an application of such tested propositions.

Mary Kellogg

Assistant Professor of Mathematics, BMCC of CUNY, New York.

Mathematical Methods in Isaac Newton's *Philosophiae Naturalis Principia Mathematica*.

The disappointment faced by anyone of our century who knows a little mathematics on looking into Isaac Newton's *Philosophiae Naturalis Principia Mathmatica* is legendary. The speaker will discuss reasons for the difficulties of the modern reader, suggest an analysis of why Newton worked the way he did, and point out ironies of the situation.

Alan Gabbey

Head of the Department of History & Philosophy of Science,
Queen's University, Belfast

On Euler's principles of mechanics

On 3 September 1750 Euler presented to the Berlin Academy of Sciences a memoir bearing the title "Découverte d'un nouveau principe de mécanique". In his 1967 article on "The reception of Newton's Second Law of Motion in the Eighteenth Century" (<u>Archives</u> <u>Int</u>. <u>Hist</u>. <u>Sc</u>., vol. 20, pp. 43-65), Thomas Hankins showed some justified puzzlement on the question of the identity of the "nouveau principe" that Euler claimed to have discovered, but assumed that it was the generalized analytical formulation of "Newton's Second Law" that Euler presented in Arts. 20-24 of his memoir. After a re-reading of Hankins' paper, Euler's memoir and cognate writings, and Truesdell's comments on this aspect of Euler's mechanics, the puzzle of the precise relation between Euler's title and the content of the memoir remains. An attempt is made to resolve the puzzle by noting that the express purpose of the memoir was to establish "nouveaux principes" that would solve the problem of the motion of solid bodies rotating about variable axes of rotation, a general problem that Euler claimed could not be solved by means of the principles of mechanics known up to that time. Some consideration is given to the question of what Euler understood by a "principle" of mechanics.

William R. Shea

Professor, McGill University, Montreal

The Evolution of Johann Bernouilli's Cartesianism

At the time of Newton's death in 1727, the year following the publication of the third edition of the *Principia Mathematica*, the vast majority of Continental scientists still felt that his objections to Descartes' theory of vortices could be countered. Johann Bernouilli, who won the prize for the 1730 Essay Competition of the Académie Royale des Sciences, questioned Newton's mathematics and redid his computations of the velocities of the layers in the vortices. These new calculations, according to Bernouilli, were in no way hostile to the Cartesian theory, and he proceeded to explain the elliptical path of the planets but without supposing, as Descartes had done, that the vortex that carried the planet round the sun was compressed into an oval by the pressure of neighbouring vortices. Bernouilli supposed that the planet had an oscillatory motion to and from the centre that resulted from its being originally above or below the point at which it would float in equilibrium in the vortex. The problem was to show that the oscillatory motion resulted in an elliptical path. Bernouilli realized that he could not claim more than that the oval might be an ellipse.

Johann Bernouilli was again awarded the prize for the Essay Competition of 1734. His essay, entitled "Essai d'une nouvelle physique céleste", shows important developments in his thought since 1730, and this talk will focus on some of them. Bernouilli admitted attraction as a fact but he still found the Newtonian concept or principle of attraction opaque, indeed unintelligible. He abandoned his earlier explanation of the elliptical orbits of the planets and replaced it by the hypothesis of a perpetual stream of matter tending to the centre from all sides and carrying bodies with it. He hoped in this way to account for all the phenomena that Newton derived from his theory of universal gravitation. His conjecture was not unattended by difficulties in the application of the principles of mechanics but it was successful in several ways. It gave rise to objections of a more philosophical than scientific nature, and it goes a long way in helping us understand the slow demise of Cartesianism.

Andrew Ian Dale

Associate Professor, University of Natal, Durban, R.S.A.

BAYES'S THEOREM: A QUESTION OF PRIORITY.

A recent assertion by Stigler that Bayes was perhaps anticipated in the discovery of the result that today bears his name is exposed to further scrutiny here. The distinction between Bayes's Theorem and the Inverse Bernoulli Theorem is examined, and pertinent early writings on this matter are discussed. A newly-discovered result by Bayes is also presented.

Theodore S. Feldman

Assistant Professor, University of Southern Mississippi

THE QUANTIFICATION OF EXPERIMENTAL PHYSICS IN THE LATE 18TH CENTURY

Physics before the last third of the 18th century was not a mathematical but a qualitative, literary discipline. Theories of physics relied on concrete, pictorial imagery that successfully resisted quantification. Experiments, though they included quantitative measurements, were not intended to generate the systematic, precise data that alone can support a proper mathematical physics.

Around the middle of the 18th century natural philosophers began to abandon pictorial imagery in favor of more abstract concepts such as latent heat and electric charge. These concepts were quantifiable, and during the second half of the century they were employed in increasingly sophisticated mathematical reasoning. After 1770 natural philosophers constructed more accurate instruments and employed them in systematic experimentation, going to great lengths to insure reliable results. This "exact experimental physics" provided, for the first time, precise, systematic measurements of physical quantities. The combination of exact experimental physics with abstract, quantifiable concepts resulted in the emergence, after 1790, of a recognizably modern mathematical physics.

These developments occurred in all the subject areas of experimental physics and in all countries that cultivated natural philosophy. A more austere manner of philosophizing may have encouraged the shift to abstract concepts in physical theories after mid-century. Socio-economic factors favored the new experimental techniques. Improvements in materials technology and precision machining made possible more accurate instruments and means of calibrating them exactly. The civil and military arms of governments of the late Enlightenment required precise measurements for cartography and hydrography, as did some of the new industries. Agricultural, medical and meteorological societies of the period emphasized systematic, accurate observation. Experimental physics both benefited from and contributed to these developments.

The history of quantifiable concepts and exact experimental physics has been examined in electricity (J.L. Heilbron, <u>Electricity in the 17th and 18th centuries</u>, 1978) and in parts of meteorology, which belonged to experimental physics at the time (Feldman, <u>History of meteorology, 1750-1800</u>, Ph.D. Berkeley, 1984 and "Applied mathematics and the quantification of experimental physics," <u>HSPS</u> 15:1, 1985). Historians have also investigated the technology of precise instrumentation, especially the instrumentation of astronomy and surveying. It remains to coordinate this information with studies of other parts of experimental physics to define the broad movement of quantification.

June Z. Fullmer

Professor of History, Ohio State University, Columbus, Ohio 43210

PRIDE AND SENSIBILITY: THE LADY'S DIARY, 1774-1817.

While The Lady's Diary: or, the Women's Almanac was published continuously from 1702 to 1841 (when it married The Gentlemen's Diary; the conjugal publication survived until 1871) it reached its apogee of fame as Britain's chief mathematical outlet only under the editorship of Charles Hutton (1774-1817). Some attention has been paid to its mathematical contents and to its mathematical contributors for that period, yet, the British mathematical community remains largely obscured from view. The non-mathematical parts of the publication have also been ignored. However, it is possible to gain some insight into the mind of that community, for Hutton used his authors and readers to present a seemingly integrated view of the world. Analysis of the sections of The Lady's Diary devoted to natural philosophy, for example, and of those sections devoted to linguistic puzzles permits one to say something about the intellectual and social structure of the mathematical community in Britain during the Regency. While the community's individual members remain relatively unknown they can, indeed, be characterized.

Pb

Wu Yi-yi, Princeton University/East China Normal University

MOHIST OPTICS: OBSERVATION AND OPERATION IN B 17-21, A REINTERPRETATION

The *Motzu*, or *The Books Of Mohism*, the representative work of a philosophical school in ancient China circa the 5th century B.C., is not only prominent philsophically, but scientifically as well. Among many entries dealing with scientific topics in the book, the five entries B 17-21 (numbering system follows A.C. Graham) are believed related to optics. Mistranscribed and ill-emended as it is, the text sees quite obscure. Discussing entry by entry the emendations and explanations put forth by many Chinese scholars and Western researchers, this paper argues for a plain and natural interpretation. The paper proposes that these entries are records of phenomena concerning shadows and instructions on how to cause them to arise. Briefly, these five entries can be read as:

B 17 -- To have shadow follows. By remarking (the light source?)

B 18 -- To have two shadows. By doubling (the light source.)

B 19 -- To have the shadow upside down. By (using) a pinhole.

B 20 -- To have the shadow cast in the direction of the sun. By circling the object.

B 21 -- To change the size and/or the form of a shadow. By (moving?) near or far away and/or (having the object) upright or tilted.

Syntactically, each entry consists of two parts. The first part is a description of the phenomenon, which is made out on the grounds of careful observation. The second part is a kind of operational instruction.

A detailed instruction of the various emendations gives evidence that the observation-operation interpretation is in accord with the original text of one of the earliest versions, named after its editor, Pi Yun. Especially for B 20, both literal and technological explication can be found convincing. This entry is concerned with a demonstration that a man with a mirror circling an object in order to find the propitious position wherein by the reflection of the mirror, a shadow is cast towards the sun. Epistemologically, this is among the first steps in the recognition of Nature.

C. Hakfoort

University of Technology, Eindhoven/ University of Utrecht, Utrecht
(The Netherlands)

TORN BETWEEN THREE LOVERS:
ON THE HISTORIOGRAPHY OF 18th-CENTURY OPTICS

Historiographical and historical studies by T.S. Kuhn, J.L. Heilbron and others, have learned historians of physics to look for *two* main scientific traditions — experimental and mathematical —, associated with two clusters of physical sciences — classical and Baconian. It has been remarked, by G.N. Cantor, that optics does not fit neatly into either of these categories. Optics seems to be an anomaly in Kuhn's two-fold division. In a study of optics in the 18th century, especially in Germany, I have argued for a division into *three* traditions and related fields (1). These traditions are: experimental, philosophical and mathematical. The corresponding fields are: experimental parts of natural philosophy, theoretical parts of natural philosophy, mathematics. My thesis is, that 18th-century optics, wich was represented in all three fields, can be better understood, when the changing relationships between optics and the three fields are taken into account.

I will substantiate this thesis by showing that it helps to reformulate and solve historiographical problems in the study of Enlightenment optics. For example: (a) What prevented the full mathematization of physical optics during the 18th century? (b) Why was the mathematical part of Euler's wave theory of light neglected by German scientists, while at the same time his general theoretical position — light is a wave in a subtle medium — was widely adopted? (c) How should the remarkable turn to qualitative chemical arguments in German optical debate around 1790 — in contrast to the mathematization until that time — be characterized and explained? (Note that this shift is, in important respects, opposite to the one in 18th-century electricity.)

In conclusion, the implications of a three-fold division for the historiography of physical sciences other than optics, will be discussed.

(1) C. Hakfoort, *Optics in the age of Euler* (in Dutch, with a summary in English; Amsterdam: Rodopi, 1985).

James Evans

Assistant Professor, University of Puget Sound, Tacoma, Washington

RUMFORD AND THE SWISS / PICTET AND THE BRITISH

Marc-Auguste Pictet, professor of natural philosophy at the Academy of Geneva, conducted a remarkable series of experiments on radiant heat in the 1780's. Through the publication of his Essai sur le feu in both French and English language editions, his results became widely known and had an important influence on the British investigators, most notably on the Anglo-american Count Rumford. Moreover, Pictet, through his editorship of the periodical Bibliothèque britannique, played a significant role in making British discoveries, including those of Rumford, better known on the Continent. Using both printed sources and unpublished letters, we shall explore the personal and professional links that existed between Rumford and the Genevan physicists on the one hand, and between Pictet and the British investigators on the other, with special attention to the personal relations of Pictet and Rumford.

Burghard Weiss

Technische Universität Berlin, Berlin, Federal Republic of Germany

THE PRINCIPLE OF MOVABLE EQUILIBRIUM IN EARLY 19TH CENTURY PHYSICS

In 1791 the Genevan natural philosopher Pierre Prevost (1751-1839) published his first account of the principle of movable equilibrium: That all bodies regardless of temperature are constantly radiating heat and that an equilibrium of heat between two bodies consists of an equality of exchange. The historical analysis of Prevost's unpublished scientific papers kept at the University Library of Geneva has shed light on the origins of his principle of movable equilibrium. Based on earlier ideas about the corpuscular-mechanical nature of heat which had been brought forward by Nicolas Fatio de Duillier, Daniel Bernoulli and Georges-Louis Lesage, his new approach to thermal equilibrium had been specially stimulated by problems of magnetic theory, about which he had published De l'origine des forces magnétiques in 1788 (cf. B. Weiss: Zur Entstehung des Begriffs *dynamisches Gleichgewicht*: Pierre Prevosts physikalisches Skizzenbuch aus den Jahren 1788 bis 1792, Sudhoffs Archiv 68 (1984), 130-142).

Despite its importance for the mathematisation of radiation theory executed by Joseph Fourier in 1807 to 1809 (cf. R. M. Friedman: The creation of a new science: Joseph Fourier's analytical theory of heat, Hist. Stud. Phys. Sci. 8 (1977), 73-99), the new principle was only reluctantly accepted in early 19th century physics. Conceptual difficulties and the speculative method used in deriving the principle hampered the recognition of its positive heuristic value in the contemporary scientific debate. So the concept of movable equilibrium had to be reinforced in 19th century thermodynamics where its value was finally fully recognized by Gustav Robert Kirchhoff (1859), Balfour Stewart (1861), and James Clerk Maxwell (1870).

Prof. Eizo YAMAZAKI

Meiji University, Tokyo, Japan.

J.B. BIOT ET A. FRESNEL: LES PREMIERES THEORIES DES POLARISATIONS CHROMATIQUE ET ROTATOIRE. (1812-1822)

En observant par l'analyseur la lumière blanche polarisée traversant perpendiculairement une lame cristallisée, Biot pense qu'il faut décomposer par la pensée en deux parties la lumière incidente: l'une E est celle qui éprouve une action de la part de la lame; l'autre O est celle sur laquelle la lame n'agit point. En appliquant les lois de Malus aux faisceaux E et O, quand la section principale de la lame forme un angle i avec le plan de polarisation de la lumière incidente et celle de l'analyseur, un angle α, il propose les formules suivantes pour les intensités des images ordinaire et extraordinaire, F_o et F_e :
$$F_o = O\cos^2\alpha + E\cos^2 i \cos^2(\alpha-i) + E\sin^2\alpha \sin^2(\alpha-i) \quad F_e = O\sin^2\alpha + E\cos^2 i \sin^2(\alpha-i) + E\sin^2\alpha \cos^2(\alpha-i)$$
Biot pose ensuite $E\sin i \cos i \sin(\alpha-i)\cos(\alpha-i)$ comme terme de correction, et parvient finalement aux expressions suivantes:
$$F_o = O\cos^2\alpha + E\cos^2(2i-\alpha) \qquad F_e = O\sin^2\alpha + E\sin^2(2i-\alpha)$$
On peut penser que les coefficients de terme de correction ont été déterminés pour simplifier ses expressions et satisfaire la condition $F_o + F_e = O + E$. Il montre par les expériences que les épaisseurs auxquelles les diverses teintes E se forment dans les lames d'un même cristal sont proportionnelles à celles qui sont marquées dans la table de Newton pour les lames minces d'une même substance.

Une fois que Biot eut trouvé ses formules empiriques en 1812, il fit tout son possible pour les déduire analytiquement. Il dut fixer son attention sur ce que l'azimut i n'intervenait dans ses formules que sous la forme de $2i-\alpha$, et il arrive à l'idée de la polarisation mobile.

Biot semblait s'intéresser bientôt à l'étude de la polarisation rotatoire qui lui paraissait être un phénomène plus élémentaire que la polarisation chromatique. Si celle-là devait être expliquée par la rotation du plan de polarisation, il pouvait croire que celle-ci aussi devait être expliquée de la même manière.

L'auteur veut exposer les travaux de Fresnel en comparaison avec ceux de Biot. En 1816, Fresnel expliqua l'une des teintes produite par la polarisation chromatique, du point de vue de la théorie des vibrations. La transversalité de la lumière polarisée fut suggérée alors par Ampère. Comme il ne put adopter cette hypothèse en 1816, il ne put compléter ses formules. Aprés sa première étude sur la polarisation, Fresnel, au lieu d'étudier la polarisation rotatoire, fit l'étude très élémentaire: poursuivre les modifications que produit la réflexion de la lumière polarisée. Cette étude fut publiée en 1817. Fresnel aurait eu de la chance dans ce mémoire de déclarer que la lumière se composait d'ondes transversales. En effet, dans le supplément de ce mémoire, il pense que deux rayons lumineux polarisés en sens contraires doivent être assimilés à deux forces perpendiculaires entre elles. Il trouve aussi dans ce mémoire le moyen d'imiter le phénomène de polarisation rotatoire. Mais pour perfectionner en 1822 la théorie de ce phénomène, il devait être le promoteur résolu de la théorie des vibrations transversales de la lumière. Il y parvint en 1821, en déduisant les mêmes formules que celles de Biot.

David Gooding

Lecturer, Humanities and Social Sciences, University of Bath, England

HOW PRACTICES SHAPE THEORIES IN PHYSICS

The stimulus of new instruments to science is well known. The microscope opened up new observational domains; the voltaic battery and the cathode ray tube were used to develop new experimental possibilities. The phenomena produced with them called for new theoretical approaches. They also demanded new experimental techniques. This paper argues that new theories are shaped by such techniques. This is not a novel thesis, but it remains controversial because it implies that new ways of <u>thinking</u> about the natural world are implicit in new ways of <u>doing</u> things in the world.

I examine experimental responses to Oersted's discovery of electromagnetism to show that early practitioners developed a repertoire of techniques to observe, manipulate, replicate, communicate and interpret the new phenomena. These practices became an important resource for, and influence upon, later developments in electromagnetic science and technology. I discuss: (1) the use of 'magnetic curves', adapted from existing methods of representing terrestrial magnetism and applied by Peter Barlow and Michael Faraday to the exploration of minute, local magnetic effects in their laboratories, (2) the convergence of interests underlying the transfer of these practices from the global to the local domain, and (3) the articulation of physical possibilities implicit in experimental and representational practices. I argue that much of Faraday's later work on magnetic lines of force was an attempt to develop the physics implied by practices he had learned during the 1820s. I also show that there was no clear differentiation of interests motivating the early electromagnetic work and that this was important to the rapid development of experimental problem-solving techniques.

Finally I explore some implications of the influence of representational and experimental practices on the emergence of electromagnetic field theory. This development cannot be viewed as a purely theoretical process. Methods of representation, observation, communication and argument must be placed alongside intellectual, personal, institutional and ideological sources and influences. More generally, to situate devices and techniques within the context of exploratory work draws attention to the wider significance of scientific instruments in the history of scientific theories.

Jinguang Wang

University of Hangzhou, People's Republic of China

Fuguang Zhen and his Optical Achievements

(1) A brief account of his life
Fuguang Zhen (1780 - ?) was a native of She Xian, Anhui Province. He mastered mathematics and physics and wrote many works, such as "Jingjing Lingchi" 镜镜诊痴 and "Feiyin Yuzhi Lu" 费隐与知录
(2) "Jingjing Lingchi"
This book is the first definitive optical work in China. It deals with the optical concepts, principles and instruments very clearly and brilliantly.
(3) "Feiyin Yuzhi Lu"
It contains knowledge of physics, meteorology, astronomy, biology, medicine and etc. There are 225 items altoghether in it, among them over 20 items are about optics. We'll select some of his representive tiems in this book and discuss them in three groups: (a) Pinhole, (b) Convex Lens and Lens of Ice, (c) Nearsightness, Presbyopia and Accommodation of the Eyes.

Yasuo Nakagawa

Associate Professor, Kobe University, Japan

HISTORICAL REEVALUATION OF THE STUDIES ON ATOMIC BOMB CASUALTY AT HIROSHIMA AND NAGASAKI

August 6 and 9, 1985 are the 40th anniversaries of the atomic bombing at Hiroshima and Nagasaki. It has been stressed that the Atomic Bomb Casualty Commission has been conducted in the high traditions of scientific inquiry (R.K.Cannan, Bullet. Atomic Scient., 19, No.8, 1963).

However, the early results surveyed by the former organization of ABCC, the Armed Forces Joint Commission for Investigating the Effects of the Atomic Bomb in Japan, in cooperation with Japanese scientists, were compiled in The Effects of Atomic Weapons edited by the U.S.DOD and the U.S.AEC in 1950. The early medical survey by the Joint Commission and ABCC was concentrated upon the standard killed and casualty rates, which were the important military indicators presenting the destructive power of this new type bomb. Their early studies on radiation injury were also performed primarily on fatal injuries and acute radiation injuries. Both of the surveys were essential to the civil defense program (A.W.Oughterson and Shield Warren, eds., Medical Effects of the Atomic Bomb in Japan, 1956). Moreover, ABCC had the purposes to survey the injuries by the military use of atomic bombs, to survey radiation effects on man and his posterity, and to use their therapeutic data for people suffered from other atomic bombing (H.G.Taylor, Kagaku, 22, No.6, 1952).

On the other hand, most of the Japanese surveying groups were composed of the atomic scientists who had worked for the development of the Japanese atomic bomb during the Japanese aggressive war, and of medical scientists who served for the Japanese militarism. The primary purpose of this survey under the Japanese Imperial Government was to confirm that the effects had been from atomic explosions. So their early medical survey was devoted wholly to the dead (M.Tsuzuki, Jap. Safety Forces Med. J., 2, No.9, 1955), and overlooked the injuries of the survivors caused by low level radiation, although they observed acute symptoms more than 2 km. from the ground zero (C.I.Dunham, et al., J.A.M.A., 147, No.1, 1951).

Furthermore, the subsequent medical survey by ABCC was biased by their military and social purposes. For example, many radiation deaths from leukemias and cancers before 1950, at the area more than 2 km., or under the radiation dose of 25 rems were underestimated by ABCC. The discarding of the survivors was due to their principles that people exposed to radiation under 25 rems dose were free from any radiation injury, and that many of the excess deaths of the survivors resulted from their injuries by blast and heat, from their poor economical-physical conditions at the wartime and after the bombing(K.G.Johnson and A.Ciocco, J. Hiroshima Medical Ass.,22, No.12 1969). The radiation survey on Hiroshima-Nagasaki survivors and the radiation protection standards based on it should be reexamined.

Pd

Regis Cabral

Department of History, University of Chicago, Chicago, IL, USA

The Peron-Richter Fusion Program, Argentina, 1948-1953

The atomic bomb did more than contribute to the termination of the Second World War. It also escalated the victors' into a Cold War conflict. Moreover nuclear energy promised cheap energy for industrialization and transportation and, for those who could discover its "secrets," independence and a place among the great powers. For these reasons many nations seriously considered their possibilities in the nuclear field as early as 1945. Argentina was no exception.

Peron's nation had the mineral resources, population sufficient enough for extensive projects, a large number of educated citizens, despite the greater percentage of illiterate individuals, and a reasonable, for an underdeveloped country, scientific tradition. In addition Argentina received European immigrants with technical or scientific training before, during, and after the war.

Counting on these resources, Argentinean scientists, such as Enrique Gaviola and Teofilo Isnardi, proposed to their government several nuclear projects. Nevertheless, because of the conflict between academia and Peronism, the government rejected these ideas. Stranded from Academia, the Peronist government fell prey to a minor nazist scientist, Ronald Richter. Richter's program led eventually to Peron's announcement of 24 March 1951 that Argentina had controlled nuclear fusion. It took eighteen months for the government to dismiss the program and to close down its laboratories at Huemul Island. Peron finally responded to international criticism, domestic political attacks, and the scientific evaluation of the Argentinean scientists.

In this paper we will present the major features of the Peron-Richter project, including Richter's "theories." We will also address the reactions the events at Huemul Island generated. We will conclude with an evaluation of the program's impact on the development of nuclear research in Argentina.

Day, Anne L.

Professor of History, Clarion University of Pa.   Clarion, Pa. 16214

THE U.S. POLICY OF CONTAINMENT AND THE TRANSFER OF NUCLEAR TECHNOLOGY

The aim of this paper is to provide a linkage between the U.S. policy of containment and the transfer of nuclear technology since World War II.

The design which will be outlined here will illustrate that in the first phase, 1943 to 1953, the U.S. nuclear policy was closely aligned with containment, was determined by military superiority and functioned in extreme secrecy. From 1954 to 1963, the policy of atoms for peaceful use, introduced by Eisenhower, expanded the walls of containment, with transfers encouraged under U.S. safeguards. From 1963 to the present, transfers of nuclear technology continued with greater amounts of agreement uner U.N. international safeguards. Thus, containment took on global dimensions.

While the world is faced with competitions and proliferation in nuclear technology, the paper argues for increased reliance upon existing treaties, agreements and U.N. safeguards. The paper should provide insight into the shapes given global technology by diplomatic policies as well as a look at the large scale organized networks of technology.

Sang-yong Song

Hallym University, Chunchon, Korea

CENTRAL INSTITUTE OF INDUSTRY, 1946-1961

The Central Institute of Industry, which had started as the Bureau of Mint in 1883, was reestablished after World War II. It was virtually the only institute of science and technology in Korea at that time. It not only was the center of research, but also played a leading part in science education and the popularization of science. Later its function was partly shared by the Army Institute of Technology and the Korea Atomic Energy Institute. Its influence, however, was decreased gradually until it was downgraded to a mere testing laboratory with the change of government.

Lashchyk, Eugene

Associate Professor, La Salle University

THE ROLE OF MODELS IN THE CREATIVE PROCESS IN SCIENCE AND LITERATURE

Heuristice for creativity and discovery in science and literature will be explored which go beyond Einstein's suggestion that "all concepts are...freely chosen conventions" or that scientists simply arrive at interesting theories by the creative leap of the imagination. Models, analogical reasoning, and metaphors can provide useful strategies for a heuristic of creativity in science and literature. The role that models and analogical reasoning play in scientific discovery and creativity is relatively well know. Mary Hesse, among others, has shown how scientists exploit neutral analogies and arrive at potentially fruitful hypothotheses and theories.

I want to suggest that models of the world, of nature, and of man's place in nature plays a similar role as a stimulus for the creative outpouring of the poet, novelist, or dramatist. One can go beyond the claim that creativity is inscrutable, that it is simply the outpouring of the poet's imagination. Without such models the author is totally cut off from a perspective or an interpretation. He or she is in short confronted with an infinity of possibilities and it is natural to wonder how one could make sense of the creative process. Joseph Conrad describes how he came upon the basic model of man that guided his writing of a novel. He describes how he once heard a story about an "unmitigated rascal" who had stolen a large quantity of silver. Conrad says "I did not see anything at first in the mere story." Then: "It dawned upon me that the purloiner of the treasure need not necessarily be a confirmed rogue, that he could even be a man of character." This idea that a man or a country could be both good and evil Conrad developed in his novel. "Such," say Conrad "are in very truth the obscure origins of Nostromo - the book".

The form of a poem or other literary work provides another example of how models, in another sense, can function to provide a strong direction for writer's creative energies. Robert Lowell describes how his poem "Skunk Hour" was modeled on Bishop's "The Armadillo". He says "The dedication is to Elizabeth Bishop, because re-reading her suggested a way of breaking through the shell of my old manner. Her rhythms, idiom, images, and stanza structure seemed to belong to a later century. "Skunk Hour" is modeled on Miss Bishop's "The Armadillo," a much better poem and one I had heard her read and had later carried around with me. Both "Skunk Hour" and "The Armadillo" use short line stanzas, start with drifting descriptions and end with a single animal."

A similar role was played by Newton's Principia. Paragons of science such as this functioned as heuristic for research.

Morris L. Shames

Associate Professor, Concordia University, Québec, Canada

LITERARY THEORY, ANALYTICAL PSYCHOLOGY AND THE SCIENTIFIC IMAGINATION

Science cannot claim for itself epistemological primacy owing to the historical and persisting problem of induction, a permeating ideological substrate, and a prototypical analysis which tends to focus on the "yeoman" side of science, that is, the science of Thomas Kuhn (1970) which focusses on the collectively articulated and practiced methodology of a highly conventional discipline. Woven into this, however, is the provenient cognitive act born of the creative imagination which is not addressed by the usual sociological-epistemological analysis.

The creative side of science is a hermeneutic affair which is founded on an appreciation of language, semiotics and literature and which is mediated by metaphor--sytematically understood. Furthermore, this process is psychologically underwritten by Jungian *Analytical Psychology*, in particular, the *collective unconscious* and the psychologically significant, inherited symbolism --epigenetic rules, as it were--which appertains thereto. Northrop Frye's (1957) systematically contrived theory of symbols and metaphor is the *hierarchical* model--based upon an exhaustive assay of Western literature --on which the epistemological analysis dealing with the scientific imagination turns.

Evidence for this analysis is grounded in the treatment of *analogy* and *metaphor* both and the crucial role they are presumed to play in science (MacCormac,1983; Sacks,1979). Coupled with *archetypal psychology* and the presumption, even among behaviorists, that there is an epistemic drive in man underpinning *symbolizing activity*, the argument for metaphor mediating the provenient *creative*, cognitive moment in the scientific enterprise, is a compelling one indeed.

Frye, N. (1957) Anatomy of criticism. Princeton, N.J.: Princeton University Press

Kuhn, T.S. (1970) The structure of scientific revolutions. Chicago: Chicago University Press.

MacCormac, E.R. (1983) Scientific metaphors as necessary conceptual limitations of science. In N. Rescher (Ed.) The limits of lawfulness:Studies on the scope and nature of scientific knowlege. Lanham, Maryland: University Press of America.

Sacks, S. (Ed.)(1979) On metaphor. Chicago: Chicago University Press.

Christopher B. Burch

Visiting Assistant Professor, University of Pittsburgh

MAGIC, METAPHYSICS AND ELECTROMAGNETIC METAPHOR IN MOBY-DICK

Electromagnetic phenomena serve both naturalistic and metaphorical functions in Moby-Dick. Melville's attitude towards science and philosophy (particularly natural philosophy) is disparaging. Instead, these phenomena are mysterious and in principle unexplainable, metaphorically betokening the more fundamental metaphysical realm of fate and destiny that gives rise to the naturalistic world of appearances. In short they are not scientific but magical. His use of this rich source of metaphor represents the poetic function, his attempt to say the unsayable, whether of nature or of man. As poet Melville bypasses the analytical toils that had for centuries beset exploration of fate and destiny, and sets forth free-flowing ideas with beguiling vagueness and ambiguity. Melville's episodes abound with metaphysical connections melded with natural appearances; and these connections are suggested metaphorically.

One leaves civilization in a ship, and enters a simpler Empedoclean world of earth, water, air and fire. Air and water are the dominant elements; but fire and earth play their roles--the focus here is electromagnetic phenomena and iron. They each have a benign and a malevolent side. The benign sun often betokens metaphysical navigation; its obverse is lightning in a storm and St. Elmo's fire at night.

Magnetism and lightning belong to the same metaphysical context for Melville. They are connected, for example, in the change of polarity of the compass needle by lightning. Like the sun the magnet is capable of providing both natural and metaphysical navigation. Zoroastrianism is frequently suggested, although fire may have either a dark or a lighter side.

From an "elemental strife at sea" Ahab is branded with a lightning shaped scar on face and neck; and his eponym "Old Thunder." His influence on his crew is often described in magnetic terms. Whales, and Moby-Dick in particular, are also associated with fire and magnetism. Ahab and the whale are drawn to each other, and the approach is attended by electromagnetic omens.

Ahab's mission is to step momentarily out of the constraints of destiny for his chance at vengeance. Again attended by omens in the form of extraordinary electromagnetic phenomena. One by one are lost the natural aids to navigation. Declining all warnings, Ahab follows his own metaphysical direction, and wrests from fate his chance for vengeance.

David A. Hollinger

University of Michigan

A FANATIC FOR VERACITY: SINCLAIR LEWIS'S ARROWSMITH AND THE ETHIC OF SCIENCE

The most widely acclaimed work of the first American to win the Nobel Prize for literature is a landmark in history of the ideology of science. This novel of 1925 domesticated to American culture the potentially alien cult of "pure science" and rendered it as American as Mark Twain. By adopting conventions peculiar to American literature and by drawing upon social myths uniquely powerful in the United States, Sinclair Lewis produced a work that was at once emblematic of an international enthusiasm for the ethic of science and symptomatic of a local anxiety about the fate of traditional American virtues in an increasingly commercial and technological society. In Arrowsmith, Lewis invoked two of the most powerful myths in all American history-- the myth of the protestant-saint and the myth of the pioneer-frontiersman-- and amalgamated them into a single figure, that of the German-inspired research scientist. At the same time, Lewis consolidated the overbearing masculinity of the scientific ethic, contrasted this ethic with an ethic of love, and brought to its apotheosis the distinctly phallocentric vision of science to which he was heir. As an attack on the female gender, Arrowsmith rivals Schopenhauer's essay "On Women."
In this respect and in many others, this novel displays a program of cultural criticism with roots very distant from the needs and interests of a scientific establishment in search of legitimation. Although Lewis was directly and extensively instructed by a zealous pro-science ideologue, Paul DeKruif, the text illustrates the dynamic, two-way process by which the ideology of science is generated and popularized: interests in the meta-scientific environment reach out toward science for resources, and project onto science images that serve these meta-scientific interests. Arrowsmith is not an unmediated product of the public relations office of the American Association for the Advancement of Science. Study of this text can enable "externalists" to escape from their porochial perpetuation in institutional-political form of the scientific determinism they condemn in its intellectual form; Arrowsmith, like so many other monuments in the history of the ideology of science, derives not only from the "impact of science," but from a dialectical discourse in which scientists and other elites create a public inventory of cultural symbols by means of which the meaning of the scientific enterprise is established and sometimes contested.

Rita K. Hessley

Assoc. Prof. Chem., Western Kentucky University, Bowling Green, KY 4210

THE SACRAMENTALITY OF SCIENCE: REFLECTIONS ON THE WRITINGS OF PIERRE
TEILHARD DE CHARDIN IN RELATION TO A CONTEMPORARY THEOLOGY OF MAN

In the years since the Age of Enlightenment first identified a
disparity between the claims of religion and the achievements of
science, a number of different propositions have been made about the
severity of the division and the potential for reconciling it.
   Throughout his writings, paleontologist and Jesuit priest Pierre
Teilhard de Chardin never waivered from his belief that a person can
act forthrightly as a scientist and not shrink from embracing the
fullness of religious faith. Rather than defend an imaginary and
unapproachable superiority of religious faith over the endeavors of
science, Teilhard repeatedly refers to the intimacy between the
hunger of the probing intellect and the restlessness of the
worshipping spirit which constitutes mankind.
   In a more contemporary theological perspective on man, Orthodox
theologian Father Alexander Schmemann is also forceful in his
critique of a mentality that would either secularize Christianity so
that it can properly 'help' science advance the condition of mankind,
or which would subvert 'wordly' activities as ultimately being of no
real value in relation to eternity. Schmemann, not unlike Teilhard,
sees mankind, in the fullness of his intellectual capacity, as 'homo
adorans': a man at worship. Contrary to some current religious
notions about man, Schmemann asserts that mankind should find itself
lead irrevocably toward the realization that exploring and trans-
forming the universe are acts which, in addition to being 'scientific'
are also inately sacramental.
   It is consitent with the thought of both men that a sacrament is
not a magical act having invisible, other-world affects. Nor is a
sacrament an action by which a natural or profane object is converted
into a sacred one. Rather, a sacrament serves to reveal the nature
and the destiny of ordinary events and objects. Taken together, sac-
raments show that all aspects of life are means to a knowledge of God.
   Using selected writings of these two men, this paper will
attempt to describe a synthesis of science and sacramentality that
does not merely bridge the gap between science and religion, but
which draws them back into proper unity without compromising the
integrity of either.

Christine BLONDEL

Musée de La Villette , Paris

Propagation mechanism in a medium within Ampère's electrodynamics.

Ampère wanted to will to posterity the image of being the "Newton of electricity" , image which was in fact later applied to him by Maxwell.

His fundamental formula of electrodynamics is indeed an elementary formula of action at a distance on a newtonian model.

We want to talk here of the attempts of Ampère to explain the electromagnetic effects in the space surrounding conductors, using a mechanism of propagation through a medium. These attempts can be tracked back to his first researches to unify electricity and magnetism in 1801, twenty years before the discovery of electromagnetism. They find an unrecognized outcome in 1825 in his hypothesis on "lignes d'aimantation" which are nothing but the usual magnetic flux lines. All along his research Ampère followed the methodological advice given by Maxwell in 1855 : "It is a good thing to have two ways of looking at a subject". He also showed that he was strongly influenced by the ether model of Fresnel and unitary philosophers.

LIVIU SOFONEA University of Braşov, ROMANIA
NICHOLAS IONESCU-PALLAS, Center of Physics, Bucharest

NEW CONNECTION BETWEEN MECHANICS AND ELECTRODYNAMICS.[x] AN
HISTORICAL-EPISTEMOLOGICAL APPROACH.(Basical items only).

1. The early theories of interactions between steady electrical currents, coming from A.M.Ampère, W.Weber and H. Helmholtz, are re-investigated from the "invariantive" IM standpoint put forward by Octav Onicescu (Bucharest), in his mechanics(IM) is essentially based on the Euclidean invariants of th configuration). If the "reciprocal action" between currents we mean the potential energy then : the problem may be studied relying the following principles(familiar in the epoch: 1/Symmetry under "interchanging" of currents, 2/crossed homotety, 3/principle of "ondulated" currents, 4/homotetical(dilatational) invariance of the configuration" of currents, 5/ principle of conservatice forces. The result of this investigation is the Helmholtz general formula of the interaction between two infinitesimal currents containing two dimensionless constants whose sum is 1. Unlike the Weber's formula this formula definitely contradicts the classical mechanics of Newton but is in full agreement with I.M. On this basis can be built the $\mathcal{L}_{I.M.}$ Lagrange function ; the result is the two body electrodynamic Lagrange function $\mathcal{L}_{E.P.}$ due to Einstein and Planck : derived from Maxwell's electrodynamics and special relativity, In this way the synthesis of Electrodynamics may be fully acomplished by pursuing the mechanical "line" in the tradition of Ampere, Weber and Helmholtz. Although this research line ceased at the end of XIX-th century, owing to the high reputation of Newtonean mechanics (and to a lack of temerity: required for a deeper reform) the "viability" of this line (perhaps more logical) than the historical one) is demonstrated ipso facto, giving support to our thesis on the manysidedness of scientific approach and : to the role of psychological factors in the decisional act at branching points.

2. Bernard Riemann a forerunner of Electrodynamics. His ideas; a scenario based on this ideas only can lead us to the Maxwellian synthesis (but using an original way).

3. Connections between mechanics (in a : classical &. semi quantum manners) and electrodynamics in the "problem of the most simple atom(the atom of hydrogen): some historical and epistemological comments concerning the Pauli, and Dirac first approaches.

The manysidedness of the scientific approach is very relevant.

[x]/A detailed preprint will be delivered on the occasion of the Congress.

Dan Florian Sămărescu, Research Institute for the Electrotechnical Industry, Bucharest, Romania.

THE IMPORTANT ROMANIAN CONTRIBUTION IN THE SCIENTIFIC RESEARCH OF THE ELECTROMAGNETIC FIELD DIFFUSION IN CONDUCTORS

The electromagnetic field of the diffusion and the heat diffusion, two different phenomena, have similar equations of mathematical modellation, which made possible the mutual borrowing of results.
This type of phenomena, in the domain of electroheating, especially the preoccupations for coupled equations of electromagnetic and heat field, represented the object of two Romanian scientists, namely the lamented academician R. Răduleţ and A. Avramescu who enriched the scientifical treasury with valuable original contributions.
A. Avramescu completed in 1937 the information in the field of conductors heating at short-circuiting, taking first into consideration the quadratic variation of specific heat with the temperature, thus determining experimentally the variation of thermal diffusion up to copper and aluminium melting. Afterwards taking into account the skin effect he solved the problems of heating in adiabatic state. In 1952 A. Avramescu computed the distribution of the electromagnetic field in cylindrical conductor at voltage step excitation, in transitory state. A. Avramescu introduced the time constant of the diffusion phenomena and characterised the thermal behaviour of the objects.
R. Răduleţ paid a particular attention to the eddy currents in electric induction furnace and determined the raising of efficiency in function of constants of material of object and screen, calculating the equivalent resistance and establishing the expressions for shape factors. Approximating the form of object by a rotating elipsoid, whose symmetry axis is identical to the furnace axis, he succeeded in 1937 to establish some more accurate expressions. Such works examinated from different points of view, in collaboration with prof. C. I. Mocanu, remained the most advanced in the field.
R. Răduleţ elaborated together with Al. Timotin and A. Tugulea the general theory of transitory parameters of massive conductors.
The results were internationally recognised and represent the valuable Romanian contributions in the study and applications of the diffusion of electromagnetic and heat field phenomena.

P.M. Harman
University of Lancaster, England

THE MATHEMATICAL PHYSICS OF MAXWELL'S TREATISE

Drawing on manuscript as well as published materials, the paper will discuss the framework of Maxwell's mathematical physics in his **Treatise on Electricity and Magnetism** (1873). The paper will emphasise Maxwell's interest in new mathematical methods, and will discuss his application of mathematical ideas to physical theory. The paper will characterise the way in which Maxwell's incorporation of quaternions, integral theorems and new approaches to geometry, and his distinctive approach to Lagrangian dynamical theory, enabled him to fashion the geometrical and dynamical physics of the **Treatise**.

Salvo Salvatore D'Agostino

Professore Associato, Dipartimento di Fisica, Università di Roma

METROLOGY AND DIMENSIONAL THEORIES IN MAXWELL'S TREATISE

Quite early in his scientific careear (1855) Maxwell was impressed by the equality between Weber's conversion factor and the velocity of light, an equality which he rightly perceived as an important hint in favour of his electromagnetic theory of light.

However, he was confronted with various kinds of difficulties when he searched for a demonstration, because Weber's factor, in Weber's Electrodynamic Theory, had the precise meaning of a convection velocity of electric charges, and this meaning had to be redefined in Maxwell's field theory, in order to become that of a propagation velocity of Waves.

In his 1873 Treatise, Maxwell approached the problem of the above equality by presenting arguments based on his theory of two Absolute Systems of Units and his Theory of Dimensions.

For this approach he was later criticized by Planck, Sommerfeld and others.

Schubert, Helmut

Deutsches Museum, Germany

The Liquefaction of Permanent Gases and the Discovery of Superconductivity.

   The artificial cooling of food and beverages has been in use since the older days for people of high social rank, applying evaporation and freezing mixtures. The physical reasons for these processes remained unreflected, only the practical applications were of interest. In the early days of capitalism, when the cities grew and trade expanded, cooling of food in warehouses and for long-distance transportation became a topic for engineers and scholars(Monge, Cullen, Guericke..). They developed the first thermodynamic notion,which was enlarged in the 18th and 19th century (Carnot, Faraday, Clausius..). A new method for cooling appeared, the liquefaction of gases.
   At the end of the last century, three European laboratories competed to liquefy the last permanent gases:hydrogene and helium. To win this struggle was not only a matter of scientific and experimental genius, but the laboratory equipment was also a most important factor. J. Dewar, whose laboratory resembled a machine room,succeeded in liquefying hydrogene, leaving his poorly equipped rival at the University Krakow far behind. At the University of Leiden, H. Kamerling Onnes organized low temperatue research on a then unknown level. With its school for instrument makers, its own scientific journal and equipment, which reminded one of an industrial plant,Leiden was the first place where helium was liquefied, and remained the only one for 15 years.
   Besides the liquefaction of gases,a second stream towards the discovery of superconductivity was the interest in measuring the temperature dependence of the electrical resistance. In the mid-19th century, W. Matthiessen used this behaviour to determine the purity of metals.He worked for the British Government to solve problems which occurred in laying the first trans-atlantic cable. The low temperature-behaviour of electrical resistance disproved all classical theories of electrical conductivity.Therefore for the first time, Onnes applied the new quantum physics in analogy to Einstein's theory of specific heat. His estimations noted mercury as the only metal to be suitable for measurements at temperatures next to absolute zero. The first measurement supported Onnes' quantum Ansatz, but with greater accuracy, one problem appeared: at about four degrees above zero the resistance suddenly disappeared. Two years later, this surprising behaviour of matter was called superconductivity.

Colin A. HEMPSTEAD

Senior Lecturer: History of Science and Technology Teesside Polytechnic, Cleveland, England.

ELECTRICAL CONDUCTION IN SOLIDS: THEORY AND PRACTICE, 1900-1920

The conduction of electricity through solids, superconductivity and specific heats at low temperatures posed insuperable problems to classical physics; but with Wilson's "The Theory of Electronic Semi-Conductors" (1931) the electrical phenomena could be understood. The working out of his theory to provide our picture of solids is still in progress, but his paper clarified some very obscure properties.

During 1900-1920 the real problems of solid state electrical conduction became evident; and it appeared that combinations of classical electricity and heat could not provide a complete account. However, in 1900 a series of papers by Drude indicated a solution to the problem of electrical conduction in metals. He treated the collection of electrons as a gas, and this, combined with ionic theories and electrodynamics promised a satisfactory explanation. Yet 'fringe' studies of those materials we define as 'semiconductors' refuted a simple classification of solids.

The neat binary division of solids (c1870)-conductors/non-conductors, i.e., metals/non-metals - was confused by the fact that compound solids, selenides, sulphides, tellurides and chlorides, conducted electricity. By 1900 it had proved impossible to use existing theories of conduction to describe the behaviour of such substances. Further, studies of thermoelectricity and the discovery of the Hall effect led to the recognition that the divisions metallic/non-metallic, conductor/non-conductor, electronic/ionic, were not sufficiently precise. For elements, such as Se, Te and Ge, together with compounds like PbS, PbSe and iron pyrites constituted a new class of materials. A set of properties defined the class: photoconductivity; photovoltaicity; rectifiction; negative temperature coefficient of resistance; high thermoelectric power; +ve and -ve current carriers without ions.

Much practical knowledge but little convincing theory was found from 1900-1910. Isolated workers, but rarely eminent scientists, studied the problems; however changes began. Thomson and Lindemann in England, Hall in the USA and Pohl in Germany became interested; their theories and experiments showed that new ideas were required. Electrodynamics, for example, was inapplicable, while quantum physics could not yet be employed. It became evident that pure materials formed into good crystals would be essential; for solids were more difficult to treat with than gases. However, by 1920 a state of confusion was passing; Pohl, in post at Gottingen, would provide the empirical base for Wilson's theory, while Sommerfeld's extension of Bohr's ideas outlined the conceptual form. In the twenty years from 1900 was founded semi-conductor physics; and the electrical properties of solids became interesting in their own right.

Ernst BRAUN

Austrian Academy of Sciences, formerly University of Aston in Birmingham

MECHANICAL PROPERTIES OF SOLIDS: 1920 - 1960

What determines the strength of a solid is, in a sense, the most fundamental puzzle posed to the solid state physicist. Yet the very first realistic attempts to solve the problem had to await the advent of the technique of growing single crystals and X-ray diffraction instrumentation.

Thus practical metallurgy and engineering managed on empirical results - to good effect - without basic understanding of the underlying mechanisms.

The first hypotheses of atomic mechanisms of plastic deformation were fanciful and could not interpret experimental results in detail. All hypotheses assigned a role to crystal imperfections, but it was only in 1934 that a plausible defect structure - the dislocation - was postulated. The theory emanated from two centres - Berlin and Cambridge - simultaneously and independently and gained gradual acceptance among physicists. Metallurgists remained sceptical for a long time.

The war disrupted all work on dislocations, but the post-war period saw rapid developments, first in extending the theory to the explanation of further observed facts and general elaboration, later to experimental proofs of the existence of dislocations. The electron microscope made this defect effectively visible and thus dispelled all remaining doubt about it.

Teichmann, Jürgen

Abteilungsleiter, Deutsches Museum, München

The F-center as a crystal defect from 1925 til 1940
(lecture is given in English)

In seiner 'Introduction to Solid State Physics' 1953 stellte Kittel fest, daß eine 'number of facts support this identification' - nämlich das Konzept des F-Zentrums als ein Elektron, das an eine Halogenionenleerstelle in einem Alkalihalogenidkristall gebunden ist. Er wählte drei Beispiele aus, die in den späteren Ausgaben geändert wurden. Wie typisch für Lehrbücher, stimmt diese Auswahl nicht mit der wirklichen historischen Entwicklung zusammen. In diesem Fall wirft sie ein interessantes Licht auf die rasche Verselbständigung des jungen Bereichs Festkörperphysik. Noch im Diskussionsbericht der für das Farbzentrenkonzept entscheidenden Konferenz in Bristol 1937 kann man die historische Situation gut komprimiert ablesen. Die experimentellen Forschungen, insbesondere in Göttingen bei R.W. Pohl, die am wesentlichsten zu dieser Konferenz beitrugen, sollen im Wechselspiel mit den zögernden Interpretationen dazu vorgestellt werden.

Spencer Weart

Center for History of Physics, American Institute of Physics, New York

THE SOLID COMMUNITY: EMERGENCE OF SOLID STATE PHYSICS IN THE 1940S

In 1930 "solid state physics" did not exist; there was neither a unified field of inquiry nor any set of social institutions to sustain such inquiry. By the 1960s the field was the largest sector of the physics community. This remarkable growth is an example of how science in the second half of the twentieth century has reversed its trend toward dividing into isolated specialties, and is reassembling itself into confederations of research interests. The example of the rise of the Solid State division of the American Physical Society shows some of the pressures that created the new field and some of the social forces that it had to overcome. During the Second World War, workers concentrating on applied problems, particularly workers in industry, sought a common forum and also a professional identity as specialists in the solid state. Some academic and theoretical physicists feared this would split and dilute their community, while others feared that metal specialists would separate from physics altogether. In the end the Physical Society remained intact by permitting the formation of broad internal divisions. Not only the APS division but other important institutional inventions, such as a solid state physics panel that advised the United States government during the 1950s, reflected the birth of a new discipline, not separate from the traditional physics community but near its center.

Silvana Galdabini and Giuseppe Giulliani

Dipartimento di Fisica "Volta" dell'Università, Pavia, Italy.

THE ORIGINS OF SOLID STATE PHYSICS IN ITALY: THE INSTITUTIONAL CONTEXT AND EARLY LINES OF RESEARCH BEFORE THE SECOND WORLD WAR.

In this communication we analyse the structure of the university, its operation, its relations with industry, the amount and distribution of funds devoted to research.

Among the contributions of Italian physicists, we will discuss those related with the magnetic properties of matter, the electrical conductivity of solids perturbed or not by a magnetic field, the contact potential, the dislocation theory (Volterra's contribution).

The links between the institutional context and the features of the Italian contributions will be outlined.

A.J. Kox

Institute of Theoretical Physics, Un. of Amsterdam, Netherlands

H.A. LORENTZ' CONTRIBUTIONS TO THE GENERAL THEORY OF RELATIVITY

Soon after the publication by A. Einstein and M. Grossmann of their 'Entwurf einer verallgemeinerten Relativitätstheorie und einer Theorie der Gravitation' (1913), Hendrik Antoon Lorentz took a lively interest in this attempt at a theory of gravitation. Both his scientific notebooks and his correspondence (especially with Einstein and Paul Ehrenfest) show that he spent much time mastering the formalism and understanding the scientific content of the theory. His efforts led to the formulation of a variational principle from which the field equations of Einstein and Grossmann could be derived (January 1915).

When Einstein's final version of the General Theory of Relativity was published in November 1915 Lorentz again took a deep interest. It is fascinating to see from his correspondence and his notebooks how he had to retrace Einstein's steps and duplicate some of his mistakes (e.g. in connection with the demand of general covariance) before he fully understood and accepted the theory. But once he had reached that point he showed his mastery in a series of four papers devoted to Einstein's theory, the first of which was submitted in February 1916. Using an almost purely geometrical approach, he again formulated a variational principle from which he derived a number of interesting and clarifying results.

In later years Lorentz both supervised work by others in the field of General Relativity and made his own contributions. Despite his epistemological objections against the Special Theory of Relativity, Lorentz showed himself a staunch supporter of Einstein's General Theory of Relativity.

Desiderio PAPP - Dr. Phil. & Dr. honoris causa
Membre effectif Académie Internationale d'Histoire des
Sciences (Paris) - Membre Honoraire Académie de Médecine
d'Argentine - Membre Honoraire Académie de Médecine du
Chili, etc.

LE PARADOXE QUI DONNA NAISSANCE A LA THÉORIE DE LA RELATIVITE.

Une version très connue relie l'origine de la Théorie de la Relativité à la célèbre expérience de MICHELSON et MORLEY de 1887, qui essayait de mettre en évidence la translation de la Terre par rapport à l'éther cosmique - c'est-à-dire par rapport à l'espace absolu. -

Cependant EINSTEIN a affirmé explicitement qu'il "ne connaissait ni l'expérience, ni son résultat, lorsqu'il s'était convaincu que la Théorie Spéciale de la Relativité était valable." - En effet, les "NOTES AUTOBIOGRAPHIQUES", rédigées par le grand théoricien pendant sa vieillesse, offrent une profonde introspection des points de départ de la création scientifique d'EINSTEIN, en soulignant que ce fut la thermodynamique qui l'impressionna particulièrement au commencement de sa carrière.

Or, le principe général de la thermodynamique prescrit l'impossibilité de construire un perpetum mobile. - Mais comment trouver un tel principe universel? - se demandait EINSTEIN. - Il réussit à le trouver après avoir réfléchi sur un paradoxe qui présentait de façon graphique l'impossibilité pour un observateur de pouvoir se rapprocher - et encore moins de dépasser - la vitesse de la lumière.

Ce paradoxe permit à EINSTEIN de reconnaître la caducité de la conception de simultanéité, invalidant la conception du temps absolu. - Suit l'exposition et l'analyse du paradoxe qui, d'accord à l'indication d'EINSTEIN lui-même, offrit au génial physicien le point de départ de sa transcendentale théorie.

David C. Cassidy

Associate Editor, Einstein Papers, Boston University, Boston, MA, USA

THE EARLY EINSTEIN: HIS YOUTH, EDUCATION AND SCIENCE

   Einstein's development, education and early science are viewed within the respective contexts of his time and in comparison with some of his contemporaries. While his humanistic and extra-curricular studies played crucial roles in his development, the technological orientations of his era, his family's business and portions of his education are not neglected. From these perspectives the budding genius appears rather typical in many respects, while atypical in others. The consequences of this dichotomy are surveyed for the early years of his career.

Barbara J. Reeves

University Postdoctoral Fellow, Ohio State University

Relativity, relativism, and politics: the cultural reception of Einstein's theories of relativity in Liberal and Fascist Italy

This paper explores briefly the cultural and political reception of Einstein's theories of relativity in Italy in the years immediately following the 1919 announcement at the Royal Society of London of the results of the eclipse expeditions, presented as confirming Einstein's General Theory. Focusing on the visit of Einstein to Bologna and Padua in October 1921 and his expected attendance at the International Congress of Philosophy in Naples in May 1924, I examine the nonscientific preparations for the visits -- the publication of popular articles and polemics in newspapers and magazines, the newspaper coverage of the 1921 visit itself, and their cultural and political repercussions. The theories were interpreted and appropriated for their own purposes by many sorts of nonscientific intellectuals, from engineers to philosophers to literary critics, as well as by political publicists and by Mussolini himself. I explore their conceptions of the social and political uses of scientific knowledge and assess the symbolic value of invocations of Einstein, highlighting the tensions and accommodations between science and other areas of Italian cultural and political life during the complex transition from Liberalism to Fascism.

BERNADETTE VINCENT-BENSAUDE

MUSEE DE LA VILLETTE, PARIS

P.LANGEVIN, propagandist of the THEORY OF RELATIVITY in France.

The penetration of Einstein's theory in France was mainly due to Paul LANGEVIN. As early as 1906 he lectured it at the College de France, and in 1922 - despite the decision of the boycott of German science by french scientific institutions - he invited EINSTEIN in Paris.

This paper will emphasize the philosophical aspects of this campaign. It is remarkable that a great number of the papers published by this established physicist on the relativity theory were adressed to philosophers, in an International Conference of Philosophy or in the Société Française de Philosophie.

It was through LANGEVIN that BERGSON knew EINSTEIN's theory in 1911, which enabled him to write "Durée et simultaneité"in 1922. After talking with LANGEVIN, another french philosopher E. MEYERSON decided to write his Deduction relativiste (1925), which induced G.BACHELAR to reply, a few years later, in "La valeur inductive de la relativité" (1929). Thus LANGEVIN contributed to insert the theory of relativity in the register of History of Philosophy.

The purpose of this paper is to appreciate LANGEVIN's influence on these philosophical debates by questionning his own philosophical development on the theory of relativity.

I will discuss this point first from an epistomological point of view, comparing LANGEVIN's interpretation of theory of relativity in his addresses to philosophers in 1911 and 1922. Then I will turn to his philosophical developments on history of science which he considered as a struggle for life between rival theories.

Eva Isaksson

University of Helsinki Observatory

GUNNAR NORDSTRÖM (1881-1923) ON GRAVITATION AND RELATIVITY

The Finnish physicist Gunnar Nordström (1881-1923) is best known for two contributions to the theory of gravitation: a scalar theory of gravitation (1913) and the general relativistic metric of an electrically charged body (the Reissner-Nordström metric).
    The young Nordström became an ardent supporter of Minkowskian relativistic electrodynamics during his stay in Göttingen in 1906-1907. He was influenced by early work on gravitation by Albert Einstein and Max Abraham. Nordström formulated a special relativistic theory of gravitation in 1912-1913. Its major experimental predictions were (1) too small value in the wrong direction of the perihelion shift of Mercury and (2) the non-deflection of light rays in a gravitational field. The theory was favourably discussed on two occasions by Einstein, who, together with A.D. Fokker, gave it a generally covariant formulation in 1914, when Einstein's own general theory of relativity (GRT) had still not reached that goal.
    Nordström went on to develop his scalar theory by adding a fifth dimension to the four space-time coordinates in a manner analogous to the later Kaluza-Klein generalisation of GRT. Nordström found out however that his attempt to unify electromagnetic and gravitational phenomena did not give correct numerical values. He declared his five-dimensional scalar theory a failure in 1916.
    Nordström's reasons for giving up his own line of research abruptly in 1916, three years before the decisive experimental verification in favour of GRT, were based on his theoretical inclinations. During his stay in Leiden in 1916-1918, he was impressed by Einstein's derivation of the field equations of GRT from Hamilton's Principle. Nordström's later work on GRT, especially his derivation of the gravitational field of a charged particle, was conducted using the Hamiltonian approach.
    Nordström's contributions came to an abrupt end in 1923 when he died at an age of 42, probably as a result of his earlier careless experimental work with radioactive substances.
    The author is working on a scientific assessment of Nordström's life and work. Nordström's private papers, including letters from Einstein, seem to have been destroyed. Some letters from Nordström to Einstein exist, as well as letters to others. As a professor of physics (later mechanics) of the Helsinki University of Technology, Nordström twice nominated Einstein for the Nobel prize in physics for his special relativity theory and GRT.
    The author would welcome any existing material on Nordström and his career.

Jia Sheng

Graduate student, Ph. D candiate, History of science, Cornell
University

EINSTEIN'S THEORY OF RELATIVITY IN CHINA (1920s-1970s)

In the early 1920s, Einstein's theory of relativity was introduced into China by both Chinese and foreign people. Since then Chinese people came to know more and more about Einstein and his theory of relativity. The reception and interoretation of Einstein's theory of relativity from 1920s to 1970s in China is one part of Chinese modern history of science. However, very little has been written on this subject. This paper will deal with the reception and interpretation of Einstein's theory of relativity in China, mostly about social history of Einstein's theory of relativity in China, from 1920s to 1970s.

One will find out that there are some differences between Chinese and Western counties' interpretations of Einstein's theory of relativity. In order to make study clear, this paper will be divided into four parts which correspond to the changes of interpretion of Einstein's theory of relativity in the different periods. They are: the early propagation of Einstein's theory of relativity (1920s-1940s), dialectical materialism and Einstein's theory relativity (1949-1966), criticism of Einstein and his theory in the Cultural Revolution (1966-1976), and, Practice as the sole criterion for testing the truth and Einstein's theory of relativity after 1976 in China.

Eri Yagi

Professor, Toyo University, Kawagoe, Japan

R. CLAUSIUS'S MECHANICAL THEORY OF HEAT

Having studied Clausius's series of memoirs (1850 and others), the following results have been obtained: Firstly, in his earliest analytical expression of the first law of thermodynamics, Clausius placed dQ (the quantity of heat) on the left-hand side (1850). He showed to what extent dQ differs from a complete differential such as dU.

Secondly, Clausius began to write the first law using dU on the left-hand side (1865) when he gave his equation for <u>entropy</u> S.

Thirdly, Clausius was able to recover Carnot's results as the first approximation in the mechanical theory of heat. He further considered the heat expended as the second order in an infinitesimal Carnot cycle.

In this paper I discuss the results of my analysis of Clausius's earlier work on light-absorbtion in the atmosphere and on the intensity of sun light reflected by the atmosphere (<u>Ann. d. Phys.</u> 72 1847, 294-314) along the line of the mechanical theory of heat. Here he studied the propagation of the quantity of light over a great distance on the universal scale. A particle theory that was assumed by Clausius, namely that moving vapor bubbles exist in the air, was also adapted by him. (Feb. 14, 1985)

  Yagi, <u>Historia Scientiarum</u>, Tokyo, No.20 (1981), 77-94.

  Yagi, "Clausius's Mathematical Method" to be published in <u>HSPS</u>, Berkeley.

Martin J. Klein

Eugene Higgins Professor of the History of Physics, Yale University

A HISTORICAL APPROACH TO GIBBS'S STATISTICAL MECHANICS

<u>Elementary Principles in Statistical Mechanics</u> stands alone among the publications of Josiah Willard Gibbs. It was his last work, appearing only a year before his death, and it contains his only writing on this subject. If one adds to this Gibbs's notoriously abstract and austere style of presentation, it is not surprising that we have not yet had an adequate historical account of how his ideas on statistical mechanics developed, or how they should be set in the context of contemporary discussion.

This paper will try to deal with these two issues with the help of manuscript materials in the Gibbs Collection at the Yale University Library. These include his own scattered notes on statistical mechanics, the draft manuscript of his book, student notes on his lectures, and some relevant correspondence. Taken together with his published work, including the 1884 abstract and his obituary notice of Clausius, these manuscripts allow us to see more clearly the special features of Gibbs's approach to statistical mechanics.

Ole Knudsen

Lecturer, University of Aarhus, Aarhus, Denmark

GIBBS IN EUROPE: A PARTIAL INTRODUCTION TO ADVANCED PHYSICS

In 1866, J. Willard Gibbs left his native New Haven to spend three years in Europe, studying mathematics and physics in Paris, Berlin, and Heidelberg. His prior education at Yale had given him a good background in mathematics and mechanics, but there is no evidence that he had been introduced to the theories of optics, electromagnetism and thermodynamics before coming to Europe.

The Gibbs collection at Yale contains three notebooks which give some information as to what Gibbs learned during his stay in Europe. In my paper, I will first review this material and then go on to discuss the possible influence of Gibbs's European experience on his subsequent work in mathematical physics.

F. Bevilacqua

Ricercatore, Dipartimento di Fisica "A.Volta" Università di Pavia

PLANCK ON THE PRINCIPLE OF THE CONSERVATION OF ENERGY

Planck's approach to the Principle of the Conservation of Energy (P.C.E.) has not yet received the attention it deserves from historians.
In "Das Princip der Erhaltung der Energie" of 1887, Max Planck analysed the P.C.E. at length. The analysis extended from its history and meaning to its interpretation in the fields of mechanics, thermodynamics and electromagnetism.
In 1885 this work was submitted, in a competition, to the Göttingen Philosophical Faculty and obtained the second prize, no first prize being awarded. Planck published a second edition, with added notes, in 1908, and unchanged third, fourth and fifth editions in 1913, 1921 and 1924 respectively. In the first part Planck gave a historical survey from the 16th century to 1860. He followed Helmholtz' approach of considering two historical roots of the Principle: the "ex nihilo nil fieri" and the "nil fieri ad nihilum". In the second part he gave a theoretical interpretation of the P.C.E. which is independent of the classical dichotomies of potential and kinetic energy, of intensive and extensive quantities and of free and bound energy. In the third part he compared the different theories of electromagnetism and expressed his preference for the contiguous action theory on the grounds that this theory offers the simplest version of the P.C.E. . This opinion was given prior to and independently of the experiments of Hertz. Despite the fact that the action at a distance-contiguous action debate was not settled immediately, Planck's assessment of it played a role in the shift towards the acceptance of contiguous action of the early eighteen nineties.

Ulrich Hoyer, Universität Münster, FRG

FROM BOLTZMANN TO SCHRÖDINGER

It is well-known that Schrödinger was a fervent admirer of Boltzmann's statistical achievements, which he learnt to know from the latter's disciple Fritz Hasenöhrl. Moreover, Schrödinger approached wave-mechanics from a statistical side, which can be seen from his handbook article on specific heat, and from his papers on gas theory, published in 1925 and 1926. However, it is also true that wave-mechanics developped from arguments, which were essentially non-statistical, and that Schrödinger hesitated accepting Born's statistical interpre= tation of the ψ-function. Nevertheless, it is possible to show that wave-mechanics is intimately connected with Boltz= mann's statistics. The author will study the relation of Boltzmann's 1877 paper on the second law of thermodynamics and probability theory to the time-independent Schrödinger equation, and of Boltzmann's transport equation from 1872 to Schrödinger's time-dependent equation.

Literature:

Ulrich Hoyer, Wellenmechanik auf statistischer Grundlage, Kiel 1983;

Über eine statistische Begründung der Heisen= bergschen Unschärferelationen. Philosophia Na= turalis, Vol. 20, 1983, p. 127-133;

Über eine statistische Begründung der zeitab= hängigen Schrödingergleichungen, ibidem p. 529-535.

Ludwig Boltzmann und das Grundlagenproblem der Quantentheorie. Zeitschrift für allgemeine Wis= senschaftstheorie, Vol. XV, 1984, p. 201-210.

Ulrich Hoyer

Prof. Dr. Westfaelische Wilhelms Universitaet

Bohr's Atomic Theory and Statistical Mechanics

Bohr's early work more or less clearly deals with statistical problems. This is even valid for his two papers on hydrodynamics, dating from 1909 and 1910, insofar as Maxwell already in 1868 had shown that the Navier-Stokes equations can be derived from the kinetic theory of gases. It is more evident for his theory of electrons, published 1911,which starts from Boltzmann's statistical ideas and applies them to the electron theory of metals. His paper on absorption phenomena (1913) also refers to a statistical problem. Nevertheless, Bohr's approach to the problem of the constitution of atoms (1913) was not from the statistical, but from the mechanical side. This is a circumstance deserving explanation. It was a decision of great importance which dominated the course of physics until 1926, when Born gave the statistical interpretation of wavemechanics. Bohr's conviction that either mechanics or electrodynamics or both have to be altered, statistics remaining essentially unchanged, dates from an early stage of the development of modern atomic physics. Obviously, Einstein's ideas on the basis of quantum theory must have influenced Bohr greatly, while Boltzmann's ideas must have gone to the background.

Nadia Robotti, Università di Genova, Genova, Italy

HYDROGEN SPECTROSCOPY BETWEEN 1916 AND 1925: SOMMERFELD'S THEORY VERSUS EXPERIMENTS

Many authors claim that the 1916 Sommerfeld theory of the fine structure of the hydrogen lines still maintained a "satisfactory agreement" with the experimental data, in the years 1920-1925. This thesis is supportd, for instance, by citing the fact that the "new mechanics" with the spin gives the hydrogen atom the same energy levels of the Sommerfeld's theory. From this standpoint the new definition of the hydrogen levels given in the years 1925-26 by Uhlenbeck and Goudsmit, by Slater, by Sommerfeld himself and by others, opposed to the classification given by Sommerfeld in 1916, is considered as independent of consistency problems between the empirical data and the thoretical expectations and is essentially viewed as a theoretical need, due to the parallel development of the research on the atoms different from the hydrogen.

In the present paper I reconstruct the evolution of the empirical basis of the Sommerfeld theory between 1916 and 1925. I moreover exhibit the links between this evolution and the intervention of Uhlenbeck and Goudsmit, of Slater, etcetera on the hydrogen levels and, afterwards, on the concept of spin.

The conclusion I came to is that within the empirical basis of the Sommerfeld theory strong anomalies existed, so as to make even the Kramer's analysis of the Stark effect useless. However, these anomalies did not cause the giving up to the Sommerfeld's theory, till new theoretical perspective for the hydrogen opened up, as alter native to the Sommerfeld's theory. These perspective were revealed by the introduction of new quantal nomenclatures for the non-hydrogenoid atoms, ascribed to the single electron by Pauli in 1924 and parallely by Wentzel.

In this framework and on the ground of the anomalies found in the empirical basis of the Sommerfeld's theory, Uhlenbeck and Goudsmit in 1925 hypothesized the analogy between the hydrogen and the other elements and proposed a new scheme for the hudrogen levels based on the vettor model of Landé. An analogous scheme, based on the quantal nomenclature introduced by Pauli for the alkalimetals, was suggested by Slater. The anomalies found in the empirical basis of the Sommerfeld's theory became then the proving ground of this new theoretical scheme. The successes obtained in this field stated the validity of the new ordering that had been proposed and confirmed the complete analogy between the hydrogen atom and the other atoms.

John L. Michel

Ph.D. Candidate, University of Wisconsin-Madison

WHY MILLIKAN RECEIVED THE NOBEL PRIZE

Robert A. Millikan was ostensibly awarded the Nobel Prize for 1923 for his work "on the elementary charge of electricity, and on the photoelectric effect." Millikan was cited for his "brillant method of investigation" and "extraordinarily exact experimental technique" by the chairman of the Nobel Committee for Physics. These two experimental findings were seen as providing crucial support for the new quantum theories of matter and light formulated by recent Nobel laureates Bohr, Einstein, and Planck.
An analysis of the nominations, and reports, discussions, and conclusions of the Nobel Committee for Physics from 1915 to 1922 reveals for the first time personalities and issues that influenced this award: first, the important role Millikan's American colleagues, who were not physicists, played in his nomination; second, the delaying effect of disputes Millikan had with Ehrenhaft and Ramsauer on the Nobel Committee's deliberations; and third, the influence of a strong, but offically unmentioned, recognition of Millikan's research on ultra-violet spectra by nominators and the Nobel Committee.
These behind-the-scene stories considerably enriches our understanding of both how the nomination and deliberation process leads to the award of a Nobel Prize, and how scientists assessed Millikan's experimental accomplishments.

CHEVALLEY Catherine

Chargée de recherches au CNRS (France)

MEYERSON'S CONCEPTION OF THE QUANTUM IRRATIONALITY

On August the 13 th, 1928, Harald Høffding, writing to Meyerson, stated that he just had some very interesting conversations with Niels Bohr, especially on the question of the irrationality which the Quantum theory introduced into Physics. As is well known, Bohr was Høffding's student and friend, and Høffding was Meyerson's friend and correspondant, and also a specialist of Bergson's philosophy. These connections define a specific context of the history of philosophy, and may help to understand the true significance of the word "irrationality" when it is used by Bohr, or Heisenberg, or Pauli, in relation to their critical analysis of the concepts inherited from classical physics.
    The Irrational is a central piece of Meyerson's construction of the principles of science. It will first be necessary to give a precise caracterisation of the concept, the origin of which is to be traced to Kant's assertion that the desire of the Reason must not be taken as a transcendantal principle, which is the case when one postulates the integrality of the conditional series in the things themselves. Only after this will it be possible to ask why Meyerson thought that "l'indéterminé quantique fait figure d'un irrationnel d'un genre nouveau, plus profond en quelque sorte que celui dont nous avons relevé l'existence, puisqu'il se rapportera au légal, lequel pouvait se concevoir jusqu'à ce jour comme une caractéristique du réel lui-même". The relation which Meyerson establishes between this new kind of Irrational and the role on intuition in Quantum Physics is the most interesting part of the whole theory. Finally it will be necessary to go back to the elaboration of a new "Anschaulichkeit" in the beginnings of Quantum Mechanics.

Ge Ge

Professor, Beijing Graduate School of the East-China Petroleum Institute

ON A LECTURE OF NIELS BOHR

The author of this note studied a lecture delivered by Niels Bohr before the Royal Danish Academy on Feb. 13, 1920. The title of this lecture is On the Interaction between Light and Mather (Collected Works, III, 228-240). According to his analysis, the author made some personal conclusions:

1. In this lecture, Bohr emphasized the unity of light and matter or the material character of light, this might well lead him to some idea of "matter waves."
2. Considering the early viewpoints of nature of light, he made some inspiring assertions.
3. He properly appreciated Einstein's light-quantum hypothesis, without "putting it aside."
4. Some statements may be regarded as the prototypes of the idea of complementarity.
5. As a whole, this lecture is a concise historical retrospection, which is very brief but highly instructing.

HRISTO SMOLENOV

Bulgarian Academy of Sciences

ZENO'S "ARROW" AND BOHR'S COMPLEMENTARITY

The methodological descendants of Zeno's "Arrow" can be traced as far as Bohr's Complementarity and Heisenberg's uncertainty relations. According to them, increasing precision in the spatio-temporal localization of a moving micro-object is associated with increasing uncertainty about the corresponding dynamical characteristics (energy, momentum). Zeno is quite right in stating that the flying arrow is at rest, provided it is always localized in the instant (i.e. in the atomic "now", as Aristotle has put it). For motion and rest become indiscernible from the stand-point of a single instant in its capacity of minimal duration or span with which utmost precision in the spatio-temporal localization is associated. Let us assume that a micro-object is totally localized within the limits of such an instant. Then one can say that it is moving no more that it is at rest, for the corresponding dynamical characteristics could not be determined to the extent of differentiating between motion and rest.
This coincidentia oppositorum, or - to put it in another way - the relativity of motion and rest, is to my mind the message of Zeno's "Arrow". It reveals an aspect of the dialectical unity of opposites, bridging ancient philosophy and contemporary physics.

John S. Rigden

Professor of Physics   University of Missouri-St. Louis

The Stern-Gerlach Experiment and Quantum Mechanics: The Early Phase

    The 1922 Stern-Gerlach experiment dramatized the contradictions that were inherent in the old quantum theory. On the one hand, this experiment strikingly confirmed the idea of stationary states through the confirmation of space quantization. On the other hand, as Einstein and Ehrenfest showed, it presented physicists with intractible difficulties in their attempts to explain the behavior of atoms in a magnetic field.
    With the creation of quantum mechanics in 1925-1926, the conceptual conflict between particles and waves began its long siege. There was the individuality of elementary atomic events, but there were also interference and the superposition principle with the loss of individuality. The wave equation invited the principle of superposition while the matrix mechanics did not suggest this generalization. Once again, the Stern-Gerlach experiment was employed to contrast the differences between these two formal approaches to atomic phenomena.
    Many of the discrepancies internal to the physical interpretation of quantum mechanics were resolved in 1927 when Heisenberg developed the principle of indeterminism. Operating from the basis that the meaning of words such as "energy of an atom" comes by specifying definite experiments to measure the "energy of an atom," Heisenberg analyzed the Stern-Gerlach experiment which enabled him to demonstrate one form of his uncertainty principle. The Stern-Gerlach experiment was further used to illustrate how a purely quantal concept such as electron spin did not have a direct relation with classical quantities like angular momentum or magnetic moment. For example, Bohr showed that it would be impossible to determine the magnetic moment of a free electron by means of a Stern-Gerlach experiment. This raised the question whether it was justified to claim that the electron possessed a magnetic moment.
    In this paper, it will be shown how thought experiments of the Stern-Gerlach type were used to elucidate and to clarify the ideas and concepts that came with the creation of quantum mechanics.

Tsuyoshi Ogawa, Chiba University, Chiba, Japan

Professor of Physics, College of Arts and Sciences

HISTORICAL MEANING OF THE SCHRÖDINGER EQUATION

   It is usually taken that the Schrödinger equation is the equation which is proper to quantum mechanics. This conception of the equation is considered incorrect from the historical point of view which we develop below.
   It is true that the equation was discovered by Schrödinger in the course of establishing wave mechanics, and that it has had a position of the fundamental equation of motion in quantum mechnics which rules temporal changes of states of the microscopic world.
   The state of affairs changed, however, since the same or similar types of equations were found valid and usuful in physical theories other than quantum mechanics; for example, in statistical mechanics, in classical electromagnetism, in multi-spinned field theories. They showed that the so-called Schrödinger equation was nothing but one of similar types of equations describing temporal development of physical states, and suggested that all the equations describing temporal development might be thrown into such a form as Schrödinger's.
   We should now distinguish Schrödinger's eqation from Schrödinger-typed equations. Then the historical meaning of the former is of course in the first presentation of such a form of equations in a concrete problem. But it should be also noticed that, in general, physical change or motion makes a one-parameter group of time for any mathematical representation of states of a system under consideration, and, when any time-developing generator appropriate to the group is found, one gets a Schrödinger-typed equation. Schrödinger's historical discovery was just the presentation of a clue to the above generality.

Dr Paul K Hoch

Senior Fellow, STP Group, University of Aston, Birmingham, ENGLAND

'Transmission of a new Mathematical Physics from Germany to Britain and America'

The period 1920 to 1955 saw the transformation of the older Anglo-American mathematical physics and applied mathematics done in mathematics faculties into a new 'theoretical physics' under the stimulation of:

(i) the success of the new general relativity and quantum mechanics, for the most part developed in Germany and Copenhagen;
(ii) a deliberate emphasis on facilitating international visits and fellowships, especially by the Rockefeller International Education Board and the Guggenheim Foundation;
(iii) international conferences, professional associations and summer schools;
(iv) the post-1933 migrations of refugee 'theoretical' and 'mathematical' physicists from Germany and, subsequently, from other Nazi-occupied countries.

Each of these aspects will be discussed in some detail, and particular stress will be laid on the <u>interaction</u> between social/institutional/cultural factors in each country and the migration of foreign visitors or émigrés; resulting in a <u>differential</u> absorption of personnel and growth of these specialties in different contexts. There will also be some discussion of national/cultural attitudes facilitating (or opposing) 'theoretical' and 'mathematical' science - or prefering one to the other - in each case, and of the different situations at what were to become the main university centres in these fields.

Laurie M. Brown, Northwestern University, Evanston, Illinois, USA

Helmut Rechenberg, Max Planck Institute for Physics and Astrophysics, Munich, W. Germany

DIRAC AND HEISENBERG - PARTNERS IN SCIENCE

Paul Adrian Maurice Dirac, born in Bristol in 1902, and Werner Heisenberg, born in Munich in 1901, were two of the handful of major figures who created quantum mechanics in the 1920s. During that most creative period of their lives, and until Heisenberg's death in 1976, with the exception of the period of World War II, they maintained a close scientific relationship, visiting and corresponding with each other. Paul Dirac died last year on 20 October. This talk is based upon an article that reviews their scientific correspondence, their personal friendship, and their respective views on the course of modern physics. It is intended, in part, as a memorial to Paul Dirac. For nearly fifty years, both Heisenberg and Dirac were active in the forefront of theoretical physics. Their programs often overlapped, and each was stimulated by the other. They met first in Cambridge in July 1925, and became close personal friends in Fall 1926, at Niels Bohr's institute in Copenhagen. We deal in this talk mainly with the period 1928-32, including a trip to Hawaii and Japan that Dirac and Heisenberg made together in 1929.

We note several characteristic differences in the scientific styles of the two friends. Dirac was typically a loner, who had few students and very few collaborators. He usually limited his efforts to solving one problem at a time, and frequently claimed that the guiding principle of his research was mathematical beauty. Heisenberg, as in his quantum mechanics and his later nonlinear spinor theory of elementary particles, preferred to make a global attack on a whole class of problems, frequently with a group of collaborators. For Heisenberg it was the physical ideas that were compelling, even if their mathematical representations appeared clumsy at the outset. In spite of these stylistic differences, their views on physics were often in agreement.

Olivier DARRIGOL, REHSEIS, CNRS, Paris, France

THE RECEPTION OF EARLY QUANTUM FIELD THEORY (1925-1932)

Quantum fields were introduced very early, if not prematurely, in the history of quantum mechanics. An analysis of their mixed reception shows the criteria of the leading quantum physicists for integrating or rejecting new concepts when specific new empirical predictions do not follow immediately. I will order these criteria according to the following three headings:

.Comparison to the older theories: elucidation of past mysteries (wave-corpuscle duality, Bose counting, radiation fluctuations), adding unity, deductive power and mathematical beauty, as well as algebraic complexity, spoiling infinities and quantum relativistic paradoxes.

.Immersion in the Copenhagen spirit: emphasis on the symbolic character of quantum fields, as a further renouncement of strictly causal descriptions of phenomena in space-time.

.Conceptions of the correspondence method connecting classical to quantum theories: the a priori status of the Maxwell field in the old correspondence method, meaning of Dirac's classical "pictures" as opposed to Jordan's quantum axioms.

Manuel G. Doncel

Seminario de Historia de las Ciencias, Universidad Autónoma de Barcelona, BELLATERRA (Barcelona), SPAIN.

QUANTUM ELECTRODYNAMICS SETTLES IN AMERICA:

THE OPPENHEIMER SCHOOL

We study Oppenheimer's work, before the war, on the Old Quantum Electrodynamics. This work does not represent neither a passive acceptance of quantum physics in America, nor a pure application of the new theories to cosmic ray phenomenology. It prooves on the contrary a creative inspiration improving quantum field theory in contact with its originators, as well as a severe cirticism of this theory, then provisory and in need of experimental testing.

We analyse Oppenheimer's theoretical improvements: his elaboration on Heisenberg-Pauli quantum dynamics and the infinite "proper energy" of the electron, his rejection of the theory of electrons and protons inspiring Dirac's antielectron, his (and Ehrenfest's) theorem on the statistics of nitrogen nucleus supporting Pauli's "neutron", his attempt to harmonise Dirac and Heisenberg-Pauli theories before the positron discovery, and after this discovery his (and Furry's) magistral work on the electron-positron theory and the infinite "polarization of the vacuum". We study also Oppenheimer's experimental tests of the theory: his (and Carlson's) working out Pauli's theory of "neutrons" as cosmic ray test, and mainly the successive rescues of the ever falsified Klein-Nishina formula, by his (and Hall's) computation of the photoelectric absorption, his (and Plesset's and Nedelsky's) computation of pair production, and his (and Carlson's and Server's) elaboration of the theory of multiplicative showers including a meson prediction.

We introduce Oppenheimer's activity with a study of his scientific and personal contacts with Dirac and Pauli in Europe. We conclude with a reflection on the American school founded by him, and its role in the elaboration, after the war of the New Quantum Electrodynamics.

Karl von Meyenn

Scientific collaborator, University of Stuttgart

PAULIS VERHÄLTNIS ZU AMERIKA

Während die theoretische Physik in den deutschsprachigen Ländern während der 20er Jahre noch in höchster Blüte stand, vollzog sich unter dem verstärkten Einfluß der politischen Umwälzungen innerhalb weniger Jahre die Wende. Etwa ab 1936 ist Amerika bereits in vielen Bereichen der Physik führend, insbesondere aber auf dem Gebiete der neu aufkommenden Kern- und Elementarteilchenphysik. Dieser Wandel wurde auch von den europäischen Physikern rasch zur Kenntnis genommen.

Diese Veränderungen haben die Amerikaner nicht zuletzt durch eine geschickte Wissenschaftspolitik herbeigeführt, indem sie eine große Zahl junger Talente zur Ausbildung an die bedeutendsten europäischen Forschungszentren entsandten, eminente Gelehrte zu Gastvorträgen einluden oder tüchtige junge Gelehrte aus Europa einstellten.

Pauli war damals eine der Zentralfiguren der theoretischen Physik in Europa. Viele seiner Schuler haben später Spitzenpositionen in der amerikanischen Wissenschaftshierarchie eingenommen. Er selbst war vor dem Kriege zweimal einer Einladung nach Amerika gefolgt. Während des Krieges arbeitete er in unmittelbarer Nachbarschaft zu Einstein am Institute for Advanced Study in Princeton. Seine Vertrautheit mit Amerika und die vielen Kontakte zur amerikanischen Physik spielten nach seiner Rückkehr nach Zürich eine um so wichtigere Rolle bei dem Wiederaufbau der europäischen Nachkriegsphysik.

Als vorzügliche Quelle zur Erschließung dieser Zusammenhänge soll die Korrespondenz des Physikers aus den vierziger Jahren herangezogen werden, die jetzt für eine Veröffentlichung bearbeitet wird.

Edward MacKinnon

Philosophy Department, California State University, Hayward

CONCEPTUAL CONSISTENCY AND THE INTERPRETATION OF PHYSICS

Theories of scientific explanation may be roughly divided into two types, historicist and logicist. Both share a common problem. Science is viewed as a collection of disparate units: theories, paradigms, axiomatic systems, or research programs. For familiar reasons neither approach is geared to treating any underlying conceptual consistency shared by differing theories.

I propose a different approach to the interpretation of scientific theories, one that begins with an analysis of the underlying conceptual continuity shared by different branches of physics. To get at this I start with experiments, rather than theories, and focus on necessary conditions which must be met by the language used to describe actual and possible experiments and report their results. The most basic of these requirements is that scientific dialog must fulfill the conditions of the possibility of unambiguous reference to absent or unobservable objects or events. This requires a shared language, ordinary language, supplemented by scientific terms. By adapting work of Strawson and Davidson it may shown that the conceptual core requisite for such unambiguous reference must be a representation of the world as a collection of objects with properties existing in a common space-time framework. It also requires a sharp subject-object distinction. In more familiar terms, the point of departure is the lived world and ordinary language. The way in which scientific extensions of ordinary language develop is a matter for historical analysis.

Typically, the original introduction of scientific terms was mediated by medieval and early modern philosophies of nature. Aristotelian philosophy of nature can be considered as a systematization of ordinary language, or common sense, realism. These philosophies of nature supplied a matrix for the discussion of anomalous phenomena, the introduction of competing causal hypotheses to explain them, controlled observation, and the development of early scientific instruments. As quantitative concepts and laws emerged the conceptual scaffolding supplied by the philosophy of nature became dispensible. The new quantitative terms, used to describe measurable properties of bodies, became a part of an extended ordinary language framework, only minimally dependent on explanatory theories.

For the purposes of this paper, further developments can be reduced to two essential stages. The first is classical physics, where the extended ordinary language had a basic coherence but the related theories were unable to give an adequate account of matter and its properties. The second is quantum physics, where the extended ordinary language could no longer supply a consistent representation of reality, but where quantum theory did supply an adequate basis for explaining matter and its properties. This leads to a reinterpretation of complementarity and to constraints on theories.

Stanley Goldberg

National Museum of American History, Smithsonian Instit.

MEASUREMENT AND THE CONSTRUCTION OF THEORIES

There has been a persistent confusion in the literature of the history of science and the sciences themselves between Operationism, a philosophy in which meaning is endowed to a physical quantity by the techniques used to measure it, and the operational specification of a measuring process. As Berka has pointed out (Measurement:Its Concepts, Theories and Problems, 1983, p. 20) the empiricist philosophy of Operationism is hardly compatible with the historical practice of measurement.

What has been overlooked in analyses such as Berka's and before him, that of N.R. Campbell (Physics: The Elements, 1920, rpr. as Foundations of Science, 1957) is the role played by the notions of "direct" and "indirect" measurement in allowing Operationism to seem a reasonable approach to analysing physical concepts. According to Berka the indirect measurement of a speed, for example, using a ruler and a clock, can be transformed into a direct measurement by employing a speedometer.

In fact, careful analysis of measuring instruments shows that calculation and theory are always involved, save for measurement of primitives (mass, length, time, and charge) which, by the nature of their role, must be given operational specification. A history of the measurement of a quantity such as the speed of light reveals that over time, growing technological sophistication allows for greater precision in the operational specification of the primitives, but does not change the fact that the interpretation of the pointer readings depends on the theory the instruments are designed to test.

James T. Cushing, University of Notre Dame, Notre Dame,

Indiana 46556, USA

THE IMPORTANCE OF HEISENBERG'S S-MATRIX PROGRAM FOR THE THEORETICAL HIGH-ENERGY PHYSICS OF THE 1950's

The S-matrix program of the 1960's had a major impact upon the theoretical high energy physics of that decade. Nevertheless, several of the key contributors to that program (e.g., Chew, Gell-Mann and Goldberger) have stated in their recollections that Heisenberg's original S-matrix program of the early 1940's provided no direct motivation for or influence upon their own work in the late 1940's and early 1950's. A common perception is that the Heisenberg program encountered difficulties quite early and then quickly died out. One can easily get the impression that the original Heisenberg program was irrelevant for the theoretical developments which provided the background out of which the S-matrix program of the 1960's emerged. While it is true that Heisenberg soon abandoned his S-matrix program in favor of his nonlinear field theory of fundamental interactions, Heisenberg's ideas gave rise to a set of questions, such as how the causality requirement, suggested in the mid 1940's by Kronig as a constraint on the S-matrix, was to be implemented in the S-matrix theory and how the interaction potential of Schrödinger theory could in principle be determined by scattering data. Work on such problems by many theorists produced a series of papers eventually leading to relativistic dispersion relations, a program usually associated with Goldberger and Gell-Mann. The background necessary for Chew's own early investigations on reaction amplitudes was quite independent of any connection with Heisenberg's S-matrix theory. However, it was the confluence of Chew's and Goldberger's programs which led to the S-matrix theory of the 1960's. It is not at all clear that the latter S-matrix program could have become a viable candidate for a theory of strong interactions if it had not been for Heisenberg's seminal papers on an earlier program, one quite different from, but in the same spirit as, its successor.

Peter Galison

Program in History of Science, Departments of Physics and Philosophy, Stanford University, Stanford CA 94305

Instruments, Arguments and the Experimental Workplace in High-Energy Physics [1]

The history of modern physics has focussed almost exclusively on the development of two theories: special relativity and non-relativistic quantum mechanics. In this paper I seek to shift the balance towards experiment and to do so in a way that periodizes the discipline in a fashion that is not dictated by the theorists´ retrospective vision. In particular, one is not obliged to break up historical epochs into unit discoveries, e.g. the discovery of particular particles or effects. For the experimentalist, the continuity of work is much more frequently determined by common instruments, techniques, and modes of data analysis.

When one looks at the history of experimentation through the experimentalist´s eyes one sees continuities, for example, between the cloud chamber, emulsions, and the bubble chamber. One can understand how natural it was for experimentalists to progress from counter to spark chamber and then to wire detectors work. It proves useful to break up the history of physics instrumentation into two traditions, which we will label traditions of image and logic. As used here, "tradition" designates a skill group -- for example image devices have certain scanning and analysis techniques in common, logic devices demand familiarity with electronic switching, counting and logic circuits.

Instrumentation must also be understood in the context of the work organizations in which they were used. We cannot understand either the origin or use of instruments without distinguishing devices used in table top craft work from those associated with hierarchical line work. Finally, we find that skill groups and work organizations are frequently tied to particular modes of argument: with image-producing devices one can invoke "golden events" as evidence. "Logic" type devices rely on statistical demonstrations. In the conclusion of this paper I will discuss how these two traditions of experimentation have come together in the digitized-visual instruments of the 1980´s.

[1] P. Galison, "Bubble Chambers and the Experimental Workplace," in O. Hannaway and P. Achinstein, Experiment and Observation in Modern Physical Science (MIT-Bradford, 1985)

John E. Chappell, Jr.

1212 Drake Circle, San Luis Obispo, California 93401, U.S.A.

RE-INTERPRETATION OF EXPERIMENTS AND REASONING IN MODERN PHYSICS:
A NECESSARY RESPONSE TO CULTURAL CONCEIT AND INVALID LOGIC

History of Science only rarely fulfills one of its potentially most useful roles: to double-check contemporary science for errors in its principles and practices. Serious errors can be found, illustrating the modern phenomenon of "cultural conceit" (See my contribution in Social Sciences at this Congress), whereby scientists have developed theories divorced from experience and valid logic. This is obvious from considering the dozens of papers discussing "tachyons"—particles of superluminary speed which have never been found, but the existence of which can be "deduced" if time and causation can flow backwards.
   The main problem is in physics, as claimed by M. King Hubbert's 1962 presidential address to the Geological Society of America (GSA Bull., 1963). He found appalling shabbiness, attributable to abandonment of classical physics. To further this critique one may properly consider the original papers relating to special theory of relativity (STR).
   At the 1971 ICHS I emphasized the fact that no unequivocal proof of relativistic mass increase had ever been found. None has even today. It has only been shown that force on a moving charge decreases as velocity increases, without at all proving which component of force or charge changes. Hence the most convincing "proofs" of relativistic time dilation, those involving meson lifetimes, all rest on arbitrary assumption; for all begin with measurements involving mass, and all introduce relativistic mass variation in order to yield relativistic time dilation. In this and other ways we can conclude that nuclear energy is a technological achievement lacking firm theoretical foundation. The problem is less to find more experimental facts, than to interpret properly those already on hand: to pay as much attention to rational processes and logic, as to measurements and mathematics. We must look for every last nuance in an observation, in the spirit of phenomenological philosophy, and build upon it with impeccable logic.
   The initial mis-step in the illogical labyrinth of STR was revealed in 1962 by American philosopher M. Evans (in journal Dialectica, publ. in Switz.). Evans showed that simultaneity cannot possibly be relative to motion, unless the Law of Contradiction is broken. Einstein broke this law early in his 1905 paper by assuming two different velocities for the same light beam. The error is disguised by omitting one of the last steps in the argument. When Einstein said the light reached the observer on the hypothetical train at a speed modified by the speed of the train, he was in effect opting for a Ritzian, additive interpretation of the 1887 Michelson-Morley experiment; when he later implied that the same light beam must have travelled at constant c, he was opting for the DeSitter double star argument. The first option accords with the Galilean relativity principle; the second, with Maxwellian optics. Together they represent the two basic postulates of STR, and are equivalent to A and non-A. The "achievement" of STR was simply to declare both A and non-A true at once—i.e., to abrogate the rules of logic. But there are logical ways to resolve the dilemma (See, e.g., J. Chappell, Jr. in Specs. in Sci. and Techn., 1979,1980).

Naohiko Hiromasa

Tokai University, Kanagawa, Japan

INTRODUCTION AND DEVELOPMENT OF THE MODERN PHYSICS IN JAPAN (1868--1912)

After the Meiji Restoration(1868), the modern physics was regularly and systematically introduced into Japan. Before that time, western science had been only irregularly and fragmentarily introduced. The history of physics in Japan started from 1868. In this paper, as a first step of my investigation on the history of physics in Japan, I would like to discuss the history of the modern physics in the Meiji Era(1868--1912).
 In the early Meiji Era, foreign teachers, who were employed by the Japanese goverment and came from Great Britain, America, Germany, France and so on, played an important role in the introduction of the modern physics. Especially, Scottish teachers, who had learned physics or engineering at the university in Scotland, exerted a great influence on the history of physics in Japan. They lectured on physics, mainly elctromagnetics and thermdynamics which had been made progress in the 19th century and closely connected with engineering, and investigated chiefly the electromagnetic properties of matters and seismology at the University of Tokyo and the School of the Ministry of Engineering which established at 1877.
 At 1886,these two organs of education were combined and became the Imperial University. After that time, almost foreign teachers returned to their country. At the latter half of the Meiji Era, Japanese physicists, who were educated by foreign teachers, played an important role in the development of the modern physics in Japn.
 There are some investigations of the hisory of physics in the Meiji Era, mainly from the viewpoint of external history. I will discuss it from the viewpoint of internal history.

Jan Oosterhoff, Erasmus University, P.O.Box 1738, 3000DR Rotterdam

lecturer, History Department, Erasmus University

The Rise of a "National Physics" at Leiden University at the end of 19th Century

In 1859 a new laboratory building was founded at Leiden University (Netherlands) for chemistry, physics, anatomy and physiology. Its purpose was mainly for education support. When in 1874 Van Bemmelen, the new appointed professor in anorganic chemistry, intended doing research, the laboratory was soon too small. Van Bemmelen had been a teacher at several secondary schools. These were the in 1863 founded "Hogere Burgerscholen", with an extended science program according the German "Realschule". There he had met as pupils Lorentz and Kamerlingh Onnes, who became later both professor in physics at Leiden.

These new secondary schools had small but efficient laboratories, where pupils guided by teachers did research in chemistry and physics. These laboratories were in most cases better equipped than the old ones at the universities. Pupils, who came in general from other strata than those who visit the traditional Latin Schools included later Nobel Prize winners as Van der Waals, Van 't Hoff, Zeeman, Lorentz, Kamerlingh Onnes and Einthoven. From this new spirit emerged a prosperity of physics in the Netherlands, called by Lorentz a "National Physics". This phycics advocated a going hand in hand of theory and practice against the old tradition of practical and empirical work.

The program of Kamerlingh Onnes on low temperatures required a new laboratory. The enlargement of the old laboratory and the construction of a new one, the high costs of instruments and machinery and the education of skilled instrumentmakers means at the same time the transition of 'little science' to 'big science' as the change of the "old professor' into a manager.

RON JOHNSTON

PROFESSOR, DEPARTMENT OF HISTORY & PHILOSOPHY OF SCIENCE,
UNIVERSITY OF WOLLONGONG

Accountability or Expediency in Public Scientific Institutions
- the Establishment of ANAHL

The establishment of the Australian National Animal Health
Laboratory has for many years been presented as the major element in
the Australian armoury against an exotic disease incursion.

This paper examines the way in which the series of decisions to
construct the $150 million facility were made. In particular, the
arguments and evidence used and interests involved in establishing:
1. the need for a maximum security animal health laboratory;
2. the economic justification of the expenditure; and
3. the value of the proposed functions of diagnosis, training,
research and vaccine production;
are critically reviewed.

The debate about the need and nature of ANAHL was conducted almost
entirely among institutions and individuals committed to the
laboratory and their perceptions of need and judgements of value
largely determined the decisions reached. The changing scientific
and political climate led to a continuing shift in the arguments
and evidence offered, but these were all designed to mainting the
basis for the initial conception of ANAHL. These shifts, however,
eventually served to highlight the uncertainties and value
judgements in the arguments, and to an erosion of the credibility
of the decison-making institutions and the proponents of ANAHL.

Susan Davies

Research Student, Monash University, Melbourne, Australia

RUTHERFORD AS EDUCATOR

Ernest Rutherford, the physicist and Director of the Cavendish Laboratory in the University of Cambridge from 1919 until his death in 1937, may not have thought of himself as an educator, but the influence he exerted on his pupils was profound.

It was not calculated; nor was it the result of a carefully thought out teaching programme. Rather it arose out of his strongly held views on physics and other matters, and on the force of his personality which impressed itself on all who came into contact with him.

For Rutherford was a man of strong utterance, definite opinions, and a good many prejudices.

He was a man of his time in many respects, who shared current attitudes to science and to the application of research. His eminence as a scientist gave additional weight and authority to these views, and he was prone to overstatement. Typically the fervour with which he espoused a particular point of view caused him to deprecate, or to appear to deprecate, the opposite point of view.

Rutherford's views on the practice of physics may be summarized thus:

he was for <u>pure research</u> and against <u>applied research</u>;

he was for <u>experiment</u> and against <u>theory</u>; and

he was for <u>nuclear physics</u> over and above all

other branches of physical inquiry.

His pupils acquired his outlook, but lacked his genius. When they rose to positions of authority in university departments and scientific laboratories in many parts of the world, they sought to put into practice what he had taught them.

In Australia, where Rutherford's pupils became leaders of the scientific community and influential in the wider political arena, Rutherford's ideas shaped the development of physics in this country and are reflected in the pattern of our scientific and educational institutions.

Yves Gingras

Postdoctoral Fellow, Dept. of History of Science, Harvard Univer.

CANADIAN PHYSICISTS: THE GENESIS OF A SOCIAL GROUP

In this communication, I study the social and institutional conditions of the emergence of physics research in Canadian universities. By concentrating my investigation on the genesis of research as a practice, I reconstruct the "process of production" of the first generation of Canadian physicists. These physicists are more interested in research than in teaching and identify themselves with the activities of an international community of scientists rather than with their activities as teachers. Thus, the history of this scientific discipline becomes the history of a social group constituted around a disciplinary identity.

In the specific case of Canadian physics, the development of engineering education in the 1870's secures the existence of the physics department as a separate entity from the mathematics department with which it was linked. On the other hand, the introduction in the 1860's and 70's of options and honours sections in the traditional BA program gives a new visibility to physics as a possible professional career. However, the full development of this new career has to wait the emergence of the first generation of university professors interested in training students in research. At Dalhousie University, research starts with James G. MacGregor, a Canadian trained under P.G. Tait at Edinburgh University. In 1879 he obtains the first chair of physics endowed in Canada. At McGill University, research is introduced by two students of J.J. Thomson: Hugh L. Callendar from 1893 to 1898 and Ernest Rutherford from 1898 to 1907. Finally, the last "father" of physics in Canada is John C. McLennan. Also trained under Thomson, he goes back to Toronto University in 1899 and trains the majority of Canadian physicists between 1910 and 1930.

Despite the efforts of these "founding fathers", the real growth of the discipline has to wait the creation of the National Research Council of Canada in 1916. This federal agency creates the first organized system to promote research on a national scale and plays an important role in legitimizing research activities in Canadian universities. After 1920, Canadian physicists slowly build a collective identity and mode of representation that culminates in the formation in 1945 of the Canadian Association of Physicists. This institution produces and spreads an image of the physicist which portrays this group of researchers as having a dominant position in the hierarchy of social representations. This dominant position is important for the future of the group since it facilitates its reproduction by attracting new recruits.

JUDITH R. GOODSTEIN

INSTITUTE ARCHIVIST, CALIFORNIA INSTITUTE OF TECHNOLOGY, PASADENA, USA

THE DUBRIDGE ERA AT CALTECH

Caltech's history is divided into three distinct eras. The first Caltech era was created by Hale, Millikan, and Noyes in the 1920s. Thirty years later, after World War II, the job was done all over again by Lee DuBridge and Robert Bacher.

Most observers understood at the time that the war in general, and the use of the atomic bomb in particular, had irrevocably altered the enterprise of science. Henceforth, academic science, higher education, and the federal government would dance to a different tune. On the local level, Caltech found itself in 1944 headless, its chief executive officer unable either to grasp the magnitude of the federal government's financial involvement with the school's rocket research program, or to deal effectively with the wartime factory Caltech had become. The running of the school on a daily basis fell to Earnest Watson, Dean of the Faculty and de facto chairman of the physics division and James Page, chairman of the Board. By this time, Caltech had become, in a very real sense, an extension of Millikan's personality.

The new era found its leaders in the war's scientific establishment: Vannevar Bush, James Conant, I.I. Rabi, and Lee DuBridge. After thoroughly canvassing the country, Page selected DuBridge. Thrust into the national limelight during the war as Director of the M.I.T. Radiation Laboratory, DuBridge went on to become in peacetime an eloquent, articulate, and persuasive spokesman for the cause of science in America. Federal support of scientific research was one of the first items he championed. In his ability to explain science to the public, to presidents and members of Congress, and to defend the principle of academic freedom during the McCarthy period, DuBridge had few peers, in or outside of Caltech.

Kauffman, George B.

Professor of Chemistry, California State University, Fresno, Fresno, CA 93740, USA

WILLIAM DRAPER HARKINS (1873-1951): A GIFTED BUT CONTROVERSIAL AMERICAN PHYSICAL CHEMIST

William Draper Harkins, Andrew MacLeish Distinguished Service Professor at the University of Chicago, had a record of accomplishments that few scientists have equaled. Although a chemist, he was a leader in nuclear physics at a time when no American physicists were working in this field. He predicted the existence of the neutron, deuterium, and artificial radioactivity. He proposed the whole number rule of atomic masses, nuclear fusion as a source of solar and stellar energy, "packing fraction," generalizations on the stability and abundance of atomic nuclei (Harkins' rules), the compound nucleus as a step in nuclear transformation, and the nuclear-shell picture for the atomic core. He was among the first to separate isotopes and to achieve nuclear transformations using neutrons. He synthesized the first artificially produced radioactive element. He made equally outstanding contributions to modern surface chemistry, a field that he virtually created, e.g., studies of surface tension, orientation of molecules on surfaces, adsorption, film thermodynamics, and emulsion polymerization. His views were generally ahead of his time, and even when they were later experimentally confirmed, he was often denied real credit. He was called "Priority Harkins" because of his efforts to obtain recognition.

Professeur Józef Hurwic

Université de Provence, Marseille, France

ACCUEIL REÇU PAR LA THÉORIE DES QUANTICULES DE KASIMIR FAJANS

 Kasimir Fajans considérait la théorie des quanticules, créée dans les années 1943 - 49, comme sa plus grande réussite, plus importante même que ses découvertes dans le domaine de la radioactivité. Cependant le milieu scientifique, sauf de rares exceptions, la traite plutôt comme une erreur du grand savant, un essai inutile de détournement du courant de la science. Une quanticule représente un ensemble d'électrons lié électrostatiquement à un ou plusieurs noyaux ou coeurs atomiques (noyau atomique avec les couches électroniques internes). De cette façon toute molécule, aussi bien minérale qu'organique, peut être décrite par une seule formule (ne comportant pas des traits qui symbolisent la liaison covalente). Ainsi la théorie des quanticules permet d'éviter la nécessité, dans la théorie de résonance, d'utiliser l'ensemble de plusieurs formules pour une molécule. Par la déformation mutuelle des ions, provoquée par la polarisation électrique, Fajans explique les propriétés stéréochimiques des molécules. Les adversaires de sa théorie indiquent, a juste titre, que les forces électrostatiques seules ne peuvent pas donner la structure de la molécule. Mais ce n'est pas non plus le cas de la théorie des quanticules: les forces quantiques sont ici responsables de l'existence du coeur atomique comme de celle de la quanticule qui autrement devrait exploser. Donc la théorie de Fajans n'est pas du tout en contradiction avec la mécanique quantique. C'est une théorie exacte, parallèle à la théorie quantique de la liaison chimique, mais avec possibilités limitées de prévision. Dans certains cas elle peut même expliquer plus simplement le comportement d'une molécule que la chimie quantique. La théorie des quanticules est advenue une vingtaine d'années trop tard. Si elle avait été créée avant l'apparition de la mécanique quantique elle serait probablement jusqu'aujourd'hui utilisée et enseignée comme, par exemple, la théorie de Kossel et Lewis. Kasimir Fajans pendant les trente dernières années de sa vie développait sa théorie "démodée" qui n'a pourtant jamais été acceptée par la communauté chimique.

Ruth L. Sime

Instructor, Sacramento City College

LISE MEITNER IN PERSPECTIVE

Lise Meitner (1878-1968) was an exceptionally versatile experimental physicist whose work was closely tied to the development of atomic physics from radioactivity to nuclear fission and beyond. In most accounts of the period, however, the intense interest in fission and the preoccupation with Otto Hahn tend to obscure Meitner's early work and disregard her independence, although both were essential to the fission discovery. Beginning in 1920 Lise Meitner's investigation of the complex relationship between beta spectra and gamma radiation clearly demonstrates her scientific style; each experiment, every interpretation was guided by her quantum view of atomic and nuclear processes. As a result she was certain that beta decay precedes gamma emission, and that disintegration electrons are emitted with uniform energy, while C. D. Ellis, relying primarily upon experiments of Rutherford and his co-workers, came to exactly the opposite conclusion. In the end Meitner's view of the nucleus was proved right in one respect but inadequate in another; the existence of the continuous beta spectrum required a new theory of beta decay. For Meitner the work with beta spectra led directly to new experiments with gamma radiation, pair formation, artificial radioactivity, and in 1934 to the neutron irradiation of uranium. Again guided by the prevailing concept of nuclear behavior, Meitner led the Berlin team for four years, persisting as the results became increasingly difficult to explain. Had Lise Meitner not been forced to leave Germany she would have fully shared in the fission discovery. It is not surprising that she was the first to understand it.

Lawrence BADASH

Professor of History of Science, University of California,
                              Santa Barbara

The Reaction in 1939 to the Discovery of Nuclear Fission

A large body of literature exists on the scientific and political history of nuclear weapons. There is little, however, concerning the reaction in 1939 to news of the discovery of nuclear fission. This paper consists of an examination of worldwide views during the preceding four decades about "harnessing the energy of the atom," a brief survey of the scientific accomplishments of 1939, and a close look throughout that year at the thoughts, hopes, fears, and actions that fission inspired, primarily in America.

An attempt is made to explain why the discovery came as such a surprise, despite speculations of atomic energy from 1900 onward. Why the discovery of fission generated relatively little moral or ethical introspection in 1939 is also analyzed.

Lu Jingyan

Professor, Mechanical Engineering Department, Tongji University, China

A study of the Development of Frictionology in Ancient China

   This paper sums up for the first time the accomplishments in frictionology in ancient China, juch information being gathered and published for the first time, too.
   Although frictionology is a new subject, its development goes far vack in history. In ancient China, plain bearings were widely used in carts and other machines, which gave rise to the development of ancient frictionology. According to our ancient books and archaeological information, lubrication appeared in China not later than the Zhou Dynasty and there were officials in charge of lubrication during the Warring State Period (475-221 BC). The earliest lubricant was animal fat, particularly the grease from sheep, and the earliest lubricant device called "牲 畏" (pronounced "ḥuo"), a kind of oil-can. Mineral oil was used for lubrication in the West Jin Dynasty. Attention had already been paid to the reasonable selection of wood to be used for plain bearings during the Spring and Autumn Period (770-476 BC) and later, analyses of frictional behavior of wood were made in light of the heat produced by drilling wood. Metal bush first appeared during the Warring States Period and was widely used in the Han Dynasty (206-220 AD). Measures were also taken to reduce friction in the plain bearings structure of carts. In addition, rollers had already been used to reduce friction in ancient China, and roller bearings came into existence in the Yuan Dynasty (1279-1363 AD). Other topics also dealt with in this paper.

Robert Mark, Princeton University, Princeton, N.J.

Professor of Architecture and Civil Engineering

STRUCTURAL ARCHAEOLOGY: An Interdisciplinary Approach to the History of Building Technology

The nature of design methods used to create the great, long-span temples, baths and markets of the Late Roman era, as well as the light, soaring structures of the High Gothic cathedral at the end of the twelfth century, has been one of the more persistant puzzles in architectural history. But new insights into early design and construction methods have been afforded by recent studies using modern engineering analysis techniques. These have shown how the early builders produced generally reliable, and at times truly elegant structures in the absence of scientific methods. Previous buildings were used as approximate models to confirm the stability of new designs, and behavior was learned from detailed observation of new buildings in the course of construction. The approach will be illustrated by studies of the key monuments of the two eras: the Roman Pantheon and the Cathedral of Notre-Dame de Paris.

J. L. Berggren

Professor, Simon Fraser University

A MEDIEVAL ARABIC TREATISE ON BURNING MIRRORS

The Arabic manuscript BM Add. 7473, 164b - 172b begins with a 6-page summary of material from Book I of Apollonios' Conics, followed by the words "we need nothing else from the first and second treatises of the book of Datrumus on burning instruments". There follow five propositions, with proofs, about reflection of light rays by concave paraboloidal and spherical mirrors, as well as a discussion of burning glasses.

The themes of the book are familiar from such treatises as those of Diokles and Ibn al-Haytham but the proofs are different and the construction of the parabola that is given is not found elsewhere. It is based on a lemma about the parabola found in Archimedes' work. The concluding section, on burning by objects of solid glass or filled with water, contains references to the practices of ancient artisans.

The many questions this manuscript raises remind us that we still do not know the full history of burning mirrors - and hence of conic sections - in the ancient or medieval worlds, and leave us with a new puzzle: Who was Datrumus?

Robert E. Hall

The Queen's University of Belfast, Northern Ireland

SOME REVEALING THEMES AND PROBLEMS IN IBN SINA

Ibn Sīnā (Avicenna; 980-1037) is rarely studied as deeply as his work will repay. Devotedly a systematizer, he nonetheless provided detailed theoretical explication of most parts of theoretical philosophy and medicine and of many parts of the mathematical disciplines. An obvious but difficult approach to his writings is to look for their Greek and Arabic sources; but Ibn Sīnā tends to think, or re-think, for himself, and since he writes to persuade as well as to enlighten he quotes sources and adopts a style intended to attract a particular readership. Using the doctrines of both, he cites Galen for physicians and Aristotle for the philosophically inclined, or adopts mysticizing language to present largely Peripatetic doctrines. A great need that emerges from the study of his writings is the comparative editing of his works, so that identical and parallel passages from several treatises can be identified and studied together easily. A second need is to analyze Ibn Sīnā's handling of his sources - his compounding, for example, of Galen and Aristotle in physiology and psychology. A third need is to determine the shape and extent of Ibn Sīnā's intellectual development.

That his intellectual system changed but slightly was something of a point of honor with Ibn Sīnā. To see why, one needs to know his epistemology, itself largely a consequence of his anthropology (in the old, theological sense), his noetic, and his conception of individual salvation - areas where he drew variously upon Aristotle, Plotinus, al-Fārābī, and the Ismāʻīlīs, among others. To that cluster of subjects belong also the question of Ibn Sīnā's 'mysticism', his doctrine of prophecy, and, indeed, the question of the bases of political philosophy. Closely related is the theory of the functioning of the embodied soul - where revealing individual topics include the partitioning, logically and locally, of the internal senses, the doctrine of the wahm ('estimative faculty') and its origins (and the associated problem of 'mathematicals' and that of the nature and status of 'experience' (tajriba)), and the theories of the heart and of the pneuma (rūḥ). Other issues - the questions of the ḥadd and of taṣawwur and taṣdīq, for example - lead back to the epistemological and ontological topics mentioned before. Finally one may enquire about the rôle and status of intellectual systems and their creators as seen by Ibn Sīnā - and re-enter the Avicennian nexus from there.

Qa

Edith Dudley Sylla

Professor of History, North Carolina State University, Raleigh, NC

THE OXFORD CALCULATORS' MATHEMATICAL PHYSICS

    The Oxford Calculators have been of interest to historians of science primarily because of their increased use of mathematics within physics. At the same time, historians of logic have been interested in the Calculators' logical innovations. I have argued previously (e.g. in the Cambridge History of Later Medieval Philosophy) that the institutional locus of the Oxford Calculators' work caused it to be seen as a preparation for disputations, and in particular for disputations on sophismata. This locus explains, I think, the hypothetical or imaginary tone of much of the Calculators' work: in a disputation on logic or sophismata one of the main goals was to derive the logical implications of a given statement or assumed case without regard to the factuality of the case in the first place.
    I argue here, however, that the fact that the Calculators debated imaginary or non-realistic cases in their logically or disputationally related work does not imply that they believed that this was the proper way to do physics. When we look at the Oxford Calculators' mathematical physics, we find a mathematics that is tied very closely to the assumed physics of the situation. It is, indeed, just the sort of mathematical physics that one would expect of a good Aristotelian if that Aristotelian turned his attention to giving a mathematical account. Unlike an Archimedian or Galilean mathematical physics, which typically assumes principles that would apply exactly only in idealized circumstances, such as on the assumption that a body is so far from the center of the earth that the tendencies downwards of its parts can be assumed to occur on parallel rather than converging lines, the Calculators' mathematical physics sees itself as abstracting the quantitative aspects from real physical situations. Moreover, the assumption is made that mathematical operations as well as quantities will match the physical situation, so that, for instance, the mathematical addition of abstracted quantities should correspond to physical addition. In my paper I show, mainly by examples taken from Roger Swineshead's De motibus naturalibus and John Dumbleton's Summa Logicae et Philosophiae Naturalis, just what this Aristotelian mathematical physics looks like.

André Goddu

Assistant Professor, Program of Liberal Studies,

U. Notre Dame

OCKHAM'S <u>COVENANTAL</u> VOLUNTARISM AND THE STATUS OF NATURAL LAWS

Many standard surveys of mediaeval philosophy portray William of Ockham as a voluntarist (whether the term is used explicitly or not) without defining precisely what is meant by "voluntarist" and without supplying the motivation behind a voluntarist rather than intellectualist conception of the soul, God, and causality (e. g. Gilson, Knowles, the "earlier" Leff, and even Copleston to some extent).

In effect such accounts leave the impression that there is some clear and unambiguous category of voluntarism essentially different from intellectualism without nuance, subtlety, or shade of difference. Given such a view, these historians have great difficulty in distinguishing Ockham from Luther and Calvin, and, it seems, their negative judgments of Ockham reflect their inability to appreciate the intellectual contributions of Luther and Calvin to modern thought.

In spite of efforts to the contrary (e.g. Oberman, Courtenay, and Miethke), such crude portraits of Ockham and voluntarism continue to distort Ockham's philosophy of nature in general and his views on causality in particular. Above all, these accounts render Ockham's beliefs in the principle of induction and in the constancy of the laws of nature contradictory at worst and paradoxical at best.

The paper proposes to define Ockham's "voluntarism" precisely and to show how his conception is compatible with belief in the constancy and uniformity of natural laws. Ockham's supposed <u>radical</u> voluntarism is a myth which the texts read in their entirety do not support.

Covenantal voluntarism does not support God's possible and arbitrary exercise of his absolute power, but God's actual covenant with man. The justification for Ockham's theories of induction and causality resides in a critical metaphysics which supports Ockham's conceptions of physical necessity. That the necessity of the universe rests on God's will renders its existence, not its status, contingent. God's will is the secure ground and source of the constancy and uniformities of nature.

Mira Roy

Keeper Sanskrit Manuscript, Department of Sanskrit, Calcutta University

AGRICULTURE AND METEOROLOGY IN ANCIENT INDIA

In ancient India, as in other parts of the old world, all the favourable and unfavourable meteorological conditions in the context of agriculture and animal farming were studied with reference to certain astronomical aspects. Apart from celestial bodies weather predictions were also made from animal behaviour as well as from the growth of some particular vegetation foretelling the future prospects of some particular crops. And rain gauge for the measurement of rainfall became a part of agro-meteorological studies. An important aspect was the prolonged observations made on the possible correlation between rising and setting of particular stars and planets and crop prospects. To disseminate knowledge of agricultural meteorology, weather lore or sayings were composed for effective oral transmission.

The paper discusses the evolution of ideas regarding meteorological phenomena as well as the prediction of rainfall from early times to C 1200 A.D. as evidenced by literary and other sources.

Giovanna C. Cifoletti

Ph.D. Candidate, Princeton University, Princeton, NJ

De Quantitatibus: A Philosophical Treatise on Mathematics by Johannes Kepler

In 1629 Kepler tried unsuccessfully to publish a philosophical treatise on mathematics, De Quantitatibus, which was forgotten after his death in 1630. The treatise was intended to be used as a philosophical introduction to a widely distributed mathematical textbook by Dasypodius (1530-1600), the teacher of mathematics in Strasbourg and most important author of mathematical textbooks at that time. Kepler's text collects, arranges by theme and interprets all the Aristotelian passages relevant to mathematics. In choosing this Aristotelian context to present his ideas, Kepler distances himself from the sixteenth century tradition of philosophical treatises on mathematics to which Dasypodius himself belonged. Dasypodius' works on mathematics, like Ramus', were within the genre developed after the rediscovery of Proclus' Commentary on Euclid. Instead, Kepler took a position following the Aristotelian humanism of Schegk.

While the datation of De Quantitatibus is still uncertain, it is likely that Kepler wrote it not too long after his education in Tübingen, where he received instruction in natural philosophy under the influence of Schegk. In De Quantitatibus, Kepler also exhibits some very original views on mathematics within this humanist Aristotelian context.

This paper is a commentary on De Quantitatibus, situating Kepler's philosophical views, in particular his new classification of the mathematical sciences, and his definition of the role of mathematics among the sciences in general. Mathematics was for Ramus a method; Kepler presents it here as the foundation of human knowledge.

William A. Wallace

Professor, The Catholic University of America, Washington, D.C. 20064

RANDALL REDIVIVUS: GALILEO AND THE PADUAN ARISTOTELIANS

In 1940 John Herman Randall, Jr., proposed the thesis that Galileo's scientific methodology derived from Aristotelians teaching at the University of Padua, and particularly from the works of Jacopo Zabarella. The thesis has been much controverted and in recent times has come to be generally rejected by historians and philosophers of science.
Randall argued his case without knowledge of Galileo's logical questions (MS 27) and the analysis of the demonstrative regressus contained therein. Still unpublished, Galileo's question dealing with the regressus presents problems because of several lacunae in the manuscript and because of difficulties in dating it and identifying the sources on which it is based. In his recent Galileo and His Sources (1984), the author has resolved most of these difficulties: MS 27 was written after August 1588 and was based on the lecture notes of Paulus Valla, a Jesuit teaching at the Collegio Romano. Unfortunately the portion of Valla's notes dealing with the regressus is missing, though his later published treatment of it (1622) survives and explicitly cites Zabarella as the major source of its teaching. The question is: Was Zabarella used by Valla in preparing his 1588 lectures, and thus became the source of Galileo's account of the demonstrative regress at second remove?
An affirmative reply is indicated on the basis of further research into Valla's logic course. This was preceded by that of Ioannes Lorinus, S.I., whose lecture notes in manuscript (1584) and whose published text (1620) still survive. Valla's teaching is a development of Lorinus's, who explicitly based his exposition on that of Zabarella and did so in both his manuscript and his later published text. Arguing a pari, one can maintain that Valla used Zabarella for his 1588 notes. A careful study of terms in all four expositions (Zabarella's, Lorinus's, Valla's, and Galileo's) shows minor mutations from one to the other, with essentially the same doctrine being presented by each.
Zabarella, then, was the methodologist who stood behind Galileo's MS 27. That the latter manuscript, written at the same period as Galileo's MSS 47 and 71 (the last containing the De motu antiquiora), lays out a program that Galileo followed in his later researches has been argued in Galileo and His Sources. A striking testimony of this is Galileo's use of the expression progressione dimostrativa in his Discourse on Bodies on or in Water (1612). Progressione (or progressus) is one of the terms that undergoes mutation throughout the four versions of regressus methodology mentioned above.
Thus Randall was right after all. But oddly enough he erred in attributing Galileo's basic inspiration to "a secular and anti-clerical spirit," which he thought characteristic of Padua, whereas it actually came to him via the clerics of the Collegio Romano!

David K. Hill

Associate Professor of Philosophy, Augustana College, Rock Island, IL

Speed, Distance, and the Law of Chords: A New Analysis of Galileo's Work in 116v

How did Galileo achieve a satisfactory formulaic expression of the basic laws of fall? Our understanding of his thinking on this and related problems has been much advanced in the past two decades by a careful examination of his private notes, or "worksheets." Of these notes, folio 116v has been perhaps the most discussed, since it presents persuasive evidence of a precisely controlled experiment, the results of which were subjected to detailed quantitative analysis, obviously meant to test principles pertaining to free fall and projectile motion. Just which principles, and how Galileo meant to test them, has been a matter of controversy. In this paper I argue that Galileo designed the experiment to test the principle $v \propto \sqrt{d}$: speed of descent is proportional to the square root of vertical descent. I also argue on the basis of evidence from the manuscript that Galileo inferred this crucial principle from an unexpected source, the law of chords, which must therefore be regarded as one of the key principles in the development of early modern physics.

H. Floris Cohen

Twente Technological University, Enschede, The Netherlands

THE TRANSFORMATION OF THE SCIENCE OF MUSIC 1610-1640

As a rare bird among historians of science, in his celebrated 1976 article 'Mathematical vs. Experimental Traditions in the Development of Physical Science', Thomas S. Kuhn recognizes quantitative musical theory as one of the many sciences that together constitute the history of science. In this essay on the 17th century origins of modern science Kuhn distinguishes between the mathematical sciences, inherited from the Greeks and the scholastics, and the newly emerging Baconian sciences, and then goes on to argue that, with one exception, these classical, mathematical sciences were radically transformed in the course of the 17th century. This one exception, he contends, is 'harmonics' (p. 40 of The Essential Tension). The point of the present contribution is to show that, in fact, the science of music was transformed during this period in a way quite comparable to the other mathematical sciences, and thus fits rather than contradicts Kuhn's own thesis.

The argument to be presented here goes back to primary source material, which is presented and analyzed in far more detail in my book Quantifying Music. The Science of Music at the First Stage of the Scientific Revolution, 1580-1650. Dordrecht (Reidel), 1984.

The science of music deals mainly with the issue of consonance: how to explain the correspondence - discovered by Pythagoras - between the consonant intervals, and the ratios of the first few simple integers? Why is the octave given by string lengths in a ratio of 1:2, the fifth by 2:3, etc.? Dealt with from its inception as a problem in arithmetic, in the hands of Beeckman, Mersenne, Galileo and their contemporaries the problem was recast as one in the physics of sound and the physiology of hearing. Just as in other sciences at the time, the new conceptions of mathematizing and mechanizing nature and of dealing with nature experimentally were ultimately responsible for this transformation, which, among other things, gave birth to the science of acoustics.

Thus, awareness of musical sound being made up of little 'shocks' gave rise to a new, physical theory to distinguish the consonances from the dissonant intervals. This theory explained consonance by the greater or lesser regularity with which these little shocks coincide at the ear drum or inside the sense of hearing. The theory was anticipated by Benedetti (as shown by Palisca in 1961), but elaborated into a fully-fledged account of consonance by Beeckman, on the basis of his emission conception of sound globules. Through his communications to Descartes and Mersenne the coincidence theory became known. Galileo, in the Discorsi, independently conceived and supported it (though in a greatly simplified manner), and, despite Kepler's misgivings, the theory became the ruling one for the remainder of the century. Just as it had replaced the numerological explanation embodied in Zarlino's senario, it was to be replaced itself around 1730 by Rameau's and D'Alembert's overtone theory of consonance. Not until 1863 were both the latter and its predecessor, the coincidence theory of consonance, to be incorporated as 'limiting cases' in the most comprehensive theory of consonance ever conceived: the one expounded in Helmholtz's On the Sensations of Tone.

BLAY Michel

Chercheur C.N.R.S. Centre A. Koyré, Paris, France.

VARIGNON ET LE STATUT DE LA LOI DE TORRICELLI

En 1644, Torricelli, s'inspirant directement des recherches sur la chute des graves, publie à Florence dans le De Motu Aquarum, la loi qui porte aujourd'hui son nom. Elle stipule que la vitesse d'écoulement d'un liquide par le fond d'un réservoir percé est proportionnelle à la racine carrée de la hauteur séparant le trou de la surface libre du liquide, c'est-à-dire que cette vitesse est égale à celle qu'acquerrait une goutte d'eau tombant en chute libre d'une hauteur égale à l'altitude de l'eau au-dessus de l'orifice.
Torricelli souligne par ailleurs avec netteté que l'étude expérimentale de cette loi présente un certain nombre de difficultés.
Dans cette perspective, l'Académie Royale des Sciences va, dans le cadre de ses travaux sur la force des eaux courantes à presser et à mouvoir, lancer un programme de recherches soulignant, comme le dit l'abbé Picard, la nécessité qu'il y a "d'éprouver ce que dit Torricelli". Ces études expérimentales vont se développer principalement au cours des années 1668-1669.
Huygens constate, au cours de cette période d'activités expérimentales que, jusqu'à présent, la loi de Torricelli n'a pu être "démontrée par raison mais seulement prouvée par expérience". Sa validité repose en fait sur le "principe d'expérience" suivant lequel les corps liquides en sortant par une ouverture possèdent une vitesse suffisante pour remonter jusqu'à la surface libre du liquide.
C'est Varignon qui, à l'Académie Royale des Sciences va s'efforcer, à partir de 1695, de "démontrer par raison" la loi de Torricelli.
Il présente un premier Mémoire le 23 avril 1695 dans lequel il déduit la loi de Torricelli des principes de la mécanique et des lois générales du mouvement. Sa démonstration, pour intéressante qu'elle soit, reste néanmoins incertaine dans les manipulations infinitésimales en rapport avec la force et la quantité de mouvement. Varignon reprend cette question en la généralisant dans un deuxième Mémoire en date du 14 novembre 1703.
Ces démonstrations de 1695 et de 1703 sont accompagnées de considérations relatives au statut qu'il faut accorder à ce qui est "démontré" par opposition à ce qui n'est que "principe d'expérience".
Pour Varignon, déduire la loi de Torricelli des principes de la mécanique et des lois générales du mouvement, c'est la démontrer, c'est-à-dire la "mettre tout à fait hors de doute", en "trouver la raison", ou bien encore, pour reprendre Fontenelle dans l'Histoire de l'Académie Royale des Sciences, pour l'année 1703,."satisfaire la raison".

Werner Diederich

Dpt. for Philosophy, University of Bielefeld, W.GERMANY

THE STRUCTURE OF THE COPERNICAN REVOLUTION

Scientific revolutions are often regarded as replacements of paradigms. That is, a revolutionary process in science is considered to be a kind of binary relation between an old and a new paradigm. What I want to show is that at least the Copernican Revolution does not show this linear pattern. My alternative account will be, to conceive of scientific revolutions as processes involving a change of 'disciplinary structure', typically processes of branching and merging of scientific disciplines and often also of various sorts of non-scientific traditions.
I regard the Copernican Revolution as a process taking roughly 150 years, namely the years between Copernicus' De Revolutionibus, printed in 1543, and Newton's Principia, published in 1687. This revolution is by no means a revolution of astronomy alone, but also one of physics. We should go, however, even beyond physics, and consider e.g. also the development of anatomy in the 16th and 17th centuries. It looks like there is a striking parallelism between the detection of the circulation of the blood and the development of Copernican astronomy, i.e., in Renaissance terms, of the revolutions of micro- and macrocosm. At least there seems to exist a certain common source of ideas which helped both revolutions to succeed. One idea is that of a central reigning organ, namely the heart or the sun, respectively. Thus the Copernican Revolution is hardly understandable without considering the support from Renaissance Platonism.
Scientific revolutions like the Copernican one are typically accompanied by some change in the disciplinary structure of science, i.e. in the way the involved disciplines are conceived of and hang together. You learn what 'mathematics', what 'astronomy', what 'physics', and so on, means, only together by incidentally learning how these disciplines relate to each other.—I do not claim that my account of the structure of scientific revolutions gives in itself an adequate understanding of such outstanding processes. For a fuller understanding you have to say something about how a revolution comes about. For this purpose I want to outline a model, the magnet model for scientific revolutions. Individual scientists may be compared to elementary magnets, which switch their directions according to an outer magnetic field, one after the other.

Dr A.J.Pyle.

THE ROLE OF BIOLOGY IN THE SCIENTIFIC REVOLUTION.

The classical accounts of S.R. concentrate on celestial and terrestrial mechanics - if the life sciences get a mention at all, it is in separate chapters devoted to the positive achievements of Harvey et al, little or no attempt being made to integrate these into an overview of S.R. as a whole. Yet most C17 scientists had medical backgrounds, and problems we would call 'biological' dominated their scientific literature. How should this apparent discrepancy be resolved ?

In this paper we accept the now orthodox view that the Mechanical Philosophy, with its explicit rejection of intrinsic qualities, powers, and 'natures', lay at the heart of S.R. We further endorse the thesis of Hutchison (History of Science, 21, 1983, 297-333) that M.P. can be seen as 'supernaturalist' in its consequences, i.e. as invoking direct rather than mediate divine involvement in the physical universe. We insist, however, that M.P., instead of originating in mechanics and being later applied to biology, was articulated from the outset in a variety of biological contexts. To substantiate this thesis - the centrality of Biology to M.P. and hence to S.R. - we draw heavily on the work of Jacques Roger, as well as on C17 sources, some familiar but now seen in a new light.

Stephen F. Mason

King's College, University of London, WC2R 2LS, U.K.

ISOMORPHIC THOUGHT STYLES IN THE SCIENTIFIC REVOLUTION
AND THE PROTESTANT REFORMATION

The unifications of modern science combine under a common theoretical scheme disciplines with little previous connection, and with their own individual models which are either superseded or assimilated into the new synthesis. The Grand Unified Theories and the electroweak unification of recent times, and the nineteenth century unification of electromagnetism, were largely internal developments within a well-established and securely based scientific tradition, in contrast to the pioneer unification of terrestrial with celestial mechanics during the formative period of the tradition, expressed by the scientific societies with a continuous history from the seventeenth century.

The definitive unification of the Newtonian world system incorporated theological elements congruent with those of the prevailing social milieu, whether orthodox Anglican or dissident Non-conformist. It was a universe created substantially in its present form, and governed by inviolate natural laws laid down by the Deity in the beginning. All entities moved autonomously under their own inertia, subject to the guidance and control of universal gravitation. Earlier in the sixteenth and seventeenth centuries, both the natural philosophers and the protestant reformers, particularly those of the Calvinist persuasion, had eliminated the central dogma of the older world system, the concept of hierarchical government, embodied in the Great Chain of Being, wherein each entity had dominion over those below it in the scale and served those above. The triple triads of angelic motors in the heavenly bodies lost both power and place in reformed theology and natural philosophy, leaving the residual beings of the chain organised in a scale of perfection but governed by a common set of natural laws, arguably, those of mechanics.

The homomorphism of theological reformation and scientific revolution found an intimate expression, through the **Restitutio Christianismi** (1553) of Michael Servetus, in the microcosmic system of the human frame. Unitarianism in theology implied the unification of the traditional three organ-systems and the triple spirit-motors of their individual fluids, with the theory of the lesser circulation of the blood between the heart and the lungs as a consequence.

S.F. Mason, 'The Scientific Revolution and the Protestant Reformation', Annals of Science, 9 (1953) 64-87 and 154-175.

Marta Feher

Dr. Technical University, Budapest, Hungary, Europe

The place and aim of "experimentum crucis" in 17th century methodology

One of the most curious developments of 17th century scientific methodology is the emergence of the method of crucial experience, or as it was later called, crucial experiment to a central position among methodological norms. The notion of it appears under the name of "instantia crucis" in the Novum Organum of Francis Bacon who lays no special emphasis upon this methodological principle merely lists it as the 13th among 27 "prerogative instances" helping the mind in assuring the adequacy of the inductive process. It appears as a norm of selection of one from among rival scientific axplanations having a hypothetical status in Descartes' methodology/ he uses it in his debate with Harvey/, then in Boyle's and Newton's works/ first in his optical papers/, In Galileo's Dialogue the argumentation for the Copernican astronom can already be seen as construed in concluding into a crucial experience: the phenomenon of the tides.

The rise of the method of crucial experiment to a central position in the methodology of later 17th century natural philosophers was due to several epistemological as well as social factors. The most important among the epistemological factors are: /1/ the shift from a necessitarian picture of scientific knowledge towards a contingent one;/2/ the change of the conception of experience from the Aristotoelian ordinary common-sense experience to the artificial experimentally produced experience of Galilei and Newton;/3/ the reduction of the four Aristotelian causes required as middle terms for a full blown demonstration to merely one/ to the formal in Galileo and Bacon, to the dfficient in Descartes and Newton/4/ the reduction of the four types of motion to one: locomotion; and /5/ the acceptance of the clockwork metaphor as basic in the view of the physical world.

Among socially based cogintive factors I would mention the "protestant rule of faith" /Feyerabend/, and change of interpretation of the "Scriptures".

Qc

Mark Neustadt

Graduate student, Johns Hopkins University, U.S.A.

THE RELATIONSHIP BETWEEN FRANCIS BACON'S SCIENTIFIC AND LEGAL REFORM PROGRAMS

This study concerns the influence of Francis Bacon's legal thought on his program for the reform of the sciences. It suggests that central features of the Great Instauration were borrowed from Bacon's less well known program for the reform of the common law, a program modeled on Justinian's 6th century reform of Roman civil law. Among these features are the choice of format for the Novum Organum and the natural histories, the idea of a group project for the collection of the natural histories, the pragmatic thrust of the program and its lack of a metaphysical dimension. The discovery of a previously unknown manuscript by Bacon entitled Aphorismi de Jure Gentium Maiore sive de Fontibus Justiciae et Juris, containing the kernel of his philosophy of jurisprudence, makes it possible to describe the relationship between his two reform programs with greater clarity than has previously been possible.

Desmond J. FitzGerald

Professor of Philosophy, University of San Francisco

The Meaning of Coeur in Pascal's Pensées

Blaise Pascal (1623-1662), French scientist and religious genius, is a striking combination of what has been called the divided mind of the seventeenth century. The divided mind relates to the tension between materialism of the new science and the immaterialism of the tradional religious commitments.

In his Pensées, Pascal sometimes used the word coeur (heart) as in "Le coeur a ses raisons que le raison ne connaît pas" (The heart has its reasons which reason does not understand). My study is an attempt to clarify the meaning of coeur for Pascal. Though he sometimes used the term in its ordinary sense as a vital organ, more often he gives it a cognitive intelligence.

I shall try to present evidence that Pascal's coeur has overtones which anticipate what Jacques Maritain will come to call "knowledge by way of affective connaturality." This is a kind of wisdom which in the moral order corresponds to the unscientific ethical insight of the virtuous person, but in the artistic order resembles a sort of intuitive knowledge the artisan has with respect to the use of his materials.

How did Pascal, who did not receive the ordinary formal education, come to formulate this kind of knowing?

Lech Mokrzecki

Professor, Gdansk University, Poland

Protestant Scholars in Poland in 17-18th c.: supranational importance of their output in the field of exact and natural sciences

Achievements of Polish scholars of 17-18th c. are relatively little known. Protestant researchers of the period came mainly from burgher circles of Northern Poland. They maintained strict connections with educational system, and in 18th c. also with scientific societies (first called into being in Gdansk: Societas Litteraria-1720, and Societas Physicae Experimentalis-1743). Research methods based on individual work, and in 18th c. on teamwork coordinated by newly formed societies. A number of studies were published in best known magazines (Philosophical Transactions, Acta Eruditorum etc.). Among representatives of 17th c. we should name, e.g., P. Kruger from Gdansk (1580-1639), a mathematician, the first to separate logarithms of numbers from logarithms of trigonometrical functions in the tables. Referring to J. Neper, Kruger also compiled the so far most detailed logarithmic tables (1634). His disciple, J. Hevelius (1611-1687), published a work of basic importance on topography of the Moon (1647). Many names he proposed are still used today. In Podromus Astronomiae Hevelius presented a catalogue of stars based for the first time on coordinates of stars based for the first time on coordinates of stars in the equatorial system. He distinguished 12 new constellations, out of which 7 retain his original names. Hevelius discovered slow, systematic alterations in magnetic declination, and constructed a prototype of a periscope.

Among scholars of 18th c. H. Kuhn (1690-1769) was the first to accomplish geometrical interpretation of compound numbers (1750), and a physicist D. Gralath (1709-1767) improved the Leyden jar, set a battery of condensers for the first time etc. D.G. Fahrenheit, born in Gdansk (1686) became famous for his experiments on mercury thermometers which he carried out abroad. Experiments on vacuum existence in 1647 (Walerian Magni) had also wide repercussions in Europe. Out of natural scientists, J. Jonston (1603-1673) was the author of monumental Historiae Naturalis (1650), and J.T. Klein (1685-1759) is recognized as the most outstanding taxonomist before Ch. Linne, as he classified the whole animal world and introduced new terms in zoology (e.g., Echinodermata for sea-urchins). Others contributors are: J.R. Forster, a member of Cook's expeditions (1772-1775); he wrote a Voyage Toward the South Pole and Round the World, and D. Messerschmidt (1685-1735), an explorer of Siberia (Forschungsreise durch Sibirien 1720-1727).

Andrzej Biernacki (Warsaw)

William James and Wincenty Lutosławski

Biographers of William James (1842-1910) have long been aware that his many philosopher friends included Wincenty Lutosławski (1863-1954), a Pole now remembered mainly as author of The Origin and Growth of Plato's Logic (London 1897). The two first met in America in 1893. In 1907, William James got Harvard's Lowell Institute to invite Lutosławski for a series of lectures about Poland. At least on one occasion, they met in Europe. Around 1900 William James wrote the foreword to Lutosławski's The World of Souls (London, 1924). Brief excerpts from their correspondence can be found in The Letters of William James edited by the latter's son Henry. R. B. Perry's fundamental Thought and Character of William James (1935) mentions James' relations with Lutosławski.

The Harvard College Library keeps author's copies of the two latter books, each containing interesting annotations and additions. Henry James' edition has the signature AC85.J.2376.920, Perry's AC85.J.2376.S935p in the manuscript collection. They are kept together with a copy of William James' Memories and Studies (New York, 1911), which also bears his son's annotations (60J-286). At the Houghton Library in Harvard you will find Lutosławski's letter of condolences to William James' widow (bMS Am.1092-4320) with an appended clip of a Gazeta Warszawska October 16, 1910 obituary; this is supplemented by Lutosławski's letters to Mrs. James and Henry James (bMS AM.1092.10.116). Lutosławski's 1929 letter to Perry is kept in the James Papers collection (bMS Am.1092-522).

Professor Wiktor Weintraub kindly drew my attention to some yet unexplored details. Lutosławski, who was a lifelong student of 19th-century Messianism as a version of Polish national philosophy tried to get James interested in this philosophy; he even wrote Perry, "I have read to him in improvised English translation many masterpieces of Polish literature." He is also known to have sent Adam Mickiewicz' Paris lecures to James. In his obituary of James, Lutosławski maintains James read texts by Zygmunt Krasiński and was indirectly familiar with Andrzej Towiański, August Cieszkowski and Juliusz Słowacki. James, according to Lutosławski, "eagerly" reacted to "some of those ideas."

Warsaw University Library keeps (signature BVW 173151) a copy of The Meaning of Truth (1909) William James sent Lutosławski, who studied it closely. The numerous annotations show he valued his relations with James highly; he even believed theirs was a mutual influence. He gives this to understand in his autobiography An Easy Life (1935) where he refers to James' lecture "On the energies of men."

John F. Bennett

Research Associate, Office for History of Science,

University of California, Berkeley

THE FAILURE OF OSTWALD'S ENERGETIC MONISM

In the history of scientific convergences, a culmination was reached with the insight that everything, from matter to mind, is energy. This idea probably came first to Wilhelm Ostwald (1853-1932) in 1887, though others broached it thereabouts. He put much effort into its development before World War I, and continued to believe in his "energetische Monismus" till his death. But the idea never carried conviction in influential circles. Among the reasons for its failure are the facts that Ostwald defamed matter in general as well as the atomic theory in particular in promoting energy to the top spot; that his philosophical popularizations included somewhat ludicrous ideas, such as formulae for happiness in terms of efficiency of energy utilization; and that he associated his own monism with that of Ernst Haeckel. Perhaps more fundamental, however, were his failures adequately to link energy with matter at one extreme and with mind at the other. Regarding matter, the link was made firm by Einstein, as Ostwald recognized: but because of its intense concentration of energy, matter thereby became the tail that wags the dog, so to say, rather than another "mere" form of energy. Regarding mind, Ostwald seems never to have made explicit the generalization that energy has the capacity, under certain conditions, to perceive, emote, and think, etc., as well as, under certain conditions, to exert a force through a distance--i.e., that it embraces the subjective as well as the objective world, and so is more than a physical concept. There are indications that Ostwald knew this, but he did not put it over. Thus he is not credited with hitting upon the monistic basis toward which science had been pointing since the first great convergence, of celestial and terrestrial motions in the Newtonian synthesis. But perhaps he should be.

Carolyn Eisele

Prof. Emerita, Mathematics, Hunter College of CUNY
Member, Institute for Studies in Pragmaticism, Texas Tech U.
Seminar in Philosophy of Science and Mathematics, Columbia U.

Peirce's History of Science as an Exercise in Logical Methodology

Shortly there will be available to those interested in the history of science a collection of historical studies from the pen of Charles S. Peirce (1839-1914), one of the finest minds yet to have appeared on the American scene. His philosophical thought has been known and celebrated through the 8 volumes of his Collected Papers (Hartshorne, Weiss, and Burks). Beginning in 1950 this writer undertook to gather together and publish Peirce's writings on the history of science. It soon became evident that Peirce's approach to such investigations was unique. More than that his advocacy of the new approach to mathematics springing up and flourishing by the end of the 19th century had to be publicized first and was presented in the 4 volumes (5books) of The New Elements of Mathematics by C.S.Peirce. A return to the original task has brought a revision in the presentation of those writings to reveal the play of Peirce's logical methodology at every level of activity. The volume exhibiting them is entitled Historical Perspectives on Peirce's Logic of Science: a history of science, (Mouton). Biographical elements naturally appear.

The most impressive study is that of the Petrus Peregrinus manuscript found by Peirce in the Bibliothèque Nationale while on a Coast Survey mission in Paris in 1883. His interest lay in its affording an additional fact to show that there was considerable activity in the direction of physical research in the first half of the 13th century. Peirce transcribed it completely.

And, furthermore, he was engaged to lecture at the Lowell Inst. on the history of science (1892-1893). Correspondence with Cattell and Putnam's Sons brought a contract in 1898 for the production of a volume on the history of science. Contributions to periodicals and newspapers were in many cases pure studies of episodes in history or were embellished with such references. Reports on his work in the service of the US Coast and Geodetic Survey carried similar material and in several cases were pure historical statements regarding early discovery or the improvement of technique. Long studies on methodology and the logic of history were drawn up for future publication. Lectures and miscellaneous writings were associated with contacts with institutions(The Smithsonian, The Johns Hopkins University, Columbia University, The Lowell Institute, The Carnegie Institution). Activity as a scientific consultant in chemistry involved historical settings. His manuscript on the history of spelling is a semiotic treat. Peirce ends a paper entitled "How did Science Originate?" as follows: "Now the minor currents and ripples in the history of science no doubt depend upon all sorts of accidental circumstances, chief among which may be reckoned the details of man's cerebral anatomy, and the corresponding peculiarities of his mind. But advance of the deep tide of science cannot fail to be governed chiefly by the essential relations between the laws and classes of the objects of nature."

Yukitoshi Matsuo

Associate Professor, Doshisha University, Kyoto, Japan

INFLAMMABLE AIR IN THE LATE EIGHTEENTH CENTURY CHEMISTRY

Far less attention has been paid to "inflammable air" (i.a.) than to the fixed air (CO_2) and the dephlogisticated air (O_2) in the late eighteenth century. The significance of i.a. in the history of chemistry, however, was no less important than that of the latter two. The purpose of this paper is to trace some important steps and results in researches on i.a., to examine parts i.a. played during that period, and to point out the significance of it.
In the late eighteenth century, at least four different kinds of i.a. were referred to under the name of inflammable air: i.a. from metals (light i.a., H_2), i.a. from charcoal (heavy i.a., CO), water gas (a mixture of H_2 and CO), i.a. from marshes (CH_4). The lack of clear discrimination of them from each other had sometimes caused the errors and confusion, whose elucidation had sometimes produced fruitful results.
Early researches on i.a. were done by Cavendish, Lassone, Volta, and Priestley. Cavendish' study (1766) was the first detailed examination of i.a. from metals. Although he did not always identify i.a. with phlogiston, he at least made way for a doctrine that phlogiston was i.a. Volta's study of i.a. from marshes (1777) gave helpful means for distinguishing differnt kinds of i.a.: the colours of their flames, speed of combustion, and the requiring volumes of air for detonation and the intensity of sound of detonation.
Researches on the combustion of i.a. and their products were contributory to the discovery of the composition of water. Cavendish' study (1784) came mainly from the experimental examination of Kirwan's opinion about the "cause of diminution of the common air in phlogistication." He examined six processes in phlogistication, and his synthesis of water was the result of one of them, a process of exploding i.a. with common air. G. Monge's study and A. L. Lavoisier's also derived from Volta's study of i.a. Lavoisier clearly pointed out the confusion of inflammable airs at the outset of his paper (1784). J. Watt's crucial defect was the failure to distinguish between i.a. from metals and that from charcoal.
Cavendish' 1784 paper led to a controversy with Kirwan. This controversy is very significant in several ways: that controversy led Cavendish to important experimental discoveries, and Kirwan to build up his system which was to play very decisive roles in the overthrow of the phlogiston theory against his will. Kirwan's system on combustion and acidity was based on the two doctrines that phlogiston was i.a. and that fixed air was universal acid. Unlike previous phlogiston theories his system was distinguished in its being highly systematic, quantitative and, therefore, clearly revealing its inner inconsistency.
Researches on i.a. were continuously carried on by Cruickshank, Desormes and Clement, Dalton, Davy, etc. It might be said that Davy tried to modify Lavoisier's system based on oxygen with his system based on i.a. (hydrogen).

Qe

Dr Jan Sebestik

C.N.R.S. and Université de Paris XII, Paris, France

THE INNOVATIONS IN LOGIC BY BOLZANO

Bolzano created a new system of logic and established a series of preliminary fundamental concepts (validity, degree of validity, analyticity). His logic is founded on the concept of **proposition-in-itself**, in contradistinction to judgement (a mental entity), and to sentence (a linguistic entity).

Bolzano introduced the concept of **propositional form** ("Satzform") with one or more variables ("variable ideas"). His method involves the successive substitutions of appropriate ideas for the variables in a given propositional form, and the examination of the truth values of the propositions obtained. This method leads to the classification of all propositions as valid, contravalid or neutral. He then defined two other fundamental concepts: 1. **degree of validity**, which relates his deductive logic to the logic of probabilities; and 2. (logical) **analyticity**. To construct his system, Bolzano established the analogy between propositions and ideas: "For propositions, the property of having truth or not corresponds, for ideas, to the property of actually representing or not an object." The general scheme of extensional relations between ideas can then be transferred to the propositions. However, the relations do not hold between the propositions themselves, but rather between the corresponding classes of verifying ideas. In this way, Bolzano is capable of constructing a system of logical relations between propositions and obtains, among other results, the first adequate definition of the concept of deducibility or logical consequence.

Wen-yuan Qian

Zhejiang University, China; the University of Michigan, USA

THE IDEA OF THE LAWS OF NATURE
as a logical as well as a historical question

Since 1956 there have appeared a number of studies that compared traditional Chinese and Western ideas of laws, in particular the laws of nature. The most interesting lesson that we could derive from the studies is, I think, although starting from disparate socio-historical contexts, different civilizations have broached the universal concept of the laws of nature. By the laws of nature I mean those formulations that, according to the standard of the Scientific Revolution, is typically exemplified by the Newtonian mechanical laws and other physical laws. The different approaches in different civilizations were basically determined by epistemological criteria and manners of reasoning; in sum, different levels of intellectual maturity. This line of consideration leads us to the observation that science occupies a special position in the whole range of human intellectual creations. A significant question is: What characteristics distinguish science from other types of intellectuality, and put it in a position that has for about a century become more and more enviable to other disciplines? By comparing the Scientific Revolution with the early scientific growth in ancient Greece and the Hellenistic world, I found that the protagonists of the Scientific Revolution declared the independence of science by subjecting it to a process of empirical verification, mathematization, de-metaphysicalization, segregation from religion, institutionalization, popularization, application, and axiomatization. The standard of scientific truth is thus best represented by: empirical verification, rigorous logic, and axiomatic aestheticism. In pondering China's nondevelopment of modern science from a comparative perspective, I have realized the idea of a "unique" manner of historical accumulation of exact science in any culture. (It is "unique" in respect to order, but obviously not in temporal scale; therefore, it is not wholly unique.) Now, quite logically, the idea of a unique accumulative order is related to the idea of a set of universal, though evolving, scientific criteria, which is again related to the universal idea of the laws of nature, which has been stabilized for centuries. If we accept these universal ideas, we will find ourselves in a vantage point to review a number of interesting questions, such as: What is the objective degree of the development of mechanical science in traditional China? How valid is the Needhamite hypothesis that in a different historical context, traditional China might have developed "field physics" first and got around the Newtonian "billiard-ball physics"? Should it be possible to extend the idea of objectively measuring the level of development to other exact sciences? In tracing ancient thoughts, in particular their manners of reasoning, are there objective (or scientific) evaluative guides?
   It is doubtful that it is possible to make many non-scientific disciplines into sciences. This study, however, heightens our hope that we may at least push the history of science a few steps toward the epistemological positions that are occupied by exact sciences.

Chen Ke-jian

East China Normal University, Shanghai, China

An Inductive Logic System, With Nonzero Probabilities to Some Infinite Universal Statements

This paper obtains, by changing R.Carnap's λ-system, a kind of probabilistic functions, which can evaluate some infinite universal statements positive value S, then Popper's famous argument [1] is shown to be not valid.

The termimnologies and their implications are the same as those used in R.Carnap's book. [2]

Let weight of empirical factor SM/S to be S square, and let weight of logical factor W/K to θ, then set

$$c(h_M, e_M) = \frac{S^2(S_M/S) + (W/K)\theta}{S^2 + \theta} = \frac{SS_M + (W/K)\theta}{S^2 + \theta},$$

This paper proves that

1. If evidence e_M means that all of tested S indeveduals have property M, hypothesis h means that all of not tested individuals (infinite) have property M, too, then h's degree of comfirmation by e_M

$$c(h, e_M) = \prod_{n=0}^{\infty} \frac{(S+n)^2 + (W/K)\theta}{(S+n)^2 + \theta} > 0.$$

2. If $e_M^{(1)}$, $e_M^{(2)}$ mean, respectively, that all of tested S_1 individuals and all of tested S_2 individuals have property M, $S_1 > S_2$, then

$$c(h, e_M^{(1)}) > c(h, e_M^{(2)}) > 0.$$

3. If e_M means thata at least one of S individuals which has been tested has not property M, then $c(h, e_M) = 0$. The last consequence may be welcome by Popperians.

---

[1] K.R.Popper, The Logic of Scientific Discovery, 1959, p. 363.

[2] R.Carnap, The Continuum of Inductive Methods, 1952.

Thomas L. Drucker

Assistant Professor, Department of Liberal Studies,

University of Wisconsin,
Madison, Wisconsin

INTUITIONISM AND FORMAL SYSTEMS

One of the motivations for formalization of disciplines within mathematics (geometry for Euclid, most notably) has been to avoid the possiblity of erroneous statements' being accepted as true. By making the premises of mathematical arguments as explicit as possible, the likelihood of error was thereby felt to be reduced. Lakatos (I. Lakatos, Proofs and Refutations, 1976) has indicated the shortcomings of formal logic in this regard. Intuitionism has tried to explain away some of these shortcomings by critizing traditional logic and by replacing it with logics that more accurately reflect the structure of mathematical reasoning (A. Heyting, Intuitionism: An Introduction, 1956). The criticism by Brouwer of traditional formal logic succeeds equally well against the intuitionist formal logics developed by his disciples. A consistent following-through of Brouwer's programme reduces the formal aspect of mathematics to an assortment of particular results (P. Kitcher, The Nature of Mathematical Knowledge, 1983). On the other hand, if the identification of mathematics with any particular formal system is rejected, Brouwer's programme can be seen as a defense of mathematical realism.
It serves as a preliminary explanation of how mathematical knowledge has occurred and how it can continue to occur. The wanderings of intuitionists from the paths laid down by Brouwer can be explained by the role that is attributed to formal logic and that it cannot fill.

Arcangelo Rossi

Professor, Universita' di Lecce, Italy

NATURAL INTELLIGENCE AND ARTIFICIAL INTELLIGENCE: AN HISTORICAL AND CRITICAL APPROACH

It is well known that the great physicist J.C. Maxwell first studied automatically retroactive physical systems theoretically; that is systems which are able to automatically correct every variation which may occur in the course of their functioning.

Since pre-Roman times, especially in the Hellenistic period, and later, in medieval and modern times, substantial developments of such automata had taken place, mainly for devotional or amusement purposes, and only much later for the production of material goods. However, nobody before Maxwell tried to assess the essential features of automata theoretically as the ability to correct variations automatically.

We suggest that the time was not ripe before Maxwell for a sceientific xplantation of automata. The modern industrial system had to develop its intrinsic automatism before the essential features of all automata could be recognized and a way to control them found.

This is the real starting point even of attempts to simulate the functioning of human intelligence as a kind of automa. Is this reduction of natural to artificial intelligence, even though historically, economically and scientifically motivated, really possiblè? A comparison of N. Wiener's program with Maxwell's early "cybernetics" gives us the possibiltiy of answering the question.

Walter HOERING

Prof., Tübingen Univ., Dept. of Philosophy

ABSTRACT HYDRODYNAMICS AND THE VISUAL MIND - AN ANALYSIS
OF SELFIMPOSED CONSTRAINTS WHICH LED TO MAXWELL'S SUCCESS

There has been a series of interdisciplinary seminars on the genesis of Maxwell's equations at Tübingen Univ. conducted by M.Schramm (history of science), W.Wittern (physics), W.Hoering (philosophy). We have been trying to reenact the development of M.'s electrodynamical ideas. The initial situation was characterized by M.'s familiarity with the theory of surfaces and partial differential eqations on the one hand and Faraday's description of the electrical and magnetic fields in terms of lines of force on the other. M. had the definite impression that F.'s qualitative description of those fields was pretty close to a quantitative description. In order to arrive at such a description he devised a series of abstract hydrodynamical models which were to mimic the flow of the lines of force. Initially they were only conceived as a means for facilitating "exact speculation", but then some of these fluid models were seen to be interpretable as dealing with actual electrical currents.- The process of mathematical experimentation is also visible in the successive attempts to introduce the "right" displacement term into the equations.
From the philosopher's of science point of view three traits of M.'s implicit methodology seem worth to be singled out as important beyond this special, if famous, example of theory generation and amenable to detailed examination.
1. Achieving visual representation of possible situations.
2. Narrowing down the range of possible stuations by demanding that they be described in terms of an already well understood theory (in our case: hydrodynamics). This often also implies severe mathematical constraints like continuity, differentiability, etc.
3. Keeping a measure of adaptibility by considering a whole range of models, each of which contains adjustable parameters.

Attention to such methodological points can also bring about a better understanding of the historical case discussed.

Antonio Ferraz

Profesor Titular. Universidad Autónoma de Madrid

HISTOIRE DES SCIENCES ET RÉALITÉ

L'histoire des sciences a atteint de nos jours sa maturité.Au moment où la science a pénétré profondément tous les aspects de l'existence humaine,la réflexion philosophique et historique sur cette même science peut et doit devenir une tâche nécessaire pour comprendre la situation de l'homme actuel et pour commencer à construire de nouvelles bases pour sa vie.- Ce que l'on entend généralement comme philosophie de la science,se réduit de façon injustifiée à l'épistémologie et à la méthodologie.La consideration structurelle et analytique de la science la présente comme un objet fini dont il s'agit de découvrir les articulations internes.Les points de vue plus historiques accentuent les aspects processuels,mais ne conçoivent pas non plus comme tâche fondamentale de décrire et d'analyser comment la science a composé une représentation particulière de la réalité.- L'homme organise et oriente sa vie selon la représentation de la réalité qui régit une époque.La science a construit au cours de l'histoire une représentation de la réalité différente des représentations propres à la pensée mythologique et à la pensée métaphysique.Une différence très importante est la dissociation entre l'homme et le reste de la réalité.L'homme ne se sent plus gouverné par des forces obscures parce qu'il croit connaître le caractère des structures et des processus réels et ne trouve rien d'autre que la nécessité légale ou le hasard.- La situation décrite au paragraphe antérieur est la conséquence de la détermination du réel comme "objectif".Ceci a rendu possible la détection de plus en plus précise des parties qu'on peut différencier dans la réalité et des connexions entre ces parties.Mais avec ceci,on a trouvé de nouveaux fondements pour poser le problème de l'intégration de l'homme dans la réalité.Il en ressort une contradiction entre l'attitude épistémologique que la science a favorisé et la connaissance du réel que le développement de la science a atteint.Résoudre cette contradiction est une tâche philosophique urgente.

SOFONEA LIVIU University of Brașov, ROMANIA
NICHOLAS IONESCU-PALLAS, Center of Physics, Bucharest.

LOGIC OF THE HISTORICAL DEVELOPMENT OF SCIENCE AND INNER
LOGIC OF ITS THEORIES. (Basical items only).+/

I. The manysidedness of scientific approach.

1. Scientific attitude. 2. The Unity of the science, 3. Scientific theory, 4. Genesis of scientific theory, 5. Life of scientific theory, 6. Historical & perenial character of a scientific theory, 7. Organic structure of a scientific theory 8. Ramifications of the scientific itinerary.

II. Searching the evolution of scientific thinking by the means of historical-epistemological models (H.E.M.).
1. Mesage of H.E.M., 2. Types of H.E.M.

III. Modelling examples.

1. "Naturalising" the transcendent mathematical number in the Indian thought of the Middle Ages, 2. H.E.M. of the Aristotelian physics, 3. H.E.M. of the Cartesian physics. 4. H.E.M. of the aether theories, 5. H.E.M. of the connections between classical mechanics and electrodynamics, 6. H.E.M. of the energy scale, 7. H.E.M. of Kantian thought & modern physics, 8. H.E.M. of the trajectory concept, 9. H.E.M. of the Platonic physics, 10. H.E.M. of cosmological theories, 11. H.E.M. of hydrodynamic scheme, 12. H.E.M. of the concept of motion, 13. H.E.M. of the quantum wave equations, 14. H.E.M. of evolution of some theories with hidden parameters, 15. H.E.M. for measuring the astronomical time, 16. H.E.M. of evolution of some theories with "hidden parameters2", the quantum theories. 17. H.E.M. of the connection between the representative spaces and physical theories, 18. H.E.M. of the relation between traditional mechanics and (various) modern mechanichal theories (M.T.) operating at small ac actions or/and at great velocities, 19. Model of sympathetic phenomena, etc.

IV. Genesis of physical entities, Examples.

V. The necessity of promoting a scientific theory (S.T.).

1. S.T. as the unique way of organizing (and interpreting) the "flew of events", 2. The process of gathering the primary data and their further processing, 3, The role of S.T. in the orientation of research towards major objectives, 4. The becoming of a S.T. and foreseeing of surpassing it, 5. Entropy of scientific approach and the going over from potential manysidedness to historical uniqueness. 6. Can the ratio chance/necessity be influenced?
x/. A detailed preprint will be delivered on the occasion of the Congress.

Annie Petit

Université Clermont II, UER Lettres, Département de Philosophy

La critique positiviste des institutions scientifiques

Le positivisme est souvent présenté comme un hymne confiant aux progrès des sciences. Ces jugements globaux et persistants - qui basculent le positivisme vers le scientisme - méritent d'être nuancés. Nous voulons souligner certains aspects critiques de l'oeuvre de Comte: en particulier ses analyses sévères des institutions scientifiques contemporaines, et son programme des réformes radicales.

Les attaques virulentes portent sur plusieurs fronts. Mais ce sont les Académies qui sont le plus violemment condamnées: à la fois lieux d'anarchie et de sclérose intellectuelles, "compagnies arriérées" dont le morcellement traduit "la dispersion mentale", régies par "une majorité scientifique essentiellement incompétente" et par les préjugés et passions des "coteries". Y règnent "demi-portées intellectuelles" capables au mieux "d'élaboration routinière", l'"avide concurrence" des "médiocrités ambitieuses", les "sophistes et trafiquants de sciences" qui masquent leurs intérêts sous les discours de neutralité. Comte multiple ainsi la dénonciation des "pédantocrates".

Mais les "corps savants" ne sont pas les seules cibles de la verve critique de Comte. Les écoles et les plus réputées, sont aussi examinées sans indulgence: Polytechnique, qui d'après Comte est devenue monotechnique sous la dominance des géomètres; les écoles "normales" au nom bien mal choisi.

Les critiques comtiennes s'organisent sous plusieurs chefs:
- la critique d'une formation scientifique désordonnée où a dominé l'esprit de spécialité qui a induit des conduites impérialistes désastreuses.
- la critique du peu de cas fait de l'histoire des sciences; d'où les courtes vues des savants, leur manque de relativisme et leur tendance à retomber dans de vieilles ornières.
- la critique des visées pragmatiques prématurées, alors qu'un programme d'applications conséquent ne peut être envisagé sans une formation de base aux sciences théoriques et pures. Toute la liaison des sciences et des arts est en jeu ici, ainsi que le statut accordé par Comte aux "ingénieurs" et aux technocrates.

Ces critiques peuvent peut être avoir encore quelque sens aujourd'hui.

Yong Woon Kim

Professor of Mathematics, Hanyang Univ., Seoul, Korea

Pan-Paradigm in Korean History of Science

One of the most important scientific achievements in Korea was
recorded by King Saejong of Yi dynasty (1419-1452). His academi-
cians and the king himself devoted to the development of many
branches of science, i.e., linguistics, astronomy, mathematics,
music, calendar making, medicine, measurement, metaphysics, etc.
It is clear that these achievements were inspired by one principle:
their effort to adapt classical Chinese scientific methods to Korean
environment. This also applies to political philosophy of the
dynastic Korea in which orthodox confucian doctrine was creatively
adapted to the reality of Korea. Korean mathematical history, when
compared with that of Japan and China, presents peculiar characters,
which are considered to have been molded by the same principle
mentioned above.
The author defines pan-paradigm as a philosophical principle which
has served to promote cultural activity of a certain race in a
certain period in its history.
Since both Korea and Japan once resorted to isolationist policy in
the past, pan-paradigm could, the author is inclined to believe,
be observable in the structures of their cultural activities.

Azaria Polikarov, Bulgarian Academy of Sciences, Sofia, Bulgaria

CHANGE IN THE CHARACTER OF REVOLUTIONS IN CONTEMPORARY SCIENCE

Early scientific revolutions had the character of 'Copernicanische Wende' (Kant), 'Umstülpung' (Marx), i.e. of a radical change, whereby discontinuity prevaled. This finds an expression in Kuhn's conception of scientific revolutions.

On the other hand it may be stated that contemporary revolutionary changes in scientific knowledge (conceptions are different in kind from those (early) revolutions, in the sense that the element of continuity here is more clearly manifested. Almost all facts speak in favour of this statement, among which attention should be paid to the following most typical ones:

a) Firstly, there are revolutions concerning new discoveries, e.g., in modern astronomy, molecular biology, high energy physics etc. This is noticed for astronomy by M. Rees (1981), B. Lowell (1981), for molecular biology by B. Davis (1980). The neutron circuitry is conceived of as a quite revolution.

b) As to special relativity Einstein emphazised the continuity of this theory with classical physics, and with electrodynamics especially. The same is also valid for general relativity which was framed as a further elaboration of special relativity.

c) In the transition to quantum theory discontinuity comes to the fore. Nevertheless there is also a continuity, reflected in Bohr's correspondence principle. The essential element of progress in quantum field theory has been the realization that "a revolution is unnecessary" (St. Weinberg, 1977). Likewise characteristic for the breakthrough of the theory of electro-weak interactions is the successful application of well-known methods beyond their initial domain (S. Coleman, 1977).

The continuity of scientific growth has been stressed, and overemphazised, in some recent conceptions (St. Toulmin, G. Holton, I. Lakatos).

Our conclusion is that the history of modern science does not support Kuhn's scheme: there are revolutions not associated with a paradigm-shift, as well as transitions from a paradigmatic conception to another one which are altogether less drastic than in the past.

Alexandru Giuculescu

Central Institute for Management and Informatics, Bucarest, ROMANIA

CRISES AND REVOLUTIONS IN THE DYNAMICS OF SCIENCE

The purpose of this paper is to define and explain the emergence of crises and revolutions in the development of sciences, using the polar concepts of normality and pathology applied to scientific theories.
I. A state of crisis means a decisive turning point occurring in the normal evolution of a scientific theory, followed either by recovering its normality or by losing it irreversibly. The major types of crises in the history of science may be illustrated as follows:
a) For many centuries the motions of celestial bodies have been explained by the Ptolemaic theory, that finally entered a critical phase because of the discovery of a considerable amount of "irregularities". The long period of crisis ended with the replacement of the geocentric model of Ptolemy by the heliocentric model provided by the Copernican theory.
b) Some theories bear the germs of future crises either because they contain statements obviously opposite to empirical facts, as did phlogistics in the 18th century, or because they have shaky foundations, as it happened to the theory of transfinite numbers in the 19th century. Such theories are doomed to suffer from chronic crises.
c) There are theories which undergo apparent crises similar to to the so-called crises of growth. The Euclidean geometry seemed to be paralysed by the problem of parallel lines. However, in the 19th century the non-Euclidean models opened to the old geometry a new cycle of normal life.
II. A scientific revolution denotes a complete interruption in the normal development of a scientific theory. The break is to be understood as a consequence to the failure to actualize a scientific theory, i.e. to use it further under several ways and forms, like mere reproduction, or verification, corroboration, re-enactment, adaptation, reduction, generalisation, split, hybridation or integration. While the old theory is to be, consequently, repudiated as useless or even dangerous for the advancement of scientific knowledge, a new theory has to meet the major requirements of scientific standards. It must be emphasized that a scientific revolution implies a gap between the old and new theories and it is not to be derived from possible conflicts between their partisans. The reception of a new theory means the beginning of a process of assimilation, which leads after all to the same process of actualization, that builds up the scientific tradition, or, more generally, the cultural heritage. A scientific theory is paradigmatic if its actualization consists for a long time of reproduction without essential modifications. While a paradigm asserts itself owing to intrinsic qualities, its deterioration entails finally the replacement by a new theory, which emerges as the outcome of a scientific revolution.

Xu Liang-ying

Professor of the Institute for the History of Natural Science,
Chinese Academy of Sciences

On the Accumulation- Ingeritance and the Breakthrough- Revolution in the Development of Science

Science is the accumulation of crystallized knowledge throughout the whole course of history and the historical inheritance is the fundamental character of the development of science. As the knowledge accumulates to a certain extent, breakthrough will take place. In general, scientific breakthrough possesses 6 forms, they are mainly: the creation of a new theory, the establishment of a theoretical system based on the pre-existing theories, the modification of extension of old theories, etc. Scientific revolution which is only a special case of breakthrough consists of both the "setting-up of the new" and the "breaking-down of the old", the former based on the latter. It starts from the uncompromisable contradiction between the new experimental facts and the old scientific theory. This contradiction has brought "crisis" to the realm of science. The crisis can be overcome only when the theoretical system and the fundamental concepts and principles are transformed basically. In the history of science, scientific revolution does not occur frequently. Based on the above sense, the establishment of Newtonian mechanics and molecular biology are not scientific revolution but outstanding breakthrough. In addition, scientific revolution has not to be understood as break with the previous history absolutely. In fact, the accumulation and inheritance of knowledge are the indispensible factors and prerequisites of any scientific revolution.

John P. Losee

Professor of Philosophy, Lafayette College

Kuhn, Complementarity, and the Puzzle-Solving Model of Historical Explanation

In The Essential Tension, Thomas Kuhn advanced two analogies: a gestalt-figure analogy to elucidate the relationship between history of science (HS) and philosophy of science (PS), and a jigsaw-puzzle analogy to elucidate historical explanation. Neither analogy is appropriate.
Gestalt perspectives of a visually ambiguous figure are mutually exclusive. Kuhn suggested that when historian and philosopher view developments in science, one sees a "duck" and the other sees a "rabbit". A given inquirer may shift back and forth between the two perspectives, but "no amount of ocular exercise and strain will educe a 'duck-rabbit'". But historical inquiry is required within theory appraisal and the philosophical analysis of qualitative confirmation, and philosophical judgments are required in historical reconstruction.
The gestalt-figure analogy might be appropriate if HS and PS are mutually exclusive on one level but complementary on a higher level. However, it would not be plausible to hold that HS and PS conform to Bohr's Principle of Complementarity. Unlike the case of wave-particle dualism, no first-order interpretation is available that is neutral with respect to HS and PS. Moreover, the constraint posed by a specific experimental arrangement (either the particle picture is appropriate and the wave picture is inappropriate, or vice versa) is absent in the case of the relationship between HS and PS.
The jigsaw-puzzle analogy emphasizes "similarity-recognition" as a condition of successful historical reconstruction. The puzzle-solver who has recognized a pattern, and has arranged his pieces to implement this recognition, has solved the puzzle. There is no reason for the puzzle-solver to seek alternative solutions. The jigsaw-puzzle analogy suggests a finality inappropriate to the nature of historical reconstruction.

Zely E. Rivera

Historian of Science and Medicine, Univ. of Puerto Rico

INTER-AMERICAN SCIENTIFIC EXCHANGE: HISTORIC
PERSPECTIVES ON PUERTO RICAN SCIENCE

Located in the Caribbean basin, Puerto Rico is a privileged island because of its strategic geographic location (in the midst of North and South America). Common language with Latin America and its political relations with the United States have made Puerto Rico an ideal Inter-American scientific exchange center.

Three important events in the history of parasitology in the Western Hemisphere have occurred in Puerto Rico: the discovery of Necator americanus (by Ashford, in 1899) -the causative agent of parasitic anemia-, and the discovery of Schistosoma mansoni (by González-Martínez, in 1904); also, the most effective immunologic test used world-wide to diagnose schistosomiasis was developed by Oliver (1954), a Puerto Rican parasitologist. (Hillyer et al., 1979)

Sponsored by the local government, the campaign against parasitic anemia (the first leading death cause in the island at the turn of the century) served as a model to the campaigns developed in other tropical countries as well as in the U.S. South under the auspices of the Rockefeller Foundation. (Ashford, 1934)

Among the social implications of these campaigns are a dramatic reduction in the mortality rate, the improvement in national productivity because of healthier workers, and an increased governmental attention to rural sanitary issues.

Exchange of scientific knowledge has been crucial to the advancement of science. Research work in Puerto Rico has had a positive impact in the social and health conditions, not only in the American Hemisphere, but in the whole world.

Manuel Servín-Massieu

D. Sc., Professor, Universidad Autonoma Metropolitana Mexico 21 D

ORIGINS, EXPANSION AND CRISIS OF CONTEMPORARY SCIENCE IN MEXICO

The paper analyzes the origins, expansion and crisis of contemporary science in Mexico; the emphasis of the analysis describes and discusses the role and significance of scientific institutions after the 1910 Mexican Revolution. Social implications directly related to higher education are also analyzed, as this institutions have been in the past, and still are, the core of the so called "system" of science and technology in Mexico. The different transformations that education in general, and higher education in particular, have suffered in the last 70 years is discussed in detail. A parallel line of analysis is covered in this work integrating the network of institutional R&D relationships to the social, economical and policy aspects of the overall process. The backwash effect of 20 years produced after the Revolution, on education and scientific institutions is considered to be a preparatory stage for the turning point at the Cárdenas' regime 1934-40; this marks the growth of science in the country and the integration of its elements on government policies, goals and aspirations. Data pertaining to this era is presented and discussed in the framework of prevailing policies qualified as socially and rurally oriented. Within a lapse of a few years after the Cárdenas" administration a newly elected government had policies changed, industrial growth emphazised and opened doors to the influx of foreign capital. As foreign technology began to be used preferentially for production, local R&D began to drift on its own away from the local needs of society. From 1960 to 1980 the complex of scientific institutions in México expanded through a series of new policies, new instruments for the fostering of R&D and creation of coordinating entities. Due to the economical slump of the early 80's we foresee two phenomena: a growing corporativization and a cooptation of the system and a widened difference between a few institutions of the complex with a relatively healthy resource balance and a numerous lot that are dangerously falling behind and loosing momentum; the consequences of this crisis on R&D and higher education are finally discussed.

Antonio José Junqueira Botelho

Ph.D. Candidate, C.N.A.M., Paris, France & Graduate Student, M.I.T.

LA SOCIETE BRESILIENNE POUR LE PROGRES DE
LA SCIENCE ET L'EVOLUTION DE LA SCIENCE AU BRESIL, 1948-1964

L'étude des associations du type de la Société Brésilienne pour le progrès de la Science (SBPC), comme l'a remarqué MacLeod (in R. MacLeod and P. Collins, 1981), peut contribuer considérablement à la compréhension du processus d'institutionalisation des communautés scientifiques.

Fondée en 1948, la SBPC fut la première société professionelle qui chercha à défendre les intérêts des scientifiques et de l'institution de la science telle quelle. Cette société fut créée à la suite de la lutte menée par un groupe réduit de scientifiques contre l'anti-intellectualisme qui menaçait l'institutionalisation de la recherche scientifique à l'éetat de Sao Paulo. Tout au long des années cinquante, la SBPC, malgré son caractère initialement régional et disciplinaire restreint, s'est battue pour la défense de la liberté de recherche et de la profession de chercheur. Parallèlement, la Société anima et participa activement à tous les grands débats politiques, à caractère technologique ou universitaire, de l'époque: la création de la fondation de soutien à la recherche de l'état de Sao Paulo, les conditions d'accès aux postes universitaires, l'enseignement des sciences, la politique nucléaire et finalement, le projet de création de l'Université de Brasilia, qui devrait servir de modèle pour une réforme de l'enseignement supérieur.

La SBPC, en plus des organismes gouvernamentaux creés à l'époque, contribua décisivement à l'expansion et à l'institutionalisation de la communauté scientifique brésilienne. Cet fait est attesté d'une part par l'analyse qualitative des thèmes de ses réunions et de sa revue "Ciência e Cultura", d'autre part par le nombre croissant de participants à ses réunions. Il faut aussi noter que le nombre élevé de sociétés scientifiques créées autour de ses activités confirment le role catalisateur de la Société pour l'institutionalisation d'une serie de disciplines scientifiques, telles que la génétique, les statistiques et la biologie en général.

Mais sa contribution la plus décisive fut la formation d'un certain esprit de corps dans un noyau de cette communauté. Cet esprit de corps défendait l'active participation des scientifiques dans des mouvements politiques en rapport avec la défense de l'institution de la science et fût crucial pour la survie de cette communauté dans cette période, mais surtout après le coup militaire de 1964.

Sa

Alberto Mayor-Mora, Universidad Nacional de Colombia, Bogota, Colombia

Associate Professor

UNDEVELOPMENT AND MATHEMATICS: THE POLEMIC ABOUT ITS TEACHING IN COLOMBIAN ENGINEERING AT THE BEGINING OF THE XX CENTURY

After 1902 it was visible in Colombia the beginings of the new industrial society. This fact was anticipated by the establishment of the Facultad de Ingenieria of Bogotá and the Escuela Nacional de Minas of Medellín at the end of the XIX century.

The question: what kind of mathematics should be taught to the students of engineering ?, found opposite answers from the two schools. In Bogotá, profesors under the leadership of the engineer, astronomer and mathematician Julio Garavito prefered the higher mathematics courses according to the French style of the Ecole Polytechnique of Paris that favored a systematic vision in each respective course. In contrast, in Medellín, the founders of the Escuela de Minas, educated in the University of California at Berkeley, limited courses on higher mathematics and gave importance to introductory notions that sacrificated the systematic form. In Bogotá, the ideal was to produce mathematical engineers like Henri Poincaré; in Medellín, the goal was to produce practical engineer like Frederick W. Taylor, Emerson, Gant or Henri Fayol.

About 1912, the intelectual leader of the Escuela de Minas, Alejandro López, a civil engineer, suggested that the combination of mathematics and engineering that the country needed should not be result of mathematical sophistication but of applied elementary mathematics. Colombia needed a transformation of its productive life from a qualitative to a quantitative level: work productivity, costs of production, wages and benefits. Thus, the courses of Statistics and Industrial Accounting were introduced in the programs of the Escuela de Minas. Its leit motiv was to form, in the future, in a country of scarce resources as Colombia, engineers capable of managing resources either to build a bridge, a road, a railroad or to build a factory, to exploit a mine or to manage a company.

Some years later, the scientific eponymy honored Garavito's name in the moon and López's name in a small forgotten railroad station. But history confirmed López's reasoning that for the first stages of economical growth of the country it was decisive to have a knowledge of "lower" mathematics: statistics, accounting and even demography, put into work in the public and private companies by the "bussiness" engineers socialized by the Escuela de Minas of Medellín.

Maurice Schoijet

Professor, Universidad Autonoma Metropolitana-Xochimilco

THE DEBATE ON A "NATIONAL " SCIENCE IN ARGENTINA, 1968-1975

Between 1969 and 1972 Argentina experienced a wave of democratic mass struggles against the Ongania-Lanusse dictatorship, that included working class rebellions and guerrila movements. Important intellectual and professional groups became politically active in a variety of ways, including unionization and formation of political movements of researchers and academic personnel. Such movements adopted increasingly critical positions on issues of science, technology and development, and reflected and ideological displacement away from traditional liberal idelology towards nationalism and Populism. The mathematician Oscar Varsavsky played a leading role as an ideologist of a "national" science, against the scientificism of the liberal scientists. Varsavsky's influence was also important in the creation of the Consejo Technologico, an organization of the Peronist Party that would deal with science and technology issues. As the Peronists took power in May of 1973 his ideas became official policy for a brief period that lasted until 1974.
We believe that Varsavsky's ideas constitute a Populist and Third World replay of the "two sciences"---bourgeois and proletarian, theory, concocted in 1920 by the Russian ideologist Alexander Boddanov, that inspired the Lysenkoist ideological counterrevolution in the Soviet Union. Varsavsky was correctly criticized by several scientists and philosophers who followed the liberal tradition of the British scientist Michael Polanyi. They pointed at the obscurantist implications of Varsavsky's positions, and defended the universality and autonomy of science. Sociologist Eliseo Veron made an important contribution in criticizing both Varsavsky and his critics, calling attention towards their limitations on the reations between science and politics. Following Veron, we suggest that the rise of Bogdanovian science policies requires some preconditions related to the nature of the state, the conjuncture of class struggles and characteristics of a given national scientific community.

References
1. Oscar Varsavsky, Ciencia, politica y cientificismo Centro Editor de America Latina, Buenos Aires, (1969)
2. Eliseo Veron, Imperialismo, lucha de clases y conocimiento, Editorial Tiempo Contemporaneo, Buenos Aires (1974)
3. Oscar Varsavsky, Gregorio Klimovsky, Tomas Sompson et al., Ciencia e ideologia, Ediciones Ciencia Nueva, Buenos Aires (1975)

DR. V. A. Narayan

Professor in History   Patna University

ROLE OF PUBLIC ASSOCIATIONS IN THE DEVELOPMENT OF SCIENTIFIC KNOWLEDGE IN BENGAL IN THE COMPANY'S PERIOD

Several public associations like the Asiatic Society of Bengal (1784) the Agricultural and Horticultural Society, (1820) and Calcutta Medical and Physical Society (1823) the Native Medical Society (1830) played a significant role in the development of scientific knowledge in Bengal during the Company's period. Most of these associations had their own journals in which important contributions were made by contemporary men of letters. At a time when British administration in India remained apathetic to the development of science and technology in India, these associations formed by energetic intellectuals served as a beacon light for the spread of scientific knowledge. Development in science was in a formative period and more due to individual and indigeneous effort then state patronage.

Mel Gorman

Professor of Chemistry, University of San Francisco

INTRODUCTION OF CHEMISTRY AND OTHER SCIENCES TO INDIA:
ROLE OF THE CALCUTTA MEDICAL COLLEGE

At the present time the basic sciences of anatomy, botany, chemistry, physiology, and physics are studied by prospective medical students in undergraduate institutions. But in the early nineteenth century they were given in the medical schools. In colonial India at that time the most outstanding school of medicine was the Calcutta Medical College. This institution was founded by Lord William Bentinck, Governor General of India under the East India Company from 1829 to 1835. During his term of office he decided that an improvement in health care for the general population would result from medical education of Indians at a level comparable with the best medical schools in Europe, with instruction in English.

The college was opened on June 1, 1835 in temporary quarters, and with only three instructors, but it expanded rapidly. Soon it moved into a building of its own including laboratories containing apparatus from London instrument makers and suppliers. A good library was started and the acquisition of European books on the fundamental sciences and medicine was continued on a regular basis. Of course, the most important contribution to the success of any educational institution is the quality of the faculty.. The various members of this body of teachers were recruited from the medical services of the East India Company. These doctors were well qualified by their education in the universities of England and Scotland, where medical education emphasized the basic sciences. Moreover, many of these physicians and surgeons were the scientists of their day, demonstrating a marked universality and diversity of talent. Many of them contributed to the advancement of their fields by research which was published in prestigious journals of Europe. As the fame of the institution spread, students from regions of India outside of Bengal were sent by their local governments to the Calcutta Medical School. In spite of doubts by many Englishmen as to the feasibility of a medical school for Indian youths, there are various indications of medical success by graduates of this college, including the ability to understand the basic sciences to an extent equal to comparable students in English universities.

This work was supported chiefly by a grant from the American Philosophical Society; additional funding was provided by the Faculty Development Fund of the University of San Francisco.

Satpal Sangwan

Scientist, National Institute of Science Technology & Development
Studies (CSIR), New Delhi, India

Science Education in India Under Colonial Constraints (1792-1857)

This paper examines the evolution of scientific-technical
education in India under colonial constraints. What were the factors
which affected British educationpolicy in India, how science
education came in their plans and what were their imperial interests
behind the move, constitute the main theme of thepaper.

Science education develoed in three phases, viz, 1792-1813;
1814-1835 and 1836 to 1857. There was a stiff resistence to
the idea of giving education during phase first. In second phase
the British had relented the plan but favoured the promation of
oriental learning. However, some education in Western medical
science was arranged. There was a tremendous pressure for scientific
technical education during third phase which led to the opening
of a few medical and engineering institutions. The British
educ ation policy was a defective on various accounts. Total
rejection of oriental literature and science and its literary bias
were its main drawbacks. Moreover the idea behind their plans
was not to make Indians scientists or inventors but to prepare a
class of writers, interpreters and assistants in various scientific-
technical schemes.

The paper is entirely based on archival material.

Deepak Kumar

Scientist, National Institute of Science, Technology
& Development Studies, Hillside Road, New Delhi-110012

Colonial Science and Indian Response (1820-80)

British activities in the field of scientific explorations, institution building, science education (later research), etc. did evoke some response from the local populace, particularly its educated class. The response was prompt, often spontaneous, and always in direct proportion to the opportunities offered. It manifested itself in acceptance and somtimes in rejection of certain changes in the academic curricula by the students and of certain innovations in technique or machinery by local manufacturers and cultivators. Often the response found echoes in active native participation in the officially patronised scientific associations or institutions, but sometimes it made them search for a distinct identity and establish institutions, scholarships and facilities of their own.

John J. Beer

Associate Professor of History, University of Delaware, U.S.A.

A YANKEE FOR GANDHI'S SPINNING WHEEL: RICHARD B. GREGG AND THE MAHATMA'S RENEWAL OF VILLAGE INDUSTRIES

In retrospect, we recognize that Gandhi's attitude toward science and technology was more prophetic than reactionary. His campaign for homespun cotton (khaddar) was not only an early influential example of "appropriate technology" in action, but served simultaneously as an alternative non-military instrument for national liberation and internal social reform.

Among the great variety of people whom Gandhi attracted was Richard B. Gregg, a Harvard trained lawyer with experience in science teaching and seven years of work in industrial relations for New England textile companies and unions. Gregg spent 1925 - 1929 in India studying the economic and technical feasibility of reviving the rural textile crafts. His subsequent book, The Economics of Khaddar, featured original engineering insights and an informed economic analyses which Gandhi welcomed and used in answering critics.

This paper deals centrally with Gegg's contributions to Gandhi's village industry movement.

Raman Srinivasan

PhD Candidate, Dept of History and Sociology of Science, U. of Penn.

SCIENCE POLICY AND ORGANISATION IN INDIA 1947-1967

India is the land of Gods. According to Hindu mythology, the heavens are populated by 330 million divine beings. Thus, on an average, every two Indians share a God. To this vast galaxy of Gods, two impressive additions were made in 1947, when the British left India. The new Gods were science and technology.
My paper attempts to study the implantation of western science and technology in modern India. A big push was given to the science sector by the Nehru government during the first two decades of independence. This period is characterised by an explosive growth in scientific research organisation and funding.
What was the mission assingned to science? How was science related to national goals? Why and how did the ruling elite relate to the ideology of science? Who were the institution builders and how did they organise science? This paper deals with these questions on three different levels. An impressionistic survey of the cultural ecology of the Nehru era is followed by an analysis of the organisational structure of science in India.
Colonial influences on current Indian science are traced. The third and most interesting dimension of scientific development in India is the interaction between individuals and society. This paper focusses on four individuals-Nehru, Bhabha, Bhatnagar, and Saha. The social forces that moulded these men as entrpreneurs of arganisations is explored.
The distinctive mission of Saha, outside the trinity of modern Indian science isseparately considered. The section on Saha naturally leads the discussion onto the complex but intriguing relations between science and politics and between scientists and politicians. A symbiotic relation is seen between these interest groups.
The paper concludes by summarising the cultural forces that propel science in India.

JAIN, Suresh C.

Lecturer, Regional College of Education, AJMER-305 004, India.

BIOLOGICAL EDUCATION IN INDIAN SCHOOLS AND SOCIAL VALUES.

Biology deals with livings including human beings whereas education helps in promoting physical, mental, emotional, social, moral and spiritual aspects of the learner(human). The impact of present value-less science on our society and increasing inhumanity is clearly visiable in our day-to-day life. There is an urgent need of inculcating basic values of humanism, social and national including non-violence, respect for life and proper utilisation of various natural resources. School education in India runs for 12 years and the teaching of science including biology forms an integral part of first 10 years and is compulsary to all the students.

School biology provides understanding of livings including man and their interrelationship to environment. Information on human body, health and hyegine, conservation of nature and natural resources, agriculture and animal husbandary is included in the curriculum. Teaching is environmentally oriented and strengthens social values of non-violence, respect and love for life, cooperativeness in the form of interdependence. The paper analyses the usefulness of school biology in strengthening and perpetuating social values among children.

Gao, Dasheng
Associate Professor, Social Science Department,
Qinghua University

Investigation on the features and Functions of Experiments in the light of the history of the science and technology

The scientific experiments have become an independent social practice that underwent a long historical development. Basically, the ancient natural science was based on direct experiences of production perceptions of the physical world. But the sprout of scientific experiments already appeared at that time. During the Warring States Periods (770-221 B.C.) in China, the optical experiment of Mohist School was a good example. About in 16-17 centuries, the scientific experiments were separated from the production practice, and began to some extent systematically and independently. Recently the results of scientific researches depent on more and more precise and large-scale equipements and advanced experimental technology.

Compared with the production practice, the scientific experiments have many advantages and characteristics irreplaceable. They can purify and simplify natural phenomena. With the help of material means such as precise instruments and equipments, in removing accidental and secondary disturbances in natural procedures, some properties and relations we need to know would appear in pure state. Therefore we can bring to light the laws govering the natural phenomena and production processes. The instruments and equipments are the extension of man's five senses and his brain. Raising their precision and sensitivity, it would help to discover the new natural laws. The experiments arenot only the important means to know the nature, but also are the linking point of applying scientific and technical knowledge to production. In the experiments, the discovery of new natural phenomena and the formation of new experimental technology always become the growing points of the new technology. Because of above-mantioned advantages and characteristics, it is possible to make natural science overstep the limitations of prodiction pradtice and surpass it, depending on the favourable conditions of the laboratories. The scientific experiments can push the researches of natural science theory forward, and open up a broad for developing production and technology.

Therefore, treating the relations between experiments and theories correctly, paying great attention to experiments, the construction of laboratories and experimental education—all these will give an impetus to the modernization of China's science and technology.

Candice Goucher

Assistant Professor, Portland State University

TREES, TRADE AND TECHNOLOGICAL CHANGE: THE IRON INDUSTRY AT BASSAR, TOGO (WEST AFRICA), 1400-1914

This paper investigates the parameters of technological change in the iron industry of the Bassar region of Togo, West Africa. The area represents one of the largest industrial settings known in pre-colonial Africa. The extensive remains of standing furnaces and slag heaps present a unique opportunity for the study of the scale and duration of local production and the identification of technical variability within the Bassar smelting process. The technological system was characterized by tall furnaces which utilized an induced draft, and which involved a subsequent process of refinement in closed, clay crucibles, not unlike the Indian wootz procedure which produced medium-carbon steel.

An important feature of this African industry was its reliance on large quantities of charcoal fuel. The fuel requirements of large-scale iron production led inevitably to deforestation and induced technological change. Between 1400 and 1914, the industry developed a number of fuel-efficient features probably in response to conditions of increasing scarcity.

In addition to nineteenth-century photographs and written descriptions of early German travellers and colonial officials, a rich record of technological history resides in the oral traditions of elders from Bassari blacksmithing clans. Interviews with metallurgists (many of whom are now in their eighties and have witnessed or participated in ironsmelting activities prior to the close of the German colonial period) resulted in the reconstruction of a smelting furnace in the village of Banjeli (Bassar) and the experimental reduction of local iron ores. Such ethnographic studies are valuable, not only because Africa provides a unique laboratory for directly observing preindustrial technological systems (a field of enquiry not open to historians of preindustrial Europe), but also because we can enhance our understanding of technological history in institutional and economic settings which have not been frequently studied.

Stefan G. Balan* and Florin S. Balan**

*Romanian Academy; **Central Institute of Physics, Bucharest

THE DEVELOPMENT OF TECHNOLOGY IN ROMANIA IN THE 18TH AND 19TH CENTURIES

In the 18th-19th centuries it has developed, firstly, a technique, resulting from the former activities of these centuries. Before 1700 there existed remarkable salt and gold mines, different manufactures, mints, the processing of wood, an enough thriving agriculture, important constructions and the printing was in use.

After 1700 the former forms are developed and from other countries are taken advanced forms of technique, which are improved by the Romanian people power of creativity and by the all level education. At the beginning of the period it was developed mostly an artisanian and manufacturing economy, helped by the local professional schools. It is worth to mention a metallurgy professional school at Oravita (1721), some workshops for ceramics, specialized builders, miners, weavers etc. The existence of some own sources of food, minerals, power, the building of some technical schools and means of conveyance have helped the development, especially to the end of the period, of an own economy. We must also point out the attention which was given in the 19th century to the mining of gold and salt, also of coal after 1850, the extraction of oil, Romania being one of the first countries in the world in the production of oil. The constructions and architecture have developed on a popular and traditional background, emphasized by Romanian creativity: Moldavian vault decorations with enameled ceramics, the specific Romanian exterior frescas of big monatsteries etc. It is remarkable that the reinforced concrete (patented in France in 1868), in 1884, when even in other countries there were not fixed norms for calculus and execution, a Romanian builder of genius, engineer Anghel Saligny, designed and built in the harbours of the Danube, Braila and Galati, the greatest silos for cereals of that time, 25000 t each. The same Anghel Saligny built across the Danube the great steel bridge in continental Europe - 4 km total length, with spans of 190 m.

Other examples we can also find in old tradition fields of activity, as in newer ones: in the utilization of oil, we had methods with world priority: metallurgy, with the furnaces from Resita; the railroad Oravita - Bazias (1856), only 30 years after the first railroad was built (Stockton-Darlington); in aviation we had some of the greatest inventors in the world: Traian Vuia, Aurel Vlaicu and Henri Coanda.

In the meantime, superior schools were built: the Bucharest University in 1694, the Bucharest Polytechnic Institute in 1881, etc.

I have emphasized only some findings with regard to the Romanian people creativity in technologies in this period.

Boris Yossifov Miloshev

Retired M.D.

EUROPE'S CONTRIBUTION TO THE DEVELOPMENT OF SCIENCE IN BULGARIA DURING THE 1890 - 1900 PERIOD

It refers to a monograph by the author, entitled "Europe's Contribution to the Development of Science in Bulgaria during the 1890 - 1900 Period". On the basis only of Bulgarian archives sources and literary data, the contribution by Europe to the development of science and education in this country has been established.

Owing to the tremendous amount of material, the author has treated only of the period of 1890 - 1900, which in fact was the culmination point of those mutually beneficial relations.

The presentation of the report documentarily reveals the contribution of each European country to science in this country. This is the author's aim and it contributes to the specifying of the volume of the relations between Bulgaria and Europe, and to acquainting the scientists participating in the Congress with this interesting matter.

Xavier Polanco

Centre de Recherche sur l'Amérique Latine et le Tiers Monde, 35 rue des Jeûneurs, 75002 Paris, France

THE RELATIONS BETWEEN THE SCIENCE IN THE THIRD WORLD CONTEXT AND THE HISTORY OF SCIENCE.

The first topic that I wish to stress, is the historical significance of the lopsidedness of the international scientific knowledge system. The polarization of science (S) and its impressive localization in the developed countries seem to mean an asymmetry in the way of scientific development. In this respect, the very question is to know if the polarization of S would not be the general condition of scientific progress. As long as we do not settle this question, it will remain a fait accompli either for the history or theory of S and for the rationality that S means.

Let me now take into account, the process by which the problem of science and technology (S&T) is focused in the developing countries, namely, from the observation of the state of S&T, and in the prescription of their application to development, the R&D policies are proposed with the aim of the increase the indigenous capability for S&T. But in this case, the 'state of' is not a given natural event. On the contrary, it is the result of an historical process. In consequence, we must pay attention to the problem of the historical background of the present state of S&T in the developing countries. We need a deeper historical understanding of S&T in these countries.

Also, another historical aim would be to resuscitate the S&T put aside by Western colonial expansion, and the diffusion of the European S&T model. Possibly the historical reconstruction of the process by which the underdevelopment of S&T in these countries was created (generation context) is not divorced from the design of the process required today for its removal (overcoming context).

Instead of the doctrine of the universalism of S without context, history shows us the universality of S to be closely tied (at level of its own recognition and worth) to the expansion of the Western culture (philosophical assumptions), mentality (intellectual education and training) and civilization (i.e. institutions). All this represents the 'condition of possibility' of that form of knowledge that we call now 'science'. It means to take the 'transcendental subjectivity', Kant's notion, in an objective sense, as an historical and cultural construction by means of a social process like the one that anthropologists named 'acculturation'. So, what I propose to study as our problem in the History of Science, is not exactly 'the emergence of science', but 'the spread of Western science' and its World hegemony.

Luis A. Camacho
University of Costa Rica

FOUR IDEAS ON THE RELATION BETWEEN SCIENCE, TECHNOLOGY AND DEVELOPMENT.

(1) A subordination of scientific knowledge to the solution of practical problems is not only demeaning to science itself but also self-defeating in the long run. The reduction of science to its use in achieving socio-economic development is, therefore, detrimental not only to science but also to development.

(2) Any notion of human development that does not take into account at least the following four types of variables: economic, cultural, biological and political, is inadequate for a lasting solution to world problems.

(3) If by "development" we merely mean an increase in the GNP or in the per capita income in a particular country at any given time, then the existence of a local scientific activity and of technological research and production do not seem necessary conditions. History provides examples of cases where local science and technology have not been prerequisites for development in that narrow sense of the term.

(4) Although it sounds tautological and therefore trivial, the existence of an active local scientific community is a necessary condition for the scientific development of a country as a whole. It is neither tautological nor trivial to say that science is a necessary (though not sufficient) condition for human development qua human. There seems to be a dialectical relation between science, technology and development at different stages of a process: science gives rise to technology, which fosters development in a very wide sense of the term; socio-economic development, on the other hand, may help to upgrade the scientific and technological levels of a country.

Kathleen G. Dugan

Lecturer, Graduate School, Academia Sinica

Transmitting the Language and Culture of Western Science:
Teaching HPS in a Second Language, Non-Western Classroom

The teaching of the history of Western science, like the teaching of Western science itself, is spreading from Western nations, where it originated, into other parts of the world. The challenge of adapting teaching to meet the interests, abilities and needs of non-Western students in developing nations suggests a need for evaluating the role of the history of science in transmitting Western attitudes toward science. Drawing on five years' teaching experience in two developing nations (Papua New Guinea and China), the author argues that courses in history and philosophy of Western science can be particularly valuable for students in non-Western countries which hope to use Western science and technology to promote national modernization. Understanding of the social, political and cultural dimensions of technological change is necessary for effective national planning. Courses in history of science can provide a forum for a discussion of changing values. Comparison between traditional beliefs and Western attitudes may help individual students to calrify their own values and to cope with the cultural adjustments required to study Western science in a Western-syle educational institution. When taught in a second (Western) language, history of science courses may be particularly valuabke tools for language training thus helping students acquire the skills they need to join the international academic community. The content and methods of history of science teaching may, however, have to be adjusted to fit the new context. Non-Western students not only lack a background understanding of Western history and culture, they also bring their own social, political and cultural values into the classroom with them. Teachers, needing to place Western history and Western science ina broader, cross-cultural perspective, will be challenged to understand different ways of thinking about science and society. Inthe West, courses in history of Western science may serve as preservers of culture, helping students gain a better understanding of their own cultural traditions. In non-Western nations, such courses can serve as instruments of social change, bringing students an understanding of Western culture and helping them to cross cultural divisions to paricipate in an international community.

Sc

Kei-ichi TSUNEISHI

Faculty of Liberal Arts, Nagasaki University, Japan

Biological and Chemical Warefare Research in Japan between the Wars I and II.

It was at the time of W.W.I that Japan started her chemical warfare program at two foci. One was the Army Committee for Technology and the other was the Army Medical College. During the first several years the latter took charge of experimental study and production of CW munition under the direction of Chikahiko Koizumi who was a professor of the College and would become the Minister of Health and Welfare. But after 1919 the Army Committee for Technology, which established its own research institute in that year, began to take over the CW activity done by the Army Medical College. Within ten years the former was to establish a poison gas plant on an island near Hiroshima. The College and the Deapartment of the Surgeon General wanted to promote their own rank in the War Department through promotion of the CW program, but they failed. In the Japanese Army, the combatant forces were ranked higher than those like the Surgeon General.

In 1930 Shiro Ishii, who is known today as commander of the Ishii unit, came back to Japan from Europe and was appointed an instructor in the Army Medical College. In 1932 he succeeded to found his own laboratory to study Biological Warfare with the support of Koizumi. At the same time he established an experimental station in Manchuria(north-east part of China) to practice BW. The author supposes that this station was operated under the recognition of higher officials of the War Department. From 1932 to 1945 more than 2000men were sacrificed by BW "research" in that station.

Manfred Rasch

Coal liquefaction - caught between economic and political interests.
The example of Germany (1900 - 1945)

After the Arab oil embargo in 1973 which boosted the oil price considerably, the Western Industrial Nations turned again to their home sources of energy, because they did not like the idea of being exposed to political blackmail caused by economic dependency. Brazil decided to substitute methyl alcohol made of natural products for mineral fuel oils. Other countries like the USA or the Federal Republic of Germany with huge coal deposits gave preference to coal liquefaction.
Today it has turned out that the decisions of the past influenced by political as well as economic reasons do no longer bear any economic relevance - because industry is pulling out of the costly upgrading of coal refinement unless the government continues subsidizing.
In Germany where a number of coal refinement technologies originated coal liquefaction had always been caught between economic and political interests since the invention of this technology. This is shown in a short historical review covering the period from 1900 - 1945. The following aspects will be discussed:
- political and economic conditions for coals refinement technology in Imperial Germany
- State subsidies for coal liquefaction in World War I (e.g. building of a hydrogenation plant)
- expectations of the industry to make 'big profits' with coal liquefaction after the war
- considerations of the government to overcome the coal sales crisis as well as to create jobs by building a hydrogenation industry (parallels to nowadays are obvious)
- economic reasons (foreign exchange savings) and political considerations (autarky), which led to the building of hydrogenation plants as well as Fischer-Tropsch factories during the Third Reich
- steps taken by the Allies which unintentionally lay the foundations for the revival of coal technology at the beginning of the seventies.

Anthony N. Stranges

Associate Professor, History, Texas A&M University, College Sta., TX

A SPECIAL RELATION: THE DEVELOPMENT OF SYNTHETIC FUELS IN NAZI GERMANY

For fifteen years after he had invented the high-pressure hydrogenation process for converting coal into oil, Friedrich Bergius struggled to keep his small factory in Rheinau in operation. This resulted in his establishing several corporations both in Germany and abroad in order to raise sufficient operating funds. Only in 1927 when I. G. Farben acquired Bergius's patents did the high pressure process appear to have a chance for commercial development. But I. G. Farben also struggled until 1933 when Hitler's Nazi Party gained control in Germany and entered into a fuel agreement that gave the process new life. Under the terms of the agreement I. G. Farben promised to produce at least 2.5 million barrels of synthetic gasoline per year by the end of 1935 and to maintain this production rate until 1944. It set the production cost at 18.5 pfennigs per liter. The German government not only agreed to support this price, but to pay I. G. Farben the difference between the production cost and any lower market price, and to buy the gasoline if no other market emerged. Alternatively, I. G. Farben had to pay the government the difference between the production cost, which was at that time more than three times the world market price, and any higher price obtained on the market. By 1944, because of technical improvements which lowered the production cost and the increasing market price of gasoline, I. G. paid 85 million reichmarks to the government.

From the first coal hydrogenation plant constructed in Leuna in 1927 the German coal hydrogenation industry grew to twelve large plants in 1944. These plants produced 128 million barrels of synthetic petroleum in the period 1938-1945, and provided much of the German military's fuel during World War II.

At the end of the war, restrictions that the Allies imposed on Germany's wartime industries prohibited the production of petroleum from coal. When the restrictions were lifted in 1951 the coal conversion plants already had been dismantled or were refining natural petroleum, thus ending Germany's successful synthetic petroleum industry.

Robert N. Proctor

Mellon Fellow, History of Science Program

Stanford University

Medical Resistance under Fascism: the Verein Sozialistischer Ärzte

Important elements of the German medical profession expressed strong and early support for National Socialism, long before the rise of Hitler to power. About half of all German physicians ultimately joined the Nazi party; figures are even higher than this among German medical professors. German medical professionals played an important role in the initiation, administration, and execution of each phase of Nazi racial policy.

Nevertheless, it is also important to recognize that important elements of the medical profession offered resistance to National Socialism. Before 1933, the Verein Sozialistischer Ärzte articulated a program of "proletarian medicine," and contrasted this with "capitalist" medicine. In certain areas, the Verein Sozialistischer Ärzte shared common programs with the Nationalsozialistische Deutsche Ärzte Bund. And yet on the whole, the Verein Sozialistischer Ärzte, along with other socialist and communist groups, struggled against racialist ideology through organs such as *Der Sozialistische Arzt*. After being driven into exile, the Internationale Vereinigung Sozialistischer Ärzte continued to attack what it called the "barbarism" of Nazi medical policy through its new journal, the *Internationales Ärztliches Bulletin*, published in Prague from 1934 to 1939. Socialist physicians in Europe and America joined the German Verein in these protests. The Internationale Vereinigung Sozialistischer Ärzte also organized physicians in struggle against the Nazi-supported Franco forces during the Spanish Civil war; physicians also organized certain forms of resistance within German concentration camps. While it may ultimately be true that the resistance offered by German physicians was a case of "too little, too late," the example of the Verein Sozialistischer Ärzte provides an example of how medicine might have evolved in Germany, had the Nazis not come to power.

Howard L. Hall

College of Charleston

An investigation into the works and paradigms of Lewis Reeve Gibbes

Lewis Reeve Gibbes (1810-1894) served on the faculty of the College of Charleston for more than fifty years, including a short period as acting President of the College. Gibbes, a native South Carolinian, was a scientist of notable achievements in many different areas of science. Science was not especially compartmentalized during his lifetime, and Gibbes worked in areas as diverse as natural history and mathematics. Gibbes even proposed a type of periodic table in the early 1870's without knowledge of Mendeleev's work in the same area. Because of this diversity of interests, a case study of Lewis Gibbes allows one to draw conclusions about the paradigms of science under which Gibbes was operating in South Carolina.

Gibbes' work in each area of science are briefly examined, and his paradigms are explored. Various influences on his science, including the rise of American pragmatism as a legitimate philosophy are noted.

Jeffrey Escoffier

Project Director, Center for Social Research

THE EMERGENCE OF A LABOR MARKET FOR SCIENTISTS IN
ANTEBELLUM AMERICA: A PROSOPOGRAPHY

One meaning of the professionalization of science is an increase in the allocation of resources to science through the creation of employment opportunities and the development of institutions. Most prosopographic studies of scientific communities analyze a cross-section of scientific authors. In this study, I analyze the dynamics over time of an emerging labor market for scientists in the growing antebellum (1800-1860) system of higher education in the U.S. This academic labor market was plagued, by difficulties in evaluating competence, by high turnover, and it offered little room for vertical mobility. This was the result of extremely localized labor markets. The problem of evaluating competence was resolved by the use of publications as an indicator of scientific competence. Many other aspects of the reward and communication structure of the American scientific community were shaped by the need of participants (employers and scientists) to operate effectively in an emerging labor market.

Shu-ping Yao

Associate Researcher, Office of Science Policy Research,
Chinese Academy of Science, China

The Contribution and Context of American Physicists of
Chinese Descent

    The article describes in brief the history of China's beginning to send her students to study in the U.S.A. since the past century and achievements made by Ameriean physicists of Chinese descent and contributions made by some of the oustanding ones after World War II. Finally, it makes a comparison between the two different cultures and values of China and the U.S., emphasizing the importance of values to the development of science from 4 aspects----
A. Competition as a force in the development of a society.
B. Self-respect and self-confidence as the starting point to success.
C. The recognition of "privacy" as a social force.
D. The different concept of "Officialdom"
It also stresses the idea that an exploration of the Western culture will be beneficial to China's rapid development.

Dr. Ravindra N. Singh and S. Saran.

Post-Doctoral-Fellow-Deptt. of A.I.H.C. & Arch. B.H.U.

Analytical study of the Glasses of Indian Classical Age.

It is now an established fact that the technique of glass manufacturing was known to Indus people. Since then the occurrence of glass objects are a regular feature from Indian archaeological sites.

The present attempt is intended to throw light on analytical study of glass objects of Gupta Period (350-750 A.D.) which has been described as the 'Golden Age' or 'Classical Age' of the Indian History. Through it was previously assumed that this period was dark age in the history of glass making of India but recent studies have shown that this period had sufficiently developed glass technology.

Glass objects like beads and bangles are reported from many sites showing variety of colours and manufacturing techniques. Emergence of indigenous vessels are characteristic feature of this period. Samples were taken from Rajghat, Bhitari, Mason, and other sites and subjected to various analytical procedures. All the glasses are of soda-lime and silica group having colouring agents viz., Cu, Fe, Ni, Co, Mn, etc. Results are compared with other sites and probable correlations are drawn.

Lu Jingyan

Mechanical Engineering Dept., Tongji University

Huai Bing Dredged Up the Iron Buffaloes--An Outstanding Lifting Project in the Song Dynasty

Huai Bing, a Buddhist monk, was an outstanding engineer in the Song Dynasty in ancient China. It was recorded in the History of the Song Dynasty that he had successfully dredged up several iron buffaloes, each weighing tens of thousands of catties, out of the torrents of the Yellow River. This engineering work has been given the name "Huai Bing's Dredging Up the Iron Buffaloes." However, the record of this engineering project in the History of the Song Dynasty is too brief for us to realize how it was carried out. Having studied history books, visited the work site, and interviewed the local old people who know the event well, the author of this paper has now made it clear.
 The lifting project was carried out on the Yellow River, west of Yongji County, in the southern part of Shanxi Province, China. Here a great pontoon bridge had existed from the Warring States Period (475-221BC) up to the Yuan Dynasty (1279-1363AD). It was called the Pujin Bridge. In order to protect the pontoon bridge from damage which had often been caused by floods, eight iron buffaloes were cast and used to reinforce and stabilize the bridge in the Tang Dynasty. In 1063, during the Song Dynasty, the Yellow River once suddenly rose and the bridge was destroyed by the flood with the result that the iron buffaloes chained to the bridge were pulled into the river. Making use of the principles of the lever, the resultant of forces and buoyancy and other technical knowledge, Huai Bing skillfully dredged the iron buffaloes out of the water.
 The pictures attached to this paper show how Huai Bing dredged the iron buffaloes out of the river. This project was a highly important accomplishment in the history of China's lifting engineering and was a precursor of the modern pontoon lifting project.

Fang-Toh Sun

Professor, National Tsing Hua University, Taiwan, R.O.C.

The Gunpowder/Rocket Technology in Ancient China and Its Transference to the Outer World

The origin of gunpowder and rocket technology has been under intensive study in recent years by many Western scholars as well as Chinese. Though there are still some arguments to be settled, it is generally recognized that the Chinese have a long history of such technological development, tracible all the way to the 11th century, or earlier. The questions lying before us are: 1) whether such technological development in ancient China was unique or independent? or there existed some similar, independent developments in the other countries, like India, for example? and 2) assuming such early development in China was independent, how was it transferred to the other countries? These are extremely difficult questions to answer, owing to the inadequate historical information available, and the language barriers involved.

In this paper the development of gunpowder technology in China since its appearence in proto form in the 7th century in the T'ang dynasty to its application to the rocket weapons as well as fireworks in the later dynasties Sung to Ming, until the early years of the 17th century will first be reviewed. The transference of such early Chinese technology to Europe and the rest of the Asian countries will be traced through a chronological study of the sequence of world events, in which the gunpowder or rockets, or fireworks were recorded. The role which the Indians and Arabians played in such transference will be examined; and the possible routes forthe transference, discussed. Though the present study is by no means exhaustive, it is hoped that it will serve as some ground work for the further study of this difficult subject.

Lung Tsuen-ni
Senior Researcher, Industrial Technology Research Institute
Chutung, Taiwan, Republic of China

THE HISTORY OF CEMENTATION COPPER ON IRON —— THE WORLD FIRST
HYDROMETALLURGICAL PROCESS ORIGINATING IN MEDIEVAL CHINA

'Cementation' is the term used in hydrometallurgy to describe the electrochemical precipitation of a metal from solutions of its salt by another that is more electropositive. The precipitation of copper on iron from natural copper-bearing solutions is s classical example of a relatively ancient art applied successfully for centuries by employing one of the earliest known chemical reactions, specifically:

$$CuSO_4 + Fe \longrightarrow Cu \downarrow + FeSO_4$$

About 1500, Basil Valentine, in his book 《Currus triumphalis antimonii》, referred to copper cementation. Similarly, Paracelsus the Great (1493-1541), referred to the process in the 《Book Concerning the Tincture of the Philosophers》 and indicated that the process may have been a commercial means to produce copper. In Georgius Agricola (1494-1555)'s 《De natura fossilium》(Forben Press, Basêl, 1546), he also referred to the erosion of iron with mine-water to produce copper. Rio Tinto in Spain produced copper commercially by precipitation (in Spanish, cementatión) on iron in the sixteenth century. But recognition of the reaction and development it to a commercial copper-producing process in China is many centuries earlier than in Western civilization. It is believed to be world earliest hydrometallurgical process, being the result of the operations of the medieval Chinese alchemists. In China, the annual copper production which was recovered from copper-mine drainage-water by cementation reached the hundred-ton level during the period of Northern Sung dynasty (960-1126). This paper will discuss the study results of the earlier history of development of cementation copper on iron in China with reference to the voluminous amount of literature such as 《Huai-Nan Wan Pi Shu 〔The Ten Thousand Infallible Arts of (the Prince of) Huai-Nan〕》(Western Han dynasty, -2nd century), 《Mêng-Chhi Pi Than 〔Dream Pool Essays〕》 (Shen Kua, Northern Sung Dynasty, +1091), 《Sung Shih 〔History of the Sung Dynasty〕》 (Tho-Tho el al., Yuan dynasty, +1345), 《Tu Shih Fang Yü Chi Yao 〔Essentials of Historical Geography〕》 (Ku Tsu-Yü, Chhing dynasty, +1799), 《Thien Kung Khai Wu 〔The Exploitation of the Works of Nature〕》 (Sung Ying-Hsing, Ming dynasty, +1637), etc.

Zhang Zhong-jing
Research Associate, Institute for the History of Natural Science,
Academia Sinica, Beijing, China

THE SOCIAL STATUS OF SCIENTISTS AND INVENTORS IN FEUDAL CHINESE SOCIETY

Although the Chinese people had a most remarkable record of achievement in science and technology for 2000 years, the social status of scientists and inventors was low. Science and technology were completely under the control of the imperial power in Chinese feudal society. We shall discuss four aspects of this situation.

1. Refined techniques were held in low esteem in ancient China. Chinese emperors rejected "excessive technique" completely or else treated its products as mere playthings or toys. New ideas and inventions were generally suspected and feared by the people.

2. The social status of inventors was low. In many dynasties craftsmen were not permitted to occupy positions of high status. They had little if any personal freedom. Even the names of those who produced most of the inventions were not preserved.

3. Several features of ancient Chinese science indicate the degree to which it was under imperial control. Those disciplines like physics and biology whose aim is to probe the secrets of nature and determine its laws were neglected and did not develop. No scientific schools grew up in China, and only individual studies were made. Even scientific disputes were sometimes decided by the emperor's authority. Also, we discuss eight famous ancient scientists to illustrate the scientists' social status.

4. Ancient scientific books were not preserved well for some special reasons that will be discussed.

Professor Tian Changhu

Chengdu University of Science and Technology, Chengdu, Sichuan, PRC

An analytical study of Chinese ancient cast technology and their influence upon the development of society of the Bronze Age

During the development of science and technology in ancient China, there existed copper in the New Stone Age, i.e., before the Xia Dynasty (B.C. 2207 1766). In Chinese Slave Society, up to Xia Dynasty, Shang Dynasty and West Zhou Dynasty, the bronze casting had achieved its height. The articles unearthed are large in size, vary in shape and are also exquisite in craftsmanship. These bronze objects testify to the remarkable progress in the smelting and foundry technique of that time. In Xia Dynasty, the ceramics achived great successes. The bronze casting was based on ceramics in mould and furnace.

In Shang Dynasty, the ancient Chinese people developed a complete set of techniques of bronze casting, ranging from burden melting, moulding to pouring and clearing. For instance, Dings had helped to bring about a foundry technique breakthrough, in both of the foundry technology was mould and bronze alloy. The burden of bronze alloy showed evolution from binary alloy cu-sn to three component alloy cu-sn-pb such as the technology of bronze cast weapons which exercised influence on the development of the Society of ancient China. As for the technology of bronze cast mirror, the hardness of such bronze reached more than 300 HB, an advantage in abrading and polishing. The foundry of bronze mirror has given an impetus to technique of high hardness cast alloy, quench hardening, abrade and polishing.

The author proposes that the bronze castings helped to bring about a foundry technique breakthrough, transition from Slave Society to Feudal Society, and this transition period was West Zhou Dynasty, and propelled the Society of ancient China forward.

(Joseph) Cheng-Yih Chen

Professor of Physics, University of California, San Diego USA

A STUDY OF HARMONIC PROGRESSIONS IN THE UNEARTHED BRONZE SET-BELLS OF THE 5TH CENTURY B.C.

In 1978, 114 pieces of music instruments were unearthed in China from the tomb of Marquis Yi 侯乙 of State Tseng 曾 of the -5th century. Among these instruments there are, for example, zither, reed mouth-organ, pan-pipes and stone set-chimes. But the most magnificent is the set of well-preserved bronze set-bells, 64 in total. Recently, the frequencies of these bells have been measured by the Peking Music Institute and jointly by the physics department of Fu Tan 復旦 University and the bronze division of the Shanghai Museum. These measured frequencies provide a valuable opportunity for the study of the harmonic progressions in tonal systems of the Chou 周 times.

The oldest extant Chinese record on a quantitative procedure for the generation of harmonic progressions is found in the 'Kuan Tzu' <<管子>> (The Book of Master Kuan) compiled in the -4th century. In this procedure harmonic tones are generated from their 'parent' tones by multiplying either by 4/3 for the up generation or by 2/3 for down generation. The procedure is illustrated in the 'Kuan Tzu' through an example in which the generation of the pentatonic scale in the 'chih' 徵 mode is given. In the 'Lü-Shih Ch'un-Ch'iu' <<呂氏春秋>> (Master Lü's Spring-Autumn Annals) of -239, the up-and-down principle is extended to the generation of a set of twelve tones. Since the twelve tones are obtained by 7 up-generations and 5 down-generations, we have $(4/3)^7(2/3)^5 = 524288/531441$ which is the comma maxima, mistakenly associated with the name of Pythagoras. The number 262144/531441, one-half of the comma maxima, has been incorrectly interpreted as the Chinese 'octave' in a number of accounts of Chinese tonal systems (for a review see J. Needham, Science and Civilization in China, vol. 4, sec. 26,1962).

It has been pointed out (Chen, Proceedings of the 3rd ICHCS, Peking 1984) that the true concept of octave is contained in the up-and-down principle. Starting with an initial note X, two notes Y and Y_1 which are an octave apart can always be generated by applying the up-and-down generation simultaneously. Evidence for such a use is found by superimposing the example of 'Kuan Tzu' with that of 'Lü-Shih Ch'un-Ch'iu.' Thus, the procedure given in the 'Lü-Shih Ch'un-Ch'iu' is a general method easily extensible from one octave to the other. The octave in the scale is divided into two sizes of semitonic intervals given by the ratios 243/256 and 2047/2187. Such a semitonic progression is found among the unearthed bronze set-bells. A detailed comparison between the measured and theoretical frequency ratios is made in terms of the harmonic progressions in both pentatonic and heptatonic patterns. The results support the theoretical findings that the Chou 周 musicians had in their possession a versatile chromatic scale. A number of practical problems that can affect the determination of the frequencies of bells is discussed.

Tb

DR. H. C. BHARDWAJ
READER IN HISTORY OF SCIENCE
& TECHNOLOGY
BANARAS HINDU UNIVERSITY, Varanasi, India.

"INDIAN CONTRIBUTION TO THE METALLURGY OF ZINC IN ANTIQUITY

Though the use of brass is attested from different cultural areas during first millennium B.C., yet metallic zinc was not isolated even a thousand years later. First reference to metallic zinc in Europe is made by Paracelsus in 16th century and it was patented in 1738 by Champion.

India played a key role in the isolation and production of this metal. Indian alchemical works around 8th century A.D. (high period of Indian alchemy) mention the extraction of zinc from calamine by heating it with carbonaceous substances in covered crucible (Rasaratnakar of Nagarjuna, verses 31-32). The covered crucible was soon provided with tubulure (Tiryakpatana yantram i.e. distillation per descensum). Description of distillation of zinc in tubulated retorts is mentioned in Rasaratna Samuchchya and other alchemical works of 12th-13th century A.D.

Archaeological evidence of the recovery of zinc smelting retorts of medieval period at zewar near Udaipur (which are the earliest in the world) and the purity of Indian Zinc which was 98.99% pure in the Pre-industrial period have been critically examined and assessed. It is conclusively proved that William Champion who patented his distillation process in 1738 had learnt it from India.

Paper points out Indian priority over the Chinese in the matter of isolation of zinc and holds the possibility of their learning the art of distillation from India. This is borne out by the brisk intercourse in scientific and technological fields subsequent upon the spread of Buddhism into China. It must be, however, conceded that Chinese had developed regular production in zinc during 14th century A.D. when zinc coins were issued by Ming dynasty.

Zhang, Yunming

Chemical Engineer, Hechi Nitrogen Fertilizer Plant, China

SULFUR MANUFACTURING PROCESSES FROM PYRITE IN ANCIENT CHINA

While sulfur is a raw material of traditional gunpowder, it has never been satisfactorily explained how the ancient Chinese obtained sulfur for their manufacture of gunpowder. In this article, it will be shown that the Chinese knew sulfur could be made from pyrite as early as the 3rd Century A. D. At that time, sulfur was a by-product from the green vitriol manufacturing process. They had a special name, "Vitriol Liquid ", for sulfur.
In ancient China, green vitriol made from pyrite was a necessity for cloth dying and medicine. The book of Song History reported the output of green vitriol was about 200 tons in 996-997 A. D. From 618-1127 A. D. the government established an organization called Ping Yang Yuan in Jin Zhou (located in Shanxi Province) for its sale. Meanwhile, the Song Dynasty gunpowder recipe mentioned "Jin Zhou Sulfur" as a raw material of gunpowder.
During the Tang and Song Dynasties, the main production area for green vitriol and sulfur was Jin Zhou. It is near Kaifeng, which was the capital of the Northern Song Dynasty and the center of gunpowder production.
In this article, a map of sulfur and green vitriol production areas shows that China has only a few natural elemental sulfur reserves and these are located in remote areas. It was not easy to transport. Fortunately, they could get sulfur from pyrite.
Moreover, the development of sulfur manufacture from primitive indigenous furnace to indirect distillation process and vertical furnace is described. Its relationship with green vitriol manufacturing and the improvement of gunpowder is clarified.
From this article, a logical conclusion is that: With the development of gunpowder, great quantities of sulfur were required to support the continual fight against foreign countries. This requirement necessitated the manufacture of sulfur from pyrite in Shanxi. The purer sulfur obtained from pyrite made it possible to improve gunpowder from an incendiary to an explosive weapon.

T. Ko (Professor), Sun Suyun (lecturer) and Mei Jianjun (Assistant)

Archaemetallurty Group, Beijing University of Iron and Steel Technolo

Ancient Metallurgy of Cupronickel and Nickel Silver in China

Cupronickel was first mentioned in China in 4th century AD, in a geographical treatise of Szechuan Kweichow and Yunnan Provinces. Deatails in Geographical Records of Huili Country, Szechuan, where ancient cupronickel was produced, survey of ancient mine and slag, interview with old workers and laboratory studies, have enabled the reconstruction of ancient process of its smelting. A "blue ore" or Pyrrhotite containing 1-2% Nickel and a "yellow ore"(copper pyrite) or a black copper ore were mixed and smelted. The dark brittle product was fragmented and then roasted as many as nine times, then smelted with 30% addition of a sulfur-free iron-nickel ore of different origin.The product was finnally alloyed with refined copper to produce the final cupronickel.

Late in the 12th century, zinc was added to cupronickel, to form nickel silver. Nickel bearing brass originally used in Bactrian coins, and again known to be made in China in 18-19th centuries and exported to Europe. A recent finding shows that nickel silver containinging 27% Zn, 9%Ni was used in the form of cast billet, with the date and wight given, in AD 1187. It was a weight for 520 gm or 20 liang in Sung Dynasty for weighing silver in the Imperial Treasury, obviously for its rust-proof property. Based on historical records, analyses of a nickel brass coin of AD 1095-1098, and of the 1187 billets, evidences are advanced that zinc was already available in meatallic form in China in the 12th century AD. Zinc was produced from calamine in the Kweichow-Szechuan region with clay crucibels capped with a cooling device for zinc condenstation, a method still in use until recently.

Details of history of cupronickel and nickelsilver production, is geographical and geological aspects, description of ancient mines, analyses of ores, slags, and reconstructed smelting proaess are described. The ancient smelting of zinc is also discussed and illustrated.

Ranganatha S. Rao

Emeritus Scientist   National Institute of Oceanography, India

History of Harbour Engineering in Ancient India

The history of harbour engineering in India can be traced back to the days of the Indus Valley Civilaziation. The brick dock (2300 B.C.) built at Lothal is the earliest example of our ancients putting their knowledge of winds, tides and currents to practical use. The recent onshore and offshore explorations by the Marine Archaelolgy Unit of the National Institute of Oceanography, Goa have revealed construction of massive gravity walls and toe walls in the island of Bet Dwarka and at Kotada and Khirasar competence of the Harappans (2400-1500 BC) and their successors (1500-1300 BC) in designing and building antierosion walls is now proved by the actual structural remains encountered on shore and the intertidal zone . The submerged temples and port-installations in Dwarka waters have added a new dimension to the eustatic studies and also thrown welcome light on the history of coastal engineering in India.

At the pre-Roman period port ofKaveripatnam (3rd century BC) a brick wharf-cum-jetty was built for berthing ships and hauling cargo. The brick jetty at Rajbandar is an important engineering feat responding to sea-level fluctuations during the first seven centuries of Christian era.

A. G. Keller

Senior Lecturer, University of Leicester

Recent work on "Los Veintiun Libros de los Ingenios"

Since L. Reti first described the five-volume MS at Madrid, "Veintiun Libros de los Ingenios y de las Maquinas," to 1965 Warsaw ICHS, there have been important developments in work on this MS. The traditional attribution to Juanelo Turriano has been challenged, other candidates for authorship proposed, the original MS published in 1983; and a fully annotated English translation is now complete, and, it is hoped, will soon be published.

Salient points to be discussed are: a) use of Latin or Italian sources, notable L. B. Alberti, on architecture. This was more extensive than first appeared. b) the irrigation system described is that of eastern Spain--so too are the horizontal watermills, of Moslem rather than European type, which dominate the sections on machinery. c) attempted mathematisation of machinery.

José A. García-Diego

President of Asociación para la Historia de las Técnicas

ENGINEERING PROJECTS FOR RECONSTRUCTION OF
JUANELO TURRIANO WATERWORKS IN TOLEDO

June 13th this year was the date of the 400th anniversary of the death of Juanelo Turriano. He is still particularly remembered as one of the greatest engineers and clockmakers of the Renaissance.

Since 1969 I have devoted a part of my time to studying this historical figure and both his actual work and that attributed to him. I was also entrusted with three unusual undertakings. As a Consulting Engineer I was in charge of a feasibility study and of two successive detailed projets for a partial reconstruction -though on the same exact scale- of one of the mechanisms which Turriano planned and built, between 1565 and 1581, to raise water from the River Tagus up into the city of Toledo. They solved a greater problem than those of any cities, so that his achievements were famous in the whole of Europe.

A model of this sort would have been the most important reconstruction of one ancient hydraulic system in the world. Or I, enthusiastically, thought so. However owing to political difficulties first, and later to lack of funds (though the total cost was not great) my hopes of seeing my work carried out have been dashed.

I am resigned to this, yet I wish to state the case in this International Congress at which so many outstanding historians are gathered.

André GUILLERME

Professeur à l'Ecole Nationale des T.P.E., Lyon, France.

BERNARD PALISSY . LA LOGIQUE DE L'EAU ET L'HISTOIRE

<u>Les discours admirables sur les eaux et fontaines, tant naturelles qu'artificielles</u> édités à Paris en 1580 revèlent toute la vie de Bernard Palissy et les questions que se posent les "éclaireurs" de la fin de la Renaissance. L'eau est-elle à l'origine du monde ? Est-elle irrémédiablement porteuse tout à la fois de la pureté et de la souillure ? Est-elle aérienne ou souterraine ? Que sont ses vapeurs : dangereuses ou exploitables ? Entre "Théorique" et "Pratique", Palissy choisit le second et abonde dans le sens du déterminisme. Il dénonce la mode des fontaines et les charlatans qui la dispensent, critique les philosophes de l'Antiquité et les universitaires qui s'y réfèrent, condamne les médecins avides d'honoraires.

Pourtant, mort dans de tristes conditions par convictions religieuses et politiques, Palissy ne devient "immortel" qu'au début du XVIIIè siècle lorsque se posent à nouveau les questions sur l'origine de l'eau. La seconde moitié du XIXè siècle en fait un héros précurseur du positivisme pour certains, patriote pour d'autres, ou martyr du protestantisme, tandis que l'enseignement public de la Troisième République ne reconnaît en lui qu'un inventeur obstiné que tout bon Français devrait être, pense-t-on alors.

On analyse d'une part l'apport de Bernard Palissy aux sciences et techniques de son temps et d'autre part comment la société française a jugé ses écrits.

Jane Mork Gibson

Contract Historian, Philadelphia, Pennsylvania, U.S.A.

The Fairmount Waterworks in Philadelphia: Blending Science, Technology and Architectural Beauty in the Nineteenth Century for the Public Good

Philadelphia's Fairmount Waterworks in the 1830's became the prototype for other cities and was famous throughout Europe and the United States. The complex was constructed in several stages from 1814-1872, initially in response to the medically urgent need to provide sufficient untainted water and to "purify" the air by fountains and the washing of streets. There were three distinct power systems for pumping, using steam engines, breast wheels, and hydraulic turbines. The technology thus employed was housed in architecturally harmonious buildings that came to resemble a group of neo-classical temples with the surrounding landscape developed as a public park. Scientific studies documenting the need for better pollution control with filtration beds brought about the closing of Fairmount Waterworks in 1911. Today the buildings remain, with one turbine and pump in situ (1851), and are being restored as a tribute to the nineteenth-century manner of blending science, technology and architectural beauty--utilitarian structures in harmony with nature.
    In 1815 two early steam engines (high- and low-pressure) were installed in an Engine House to pump water from the Schuylkill River to a reservoir 92 feet above, at the top of Faire Mount. Water then flowed by gravity through wooden pipes to public hydrants and to houses and industries subscribing to the service. Cast-iron pipes were introduced in 1819. The need for more water and the high cost of operating the steam engines, as well as the primitive state of scientific knowledge re strength of materials and thermodynamics, brought about a change in the system to water power in 1822. A dam was thrown across the Schuylkill and a forebay and mill house constructed for eight vertical water wheels. In 1825 the Watering Committee stated that the cost of raising water was less than $4 per day. To protect the quality of the water, the city began to acquire land bordering the river for Fairmount Park. The new technology of the hydraulic turbine was successfully introduced at Fairmount in 1851. In the 1860's a New Mill House was added for three large state-of-the-art turbines, and in the 1870's the water wheels were removed from the Old Mill House to install three more turbines. Accompanying these changes, additional reservoirs and distribution systems were added, always maintaining the park-like setting and the architectural beauty.

Tc

Bernard O. Williams, University of Kansas, USA

U S Joint Effort to Define Electronic Digital Computing, 1942

In the winter of 1942, the first United States effort to jointly define electronic digital computing was organized by the National Defense Research Council, Division Seven on automatic gun fire control. At a meeting in April, projects at Eastman Kodak, Massachusetts Institute of Technology, Bell Laboratories, Radio Corporation of America, and National Cash Register were reviewed to derive a statement of potential and likely magnitude of application. After followup sharing of information among the projects, the summer months saw the decision not to develop such devices, although the capability to do so had been demonstrated.

Basic digital counting circuits had emerged from geiger-muller counters and from instruments for measuring cosmic rays and bubble chamber events in the mid-1930's. Applications to business machines and experimental scientific calculators were beginning. As war approached, the named institutions and others turned to military applications. The individual projects were all well along in the design of serviceable devices for antiaircraft gun fire control when the option to rely on an analog system, recently developed at Bell Laboratories, was identified as best fitted to the character of the guns. The electronic digital systems could have been faster and of almost arbitrary precision, but the application did not call for this level of sophistication.

The questions considered during the joint investigation and the form of the conclusions are the subject of this paper. Files in the National Archives, University of Pennsylvania, Library of Congress, and MIT Archives are the primary sources for this discussion.

Giovanni Battimelli, Università di Roma "La Sapienza", Roma, Italy

"SENZA VINCOLI UFFICIALI": THE INTERNATIONAL CONGRESSES OF
APPLIED MECHANICS IN THE TWENTIES.

Several authors have discussed the conflict, during and after the first world war, between the internationalist ideology of scientific knowledge and the political commitment of scientists, in particular with regard to the policy of the International Research Council and of its scientific Unions.

In this talk a case study is presented of an international body which was born during the twenties, with the polemic between scientists on opposite sides being at its peak, and quickly attained unpredicted success. Preceded by an informal gathering organized by T. von Kármán and T. Levi-Civita in Innsbruck in 1922, the International Congress of Applied Mechanics, first held in Delft in 1924, was at the end of the decade a much more alive institution than many of the Unions tied to the IRC.

Two factors, which seem to be especially responsible for this success, are discussed. On one side, it is to remark the programmatic refusal, by the "founding fathers", of establishing any formal connec tion between the Congress and any official body or institution tied to the IRC or to single Governments, in order to eschew the obstacles of international scientific diplomacy: this "refusal of politics" proved to be an extremely successful political act.

On the other side, the specific nature of the discipline involved has to be taken into account. The International Congress of Applied Mechanics is seen as the body which comes to identify a new sector of scientific activity, bordering with physics, mathematics and engi neering, taking shape in those years mainly in the German-speaking scientific world.

It is recalled that proposals advanced at the IRC to establish international cooperation on technical matters, tied as they were to old-fashioned disciplinary subdivisions, and overall inspired by anti-German prejudice, totally failed to materialize.

Iskender Gökalp

Chargé de Recherche au C.N.R.S., Centre de Recherches sur la Chimie de la Combustion et des Hautes Températures, Orléans et Programme Science, Technologie, Société du C.N.R.S., Paris.

A STUDY IN THE CONSTITUTION OF A FRONTIER SCIENTIFIC FIELD: THE CASE OF TURBULENT COMBUSTION.

Turbulent Combustion (TC) is a relatively new scientific research area (its origin may be dated in the years 1930-40) constituted at the interface of the Mechanics of Turbulence and the Chemistry of Combustion. These two domains have shared very little in common before the emergence of TC : they were autonomous scientific domains in the sense analysed by D. Shapere (in F. Suppe, The structure of scientific theories, 1974, pp.518-565). The process of the constitution of the TC, its further development and current autonomization constitute an interesting case study pertaining to discussions concerning topics such that relations between scientific fields or theories, unification of science, reductionism, interfield theories, etc. Moreover, the investigation of the patterns of the autonomization of this frontier field informs us about the basic mechanisms of interdisciplinarity. Indeed, the intervention of only two autonomous domains (and both from the same side of the exact sciences/social sciences barrier) without any "background knowledge" between them, designates the TC as a simple model for interdisciplinarity studies.

In this communication, the constitution of TC as a frontier scientific domain is investigated from several angles. As a first point, different stages of its growth are sketched and compared with conventional growth patterns (three or four staged patterns as identified by the sociology of science) of non-frontier domains. The pecularities of the development stages of TC are then emphasized. In the same framework, the constitution of the TC scientific community is also investigated. Namely, the time evolution of the respective shares of the scientific origins (working initially in turbulence, combustion or other fields) of the members of this community is investigated.

Besides these socilogical informations on TC, some questions concerning the topic of the constitution of a frontier paradigm are also adressed. Within this framework, the following points are raised.
*The interaction mechanisms of the intervening fields (combustion and turbulence) are investigated. The role of a shared vocabulary is emphasized. The concept of the "flame generated turbulence", proposed in the early stages of the TC is treated with detail in order to exemplify a given state of these interaction mechanisms.
* The influence of the intervening fields on their reciprocal conceptualization is dealt with, by using data from the exemplary situations given by the necessity of inclusion of a chemical source term into the Navier-Stokes equations or by the necessity of taking into account the temperature fluctuations in the Arrhénius chemical production term. The problems raised in both autonomous fields by the incursion of these new items are stressed.
*The answers of each autonomous community to new experimental exigences raised by their unification are emphasized on the example of the advent of non intrusive optical turbulence measurement techniques.

Td

ELENA AUSEJO.— GRADUATED IN MATHEMATICS —IBM TRAINEE—

INDUSTRIAL CORPORATIONS AND DATA PROCESS AUTOMATION

This paper tries to analyze the main factors that explain the trascendental influence of the industrial activity in the development of Computer Sciences.

A look at the most relevant inventions that have helped to mechanize the Calculus techniques since the Middle Ages up to the XXth, century shows important differences between the impact of these machines in the past and in the last forty years.

This new situation suggests a new proposal to study this special relationship between science and Society from the point of view of the History of Sciende.

Gail Cooper

University of California, Santa Barbara, U.S.A.

MANUFACTURED WEATHER: AIR CONDITIONING IN AMERICA

This paper traces the impact of air conditioning upon the patterns of work and leisure in three different environments: the factory, the theater and the home. The employment of air conditioning carried explicit cultural values which are reflected in advertising rhetoric, in new architectural designs and in a particular engineering style. Installations were based upon the notion that air conditioning was not merely a mitigation of unpleasant conditions but an attempt to establish a totally controlled and artificial indoor environment. The connection between the control of environment and the control of human behavior was made explicitly by Progressive era engineers who dominated the structure of air conditioning in factories. The expectation that air conditioning could provide an effective tool for the control of behavior was established between 1902 and 1922 in the workplace. This expectation carried over into comfort installations in theaters and department stores during the 1920s and 1930s, where air conditioning was utilized to promote consumption rather than production, leisure rather than work. This dual focus on maximizing work and leisure provided a confusing legacy for the post World War II drive to establish air conditioning in the home. The balance which promoters struck between the home as a workplace and as a refuge from the world of work is indicative of social values and gender ideals in post-war American society.

Henry Etzkowitz

Ass. Professor of Sociology SUNY Purchase

THE DEVELOPMENT OF SOLAR PHOTOVAOLTAICS TECHNOLOGY AT BELL
LABORATORIES: 1950-1959

This paper will provide an analysis of solar photovoltaics
research and development activities at Bell Laboratories from 1950-
1959. The focus will be upon the interaction among research scientists, engineers and management in setting research direction, identifying potential uses for the emergent technology and initiating
development work.

Research questions include 1) the effect of the theoretical
orientation toward solid state science on the invention of a
practical photovoltaics device; 2) the effect of corporate
management research goals on scientific work on photovoltaics within the development group; 3) the uses for photovoltaics projected
by the scientists, engineering staff and management; 4) how decisions
were made to utilize photovoltaics within the Bell organization; and
5) the diffusion of scientific and engineering work on photovoltaics
outside of the Bell organization. This paper is the first phase
of a research design to examine the conduct of photovoltaics
R&D under varying institution al conditions.

Barton C. Hacker, Historian, Reynolds Electrical & Engineering Co., Las Vegas, Nevada

Manufacturing a New Order: From Military Technology to Industrial State, 1850-1950

Military institutions have always interacted consequentially with other social institutions in the history of civilized societies. Dramatic changes in military technology beginning with the introduction of gunpowder decisively shaped the modern world. New weapons demanded new tactics, new tactics in turn required new organization, and military innovation ultimately tranformed polity, economy, and society. Michael Roberts has termed the last phase of this transformation The Military Revolution, 1560-1660 (Belfast, 1956) and shown how relatively narrow technical changes in battlefield tactics could in time produce far-reaching changes in every aspect of Western civilization.

Accelerating change continues to characterize Western society, military institutions not excepted. Another military revolution began little more than a century ago. Like its earlier counterpart, it started with technological innovation and ramified throughout society. This revolution has yet to find either distinctive name or systematic analysis, but the modern industrial state is as surely a product of this nineteenth-century revolution as the eighteenth-century absolutist nation-state was of the upheaval Roberts defined.

From the mid-nineteenth century onwards, Western armies underwent a remarkable, bureaucratically sophisticated transformation. Technologically, it originated in vastly expanded rates of fire and ranges of small arms and guns during the century before World War I. Other changes of at least equal import included organizational innovations like the genral staff which allowed growing armies to be controlled, production innovations like interchangeable parts manufacturing which allowed them to be equipped, and communication innovations like railroad and telegraph which allowed them to be supplied and directed. But even when individually well studied, such changes have rarely been treated conjointly, nor have their larger consequences received scrutiny commensurate with their importance.

Military concerns, and money, have shaped higher education, for instance, in ways seldom fully acknowledged, or even recognized. Civilian technology owes more to research and development originating in military purpose than is commonly realized. Managing large-scale enterprise in the late twentieth century still relies on models devised during the late nineteenth century to control ever-growing armies. The structure and operation of the modern industrial state can only be properly understood in terms of its military antecedents, and the ongoing dynamic interaction between military and other social institutions.

Edmund N. Todd, SUNY Potsdam, Potsdam, New York

THE STRUCTURE OF CHOICE: ELECTRIFYING THREE RUHR TOWNS, 1890-1900

Electrification in three Ruhr towns during the 1890s nicely illustrates the interaction of evolving political, social, and technological systems. During the 1890s, the ability to transmit electricity over distances remained the sole recognized benefit of alternating current. Some towns built alternating current systems when the power station had to be placed outside town. But choice of system could reflect local configurations in another more subtle way. When a town was the focus for the surrounding area and developed a strong, unified elite, the town might chose alternating current to serve surrounding communities. A town which did not develop as a focal point might chose direct current to gain greater economy for a narrowly conscribed area.

The Ruhr towns of Bochum, Dortmund, and Essen illustrate these connections. As administrative, economic, transportation, and cultural centers, Dortmund and Essen dominated the surrounding rural areas, while Bochum acted as an administrative, but not an economic or cultural center of the county of Bochum. The elites of the cities also differed significantly. While old, local families promoted industrialization in Rhenish cities like Essen, outsiders did so in Westphalian cities, like Bochum and Dortmund. Unlike Dortmund after the 1860s, Bochum did not produce from immigrant entrepreneurs and local families a unified industrial leadership feeling bound to the city. Conflict between Bochum's old local elite and its new industrial elite continued into the 1890s and hindered the city's development. Unified leadership in Dortmund and Essen contributed to local cultural growth, and, in the 1890s, supported strong mayors who promoted the development of their cities and looked past municipal boundaries in solving problems.

Technological choices followed local development. Bochum built in 1892 a small direct-current system to light city hall, then grudgingly expanded the system and finally in 1906 began drawing alternating current from the county. Dortmund and Essen had strong elites and powerful mayors who displayed regional interests by chosing alternating-current systems. Dortmund built a mixed system with the alternating-current component supplying its new harbor and surrounding communities, and the direct-current component providing an efficient, low-cost and secure system for its center city lighting load. Essen chose an alternating-current system in 1897 to supply customers in areas it hoped to incorporate. In each case, deployment of different kinds of technological systems reflected local political, social, and economic roles of the towns and their elites.

Sverker Sörlin

Junior lecturer, Dept. of the History of Ideas, Univ. of Umeå, Sweden.

Natural resources and national identity: The organisation of forest research and the 'New Patriotism' in the early industrial era - the case of Sweden.

Industrialization in Sweden during the period 1870-1920 was to a large extent based on the natural resources of the country herself: ore, hydro-electric power, the forests. Even more so, this was reflected in the rhetoric. Sweden was predestined to be among the world's leading powers, it was argued, because of her rich natural resources. These also became a main field of study. Extensive research was done to map and quantify natural resources. The natural sciences at the universities expanded at an enormous speed and several independent research bodies were formed to identify and utilize this material capital. New subjects were established, even in the social sciences. "Economic geography" was offered a chair at the Stockholm School of Commerce in 1907. It was enhanced through the blend of scientific interest and national aspirations. This was nothing exclusively Swedish. All over the world there occurred a scramble for natural resources and science generally served as a "super-ideology" and an alibi for expansionist policies. However, in Sweden this became an important trait in the national identity. Patriotism centered on the natural resources and first and foremost on the forests. In the national imagery these were represented variously as oasis, as the industrialized Sweden to be, as the mythical and historical roots of the nation, and as powerful tools in the "struggle for survival" among nations. The organization of forest research thus became a patriotic obligation and took on heroical overtones. The research was state-promoted. The Royal School of Forestry expanded and the Public Institute of Forest Research was established in 1902. These are some examples of what has become main features in the national concept of Sweden during the 20th century, the so-called "Swedish model", i.e. the state being very active to promote science in order to achieve economic progress. There was also rapid growth in the number of publications on forestry, both scientific and popular. It was regarded necessary with a national re-education of the people, the forest-owning peasants in the first place, via campaigns, excursions, outdoor museums, pamphlets. Public forestry boards were set up to implement the laws that often were in conflict with the old popular ideas of resource management. The value-system of industrialized society was confronted with that of peasant-economy. The scientists were among the principal supporters of the former. Science and scientists thus took part in the production of values neededto pave the way for industrialism. Forest research helped to bring about a new image of Sweden that shattered the old national spirit of historical apotheosis and stressed modernity, rationalism and productive ability.

Raymond Duchesne

Professeur, Télé-université/Université du Québec, Québec, Canada

LES INSTRUMENTS SCIENTIFIQUES DU SEMINAIRE DE QUEBEC, 1663-1900

Depuis sa création au XVIIe siècle, le Séminaire de Québec s'est efforcé d'être une des premières institutions d'enseignement du Canada. A cette fin, le Séminaire a consacré des ressources considérables, tout au long de son histoire, à l'enrichissement de son Cabinet de physique. Au milieu du XIXe siècle, le développement de l'Université Laval, création du Séminaire lui-même, oblige les professeurs à élever le niveau de l'enseignement des sciences et à doter le Cabinet d'instruments plus puissants et plus précis.

Au tournant de 1900, la collection comprend plus de mille appareils et instruments: télescopes de Gregory et de Dollond, gyroscope de Foucault, machine d'Atwood, bobines de Ruhmkorff, pont de Wheatstone, tubes de Crookes et tubes à rayons X, etc. Ces instruments sont l'oeuvre des plus célèbres artisans de Londres, Paris et Vienne: Dollond, Newman, Pixii, Ducretet, Koenig, Ruhmkorff, etc.

Fort heureusement, cette collection a été soigneusement conservée et nous est parvenue avec une masse de documents d'archives qui permettent d'en retracer l'histoire. Ces archives et les instruments eux-mêmes sont aujourd'hui conservés au Musée du Séminaire, à Québec.

Cette collection, véritable "fossile" selon le mot du Prof. Gerald L'E. Turner, permet de mieux connaître l'enseignement des sciences dans une institution périphérique du XIXe siècle, tout en enrichissant l'histoire des instruments scientifiques.

André W. Sleeswyk, Laboratorium voor Algemene Natuurkunde,
University of Groningen, The Netherlands

THE INDUSTRIAL REVOLUTION: CHINA AND EUROPE

The history of the development of science and technology illustrates particularly clearly the commonly occurring lack of causality between successive steps in their progress. The next step in the sequence of progress is often based, not on the latest prior development, but on a retarded one abandoned earlier. This process, and the concomitant constant shifting of the locus of progress and the fact that there can be an advantage in being backward, have all been discussed at some length by Romein (1935), apparently on the basis of earlier work by Kautsky (1910). The reality of the observation seems well established. It stands in opposition to the historically untenable hypothesis of causal step-by-step progress.
Needham (1964) noticed that, although Chinese civilisation had been more successful than occidental in applying knowledge of nature to practical human needs between the 1st century B.C. and the 15th century A.D., it lagged behind afterwards. His finding fits this general observation. There is no paradox involved, and it is not necessary to postulate a perennial stagnation in the sciences in China (Qian, 1985) to explain a temporary lag in the development of Chinese science and technology. Sivin's (1975) biography of the polymath Shen Kua (1031-1095) and his essay (1982) on the development of Chinese science illustrate this point very clearly.
Although the British industrial revolution was in the first place an economic and social phenomenon, it acquired its extraordinary dimension and significance as the result of the interaction between the sciences and technology. The contribution of the sciences to the economically important development of the cotton industry was negligible, but Newcomen's atmospheric steam engine was a direct application of the physical research on vacuum by Torricelli, Pascal, Huygens, Papin and von Guericke. On the other hand, the crucial improvement by James Watt, the addition of a separate condenser, could only be explained adequately by physics in the mid-19th century (Carnot, Clausius).
There was no major technological obstacle which would have prevented the Chinese from emulating the British development of textile industry, but the concept of atmospheric pressure and physical experimentation necessary for developing the steam engine would have required a revolutionary development in Chinese science.
Moreover, Hellenistic geometry, which permeates to some extent everyday technology in the West, was unknown in China. The many additional improvements of the steam engine by James Watt (Ferguson, 1962), the use of the screw thread, the development of tooth shapes for power gearing, would all have been impossible without it. On the other hand, even in the 20th century the use of bolts and nuts was unknown to a large section of the Chinese population (Hommel, 1927). The conclusion must be that ignorance of Hellenistic geometry made a major Chinese industrial revolution an impossibility.

Mari Williams

Research Officer, Business History Unit, London School of Economics

Instrument manufacture in transition: the case in Britain, 1880-1940

The instrument sector is a small but vital area of industrial activity, serving not only scientific research institutions, but also many branches of industry. The manufacture of instruments as a commercial enterprise now largely takes place within small or medium sized companies, some of which are independent but many of which are subsidiaries of much larger multiproduct or multinational corporations. This has been the case only comparatively recently; towards the end of the last century instrument makers were generally to be found working in far smaller units, often based around a family-run workshop. In Britain in this period the limited company had only recently emerged and just a small number of instrument companies were registered. It was also the period - according to many economic historians - when Britain's world-wide industrial lead was threatened if not relinquished, a situation from which the British economy never recovered.

As well as being relevant within the obvious areas of the history of science and technology, therefore, the history of instrument manufacture between the late nineteenth and mid-twentieth centuries should also be revealing about many important issues within economic and business history. The aim of this paper will thus be to sketch out a brief account of changes in the scientific instrument industry in Britain between 1880 and 1940, emphasising the origin and diffusion of new ideas and designs, interactions between makers and users (both scientific and industrial), and finally changes in the organisation of the industry. Evidence will be drawn from a variety of sources, including the records of a number of companies, published histories of other companies, instrument catalogues and contemporary scientific and industrial journals. From the account it emerges how evidence collected for business and company history can benefit aspects of our understanding of the history of science and technology, and vice versa.

William J. Wallace, Ph.D
Professor, San Diego State University

YEHUDI

Prior to World War II very little research had been done on the subject of camouflage and especially little was known as to what objects humans could see at distances particularly under various lighting and weather conditions.

One of the outcomes of the intensive research in these areas was code name "Yehudi" - the attempt to camouflage flying aircraft and ships at sea using lights solely.

Gregory Davis

Professor, College of San Mateo

Modern Technology and Modern Art--A Common Philosophical Basis?

The process of technological change is presented by some as a self-justifying activity, regardless of negative consequences like pollution on a global scale, dehumanization, and creation of the potential for nuclear annihilation. This is a nihilistic conception, where action is free from higher values and form (change) is glorified over content (effects).
Modern art also promotes form, the artist's activity or technique, at the expense of content. Examples are Pollock's action paintings, Burrough's literary "paste-ups," and Arrabal's striking film images applied to necrophilia and sexual perversion.
This nihilism in art has the same historical roots as technological nihilism. Both derive from the impact of the Scientific Revolution on philosophy and values. Descartes and Bacon uncoupled reason from the search for the Good and gave it a utilitarian focus, the mastery of nature. Technological operations, however, were guaranteed as good in advance by the 18th century theory of progress. As disillusion with the new society, based on industry, equality, and capitalism emerged in the 19th century, however, it was clear that the process of change remained intact while the benefits promised by the ideology of progress had failed to materialize. Eventually, process itself became a value.
Hume, Kant, Laplace, and Darwin were all milestones, as philosophy lurched toward the crisis proclaimed by Nietzsche's "joyful wisdom" that God was dead and nihilism had arrived. Schopenhauer had been the first to make the irrational--a collective will to live--the basis of his philosophy, Fichte had also glorified will, seeing the unrestriced activity of the ego, seeking to realize its projects in nature, as an expression of human freedom. In 1975, Samuel Florman wrote in Harpers that unrestricted technological development was justified because it allowed human will free expression.
Flaubert and Courbet in the 19th century disconnected art from the Good and made form more important than content. Flaubert wrote a stylistically perfect novel about a mediocre adultress; and Courbet painted humble subjects, building up the paint "unrealistically" with his palette knife to emphasize the primacy of the artist and his means. In 20th century painting, Mondrian "voided" content with his geometric abstractionsism, and action painting made the creative act of painting itself the focus. At art "happenings," the material used was physically destroyed at the end.
Both art and technology which indulgently take themselves as ends independent of external values have a nihilistic character. But whereas autonomous art operates on the margins of real life, autonomous technology negates the material basis for life itself. If Heidegger was right when he said technology in the modern world "discloses being," then what does it tell us about ourselves?

Kazutaka Unno

Professor, Osaka University, Osaka, Japan

THE EUROPEAN ELEMENTS IN THE JAPANESE SURVEY METHODS DURING THE EDO ERA: WITH SPECIAL REFERENCE TO THE TECHNICAL TERMS AND THE INSTRUMENTS

According to the tradition prevalent among the surveyors of the Edo Era, the European survey methods were brought to Japan in 1641 by a Dutch called Caspar. Chronology tells us that it cannot be Caspar Schamburger, a Dutch surgeon, since he came to Japan in 1649. Nor is there any record handed down to us which proves that he used to teach survey methods as well as surgery. When and by whom were the European survey methods brought to the country, then?

An attempt is made here to solve this problem by examining the origins of European vocabulary as it is used by the surveyors of the time. Hiden Chiiki Zuho Daizensho (1717) by Hosoi Kotaku and Bundo Yojutsu (1728) by Matsumiya Kanzan are examined, as they represent manuals of comparatively early date and with full contents among those that have survived until today. The European vocabulary which is common to them include such words as watarante(kuwadarantei), kompansu(kompasu), asutarabiyo(asutarahi) and piroto and the names of the twelve months. Piroto(for piloto) is obviously not of Dutch origin, while in the rest of the words it is not always clear whether Portuguese or Spanish is represented. On the other hand, the names of the months include shanero(for janeiro), hebereiro(for fevereiro), setemboro(for setembro) and nobemboro(for novembro), which are evidently Japanized forms of the Portuguese words. These Portuguese forms for the twelve months appear already in a book commonly called Genna Kokaisho, written by Ikeda Koun in 1618. Asutarabiyo(for astrolabio) and kuwadarantei(for quadrante) both name what are originally nautical instruments; as one may expect, they are illustrated in Ikeda's book just mentioned. The appearance of the word piloto in manuals of survey methods might seem surprising, but it goes to show that the European survey methods of the Edo Era had their origin in the art of navigation in Europe.

To sum up, the claim of the Dutch sourse was a mere fiction which was intended to conceal the relation with the Portuguese who brought Christianity into the country. It was the art of navigation people had learnt from the Portuguese before they were forbidden to come to the country in 1639 that provided the foundations for the subsequent use of the European survey methods in Japan.

James E. McClellan III

Associate Professor, Department of Humanities
Stevens Institute of Technology, Hoboken NJ 07030 USA

Science after the Scientific Revolution: The Mémoires of the Paris
Academy of Sciences, 1699-1790.

With an eye toward documenting the general course of science, scientific research and disciplinary development after the Scientific Revolution, this paper reports initial findings from a computerized data base being compiled of articles and papers that appeared in leading scientific and learned society journals of the 18th century. Among the most prestigeous outlets for the publication of original research in science, making public the work of the premier scientific society of the age, and conveniently spanning the century (with 93 volumes appearing for the years 1699-1790), the famous Histoire et mémoires series of the Paris Académie royale des sciences was chosen as a valuable indicator for initial development and analysis.

Preliminary findings suggest approximately 3,400 articles published as finished scientific mémoires, comprising roughly 50,000 pages. There were about 235 authors. That averages to 15 articles per author and 14 pages per article, high figures indicating an established scientific cadre publishing polished scientific pieces. Individuals publishing memoirs were among the resident élite of the Paris institution, and families of scientists formed a notable subgroup. The number of articles per volume remains more or less constant at around 37, but, unexpectedly, the length of those articles and associated volumes increases considerably (from 12 to 16 pp./article and from 300 to 600 pp./vol.) from the beginning to the end of the century. Based on a convenient, contemporary classification embedded in the Histoire et mémoires, an initial division into disciplines reveals the following distribution of research effort in the world of ancien régime science: astronomy (33%), physique générale (17%), chemistry (14%), anatomy (11%), mathematics (8%), botany and natural history (6%), mechanics (5%), and other (6%). Short observational reports predominate in astronomy, and, notably, the number of astronomy articles doubles in the second half of the 18th-century. The uniquely 18th-century category of physique générale comes in second place; that largely static category includes a gamut of disciplines that both did and did not meld into 19th-century physics: meteorology, electricity, natural history, and scientific instrumentation, to name only the top subfields. Chemistry ranks third, with significant expansion at the time of Lavoisier and the chemical revolution. Reflecting the known separation of science from medicine, the number of fourth-place anatomy articles declines by 50% in the second half of the century. Considering mathematics and rational mechanics together (at 13%), the mémoires of the Paris academy would seem the very bastion of these highly-specialized sciences. For all their importance botany, natural history, and other inquiries score low. Comparisons are made with other 18th-century scientific journals.

Bo S. Lindberg
Assistant professor, Dept for History of Science and Ideas
University of Göteborg, Sweden

"LINGUA ERUDITORUM VERNACULA" LATIN AS THE LANGUAGE OF SCHOLARS AND SCIENTISTS IN 18TH CENTURY SWEDEN

The 18th century was the heyday of Swedish science, when Linnaeus, Torbern Bergman and others became famous all over Europe. This flourish of science was accompanied by an intense debate, whether scholars and scientists should teach and write in Latin or in the Swedish vernacular. In the arguments, three main dichotomies are discernible:

1. <u>Traditional and international learning vs the demands for knowledge that serves the national interest</u>(economics and applied science in the first place, but also humanities if they deal with Swedish history, antiquities and language).This dichotomy resembles the conflicts in the preceding century between university <u>learning</u> and the new, empirical <u>science</u> of bourgeois and puritan circles.

2. <u>Theoretical vs applied science</u>. The Swedish scientists continued to write most of their theoretical works in Latin in order to reach the international audience. This passed almost unnoticed by both foes and friends of Latin.

3. <u>Science vs humanism</u>. In the first part of the 18th century, the vernacular was associated with (applied) science, while Latin was still very much the language of the humanities. Later on there was a shift, so that Latin was defended as the language of theoretical science, whereas humanities were increasingly treated in Swedish,as the national culture developed. In addition, there was a tension within the Latin camp between the scientists who regarded Latin as a merely technical language for scientific communication, and the representatives of the old humanist learning, who endeavoured to uphold a decent classical Latin, not only as means of communication but also as an expression of culture and identity.

J. Morton Briggs, Jr.

Professor of History, University of Rhode Island, Kingston, R.I.

AN ENGINEER AND THE INDUSTRIAL REVOLUTION DURING THE ENLIGHTENMENT

As early as 1756, a French mining engineer named de Genssane recognized four major problems facing industry: the need for workers with new skills, the need for government protection of fledgling industries, the crisis of fuels, and the necessity of large amounts of capital over long periods of time. (A. de Genssane, Sur les Mines d'Alsace, 1756 and 1763.) De Genssane knew of the sorry state of French miners, of their practices, poverty and disorganization. His solution was to control their practices through managers who were truly competent, and his call for the training of these men was loud and clear. Government help was needed not only to encourage industries, but to see to it that local lords did not get in the way of the large mining ventures. The crisis of wood supply is well known; de Genssane's description of it shows concern both for industry and the environment. Finally, he worried about the fact that most entrepreneurs did not have enough money to get started and to withstand losses until the profits finally began to flow. Large number of partners were required and it was not easy to keep the relations between them amicable over long periods of time. (A. de Genssane, Traité de la Fonte des Mines, 1772-1776; Histoire Naturelle de la Province de Languedoc, 1776-1780.)

De Genssane, a Corresponding Member of the Académie des Sciences, represents the union of science and techniques that was to mark the Industrial Revolution. His operations in the Vosges in Alsace and later at Villefort in the Cevennes, in Languedoc, demonstrates his style of action. Yet he came upon the scene before any of the schools (such as Ponts et Chaussées or the École des Mines) were established. His own programs were fulfilled by the work of Gabriel Jars, Duhamel de Monceau and others. Nonetheless, the frame of mind that was to produce the emphasis on new techniques and organization was present, at least in Antoine de Genssane and some of his colleagues, by the middle of the Eighteenth century.

Roger JAQUEL
                                              FRANCE
Professeur honoraire (Lycée A. Schweitzer Mulhouse)

L'HISTORIOGRAPHIE INTERNATIONALE ET PLURIDISCIPLINAIRE
RECENTE - DEPUIS UN QUART DE SIECLE - DU SAVANT POLY-
VALENT (PHYSICIEN, MATHEMATICIEN, ASTRONOME, GEOGRAPHE)
ET PHILOSOPHE JEAN-HENRI (OU JOHANN HEINRICH) LAMBERT
(1728-1777).

Pendant les deux siècles qui ont suivi les travaux pionniers de J.H. LAMBERT (1759-1764) c'est dans les pays germanophones que les études lambertiennes, aussi bien scientifiques que philosophiques, ont été- et de loin - les plus nombreuses et les plus fécondes.
Depuis 1960 l'Allemagne a gardé la suprématie dans ce domaine, mais la recherche s'est internationalisée et diversifiée. Des études novatrices paraissent aussi bien en Italie que dans les pays anglo-saxons, et des chercheurs soviétiques participent à cet essor. La France ne s'est non plus désintéressée de ce mouvement, et c'est même dans ce pays que se réunit, pour le bicentenaire de la mort de Lambert, en septembre 1977, à Mulhouse un Colloque international et interdisciplinaire qui est à la fois une résultante et un symbole de cette tendance.
Les études publiées depuis 1977, et en particulier les Actes de ce colloque (parus début 1980) semblent préluder à un renouveau vigoureux de la curiosité et de la connaissance lambertiennes, qui s'étaient assoupies pendant la décennie suivant la deuxième guerre mondiale.
Le projet d'édition systématique des oeuvres complètes de Lambert en trente volumes in-4° (envisagé par M. STECK en 1943) a été démantelé par les circonstances. Ses oeuvres complètes philosophiques- surtout en allemand- sont sur le point, grâce à H.W.ARNDT, d'être intégralement accessibles facilement (en 10 volumes in-8°).
Mais les oeuvres scientifiques de l'auteur trilingue (allemand, français, latin)- y compris ses cinquante mémoires, en français, à l'académie (francophone à l'époque de Frédéric II) de Berlin - font problème: Elles sont d'un accès assez compliqué, voire aléatoire, surtout en dehors de l'Allemagne et de la Suisse.
La bibliographie lambertienne de Max STECK de 1943, enrichie partiellement mais aussi compliquée dans sa 2e édition en 1970, est toujours indispensable. Mais les nombreuses références parues ou découvertes au cours du dernier quart de siècle rendent sa refonte plus que souhaitable.

Sven Widmalm

Office for History of Science, University of Uppsala, Uppsala, Sweden

GEODESY AND MILITARY CARTOGRAPHY AROUND 1800

During the 18th century, cartography and geodesy benefited from refinements of instruments and of physical theory. France led the way in this developement and established a tradition of cartographic and geodesic work not to be found in other countries. When geodesy-based cartography was eventually adopted in England and Sweden the context was one of nationalist scientific competition, where the accuracy of earlier measurements was defended or challenged. Knowledge and techniques imported and developed for scientific measurements were later used in the respective military cartographic surveys of England and Sweden.
Between 1784 and 1787 a triangulation was made to determine the relative positions between the observatories of Paris and Greenwich - true knowledge of the position of Greenwich having been questioned by French scientists. The English operations were supported by the Royal Society, the Board of Ordnance, and the king, who donated money to purchase a new kind of instrument - a theodolite of unsurpassed precision. In 1791 the triangulations were continued under the surveillance of the Ordnance Survey, a new organization for producing maps for military and civilian use. The Ordnance Survey employed personnel trained during the Paris-Greenwich measurement, used techniques established at that occasion, even borrowed the Royal Society-theodolite, and published their reports in the Philosophical Transactions.
In Sweden the developement followed a similar pattern. In 1801-1803 Swedish scientists re-measured the meridian that Maupertuis first had determined to ascertain the shape of the earth. This expedition was sponsored by the Swedish Academy of Sciences and by the king; further assistance came from the military. An instrument - Borda's repeating circle - was brought from France along with Delambre's instructions for its use. Just like in the British case a military survey was established a few years later which employed personnel, and used techniques and instruments from the scientific measurement.
Thus, in both England and Sweden techniques and instruments for triangular measurement were acquired in the context of a scientific measurement, and were later put to use in military cartography. In both cases the scientific measurement aspired to the highest possible accuracy, which meant that every step of the operations was scrutinized and founded on what was thought to be a sound scientific basis. In this way a "deep" technical competence was established, which later could be utilized for routinework and for propagating a tradition for this kind of work into the future.

Dr. Richard R. Yeo

Lecturer, School of Humanities, Griffith University,

Brisbane, Australia

Defining Science: William Whewell and his scientific audience in nineteenth-century Britain.

William Whewell's philosophy of science has usually been seen as the epitome of the idealist or intuitionalist position against which J.S. Mill defined and defended his empiricist account of knowledge. As a result of this concentration, there has been little study of the way in which Whewell's work was received by his scientific contemporaries. Yet Whewell was part of an intellectual community which shared his general concern about defining the nature and limits of science. In the debates relating to this issue, questions of epistemology and methodology - the subjects which interest modern philosophers of science - were often closely connected with other matters such as the organization of research, specialization and the hierarchy of disciplines, science education, and the theological implications of scientific knowledge. By investigating some of the ways in which Whewell's work was interpreted and criticized, it is possible to gain a perspective on contemporary nineteenth-century debates about the nature of the scientific enterprise, while also revealing areas of tension within Whewell's own history and philosophy of science.

As part of such an inquiry, this paper aims to show how Whewell's writings both reflected and challenged assumptions or contentions about the accessibility, singularity and transferability of scientific method. Whewell's views, which were conditioned by his epistemology and his study of the history of science, will be discussed in the context of debates on issues which included the philosophy of discovery, the division of scientific labour, the scope of scientific education, the unity of science and the relationship between different disciplines. These discussions involved men such as John Herschel, David Brewster, James Forbes, William B. Carpenter, Baden Powell and Richard Owen - commentators who did not always accept Whewell's account of the nature and progress of science.

Dr. Harvey W. Becher

Associate Professor, Northern Arizona University

"Social and intellectual considerations about Cambridge mathematics and science in the 19th century"

The preeminence of mathematics in a Cambridge liberal education had numerous ramifications, both social and intellectual. Mathematically talented middle class men gained access to Cambridge. In an intense and highly competitive environment, these men studied for honors fervently in order to achieve success, and what they studied was applied and applicable mathematics. In addition, they very often studied science with the Cambridge science professors, themselves graduate wranglers, and undertook research as undergraduates or recent graduates. Prior to the era when the sciences became paying professions, these men most often opted for careers in the established, socially prestigious, and financially rewarding professions of the time--education, the Church, and the law. Here, along with their more blue-blooded counterparts, they became leaders of society. Many of them also pursued hard, practical and/or theoretical science, making major contributions in scientific fields ranging from engineering to physics and astronomy and from political economy to biology and geology. Moreover, these graduates carried Cambridge mathematics and science throughout Britain and the Empire, and they funneled mathematically talented young men to their alma mater.

Because of the evolution of the mathematics curriculum within a liberal education, because of the difficulties in teaching abstract mathematics, and because of fundamental beliefs concerning mathematics in a liberal education shared by Cambridge men, be they radicals, Whig reformers or conservatives, Cambridge mathematics took on marked characteristics. And since Cambridge science was dominated by Cambridge mathematicians, Cambridge mathematics determined the form of Cambridge science. The Cambridge mathematics curriculum centered on problem solving, and excellence was defined in terms of elegance and simplicity. Being applied mathematicians, Cambridge men sought answers, and the validity of the answers was most often checked empirically--in the data and in the laboratory. Abstract mathematical rigor was of little concern. Science based upon physical models, the hypothetico-deductive method, and elementary physically intuitive mathematics was the ideal. In this form, mathematics dominated the curriculum, was accessible to the mass of honors students, and became the foundation of Cambridge science to the end of the nineteenth century. In the 1850s Churchmen would do hard science with elementary mathematics; at the end of the century physicists at the Cavendish would often turn out work involving mathematics more akin to that of Newton than that of Weierstrass.

Frank A.J.L. James

RICHST, Royal Institution, 21 Albemarle Street, London, W1X 4BS, England

PROFESSIONAL SCIENTISTS AND THE "SCIENTISTS' DECLARATION": THE CASE OF HERBERT McLEOD

In Britain during the early 1860s there were a number of attacks on orthodox christian theology by liberal churchmen. The most influential criticisms were Essays and Reviews and Bishop Colenso's Pentateuch. These claimed partial support from contemporary scientific work. In this paper I shall explore the reaction of orthodox professional scientists to what they regarded as the misuse of science.

This will be done by an examination of the career of Herbert McLeod (1841-1923). During the 1860s McLeod was Assistant Chemist at the Royal College of Chemistry under Hofmann and later Frankland. He kept a daily diary from 1860s until 1923 which ran ultimately to 174 volumes.

McLeod was the author of and one of the chief motivators behind the "Scientists' Declaration" of 1864-5. This argued that in the final analysis there would be no conflict between the word of God written in the Bible and in Nature; this Declaration was signed by 717 scientists. McLeod's diary for the 1860s will be used analyse the development of the Declaration. This shows that the relationship between professional scientists and their religious belief was more complex than some recent studies have indicated. In particlular I shall draw a distinction between physical scientists whose belief generally remained strong and biological scientists where it was weaker. This provides a new perspective on the relations of science and belief in this crucial decade.

Jeffrey A. Johnson, Assistant Professor of History

University Center at Binghamton, State University of New York, USA

A CULTURE DIVIDED: ACADEMIC MODERNIZATION AND THE DIVISION OF THE PHILOSOPHICAL FACULTIES IN GERMAN UNIVERSITIES 1859-1914

Mid-nineteenth-century German scholars of the natural sciences and the humanities shared in an academic culture marked by service to the classical ideal of pure scholarship or Wissenschaft, membership in a university (and especially a philosophical faculty), and a secondary education or Bildung based on the classical languages. Most scholars also shared a common social origin as sons of classically-educated men. By the advent of the First World War, however, institutional innovations and social changes were breaking up the old academic culture and often provoking hostility between scientists and humanists. The divisive issues can be expressed in the word "modernization."

Controversial though that term may be as a sociological concept, it emerges from the actual historical context in this case. "Modernity" meant matriculating students who had not received a classical secondary education; it meant giving university status to the technical colleges and weakening the dichotomy between "pure" and "applied" scholarship; it meant institutionalizing specialization, and thus dividing the philosophical faculties between humanists and scientists, or even along disciplinary lines like the technical colleges; it meant accepting into the academic elite the children of the industrial entrepreneurs and the "new middle class" of white-collar workers; and finally, it meant accepting money from wealthy businessmen and thus potentially prostituting the academic enterprise. It meant, in short, accepting the social consequences of industrialization in the universities. The central act of the modernization drama was played out between 1859, when the Prussian government sanctioned a secondary education without Greek for its bureaucrats and dissertations written in the vernacular for "modern" university disciplines, and 1914, when the University of Frankfurt am Main was opened with a wholly private endowment. This paper will examine one of the key issues, the division of the philosophical faculties.

One way of approaching the issue is to compare classical philologists and chemists as exemplars of the extremes of humanism and science. They can be placed at opposite ends of a spectrum from "traditional" to "modern" social origins, yet they sometimes behaved in apparently anomalous ways. At times classicists supported divisions of the philosophical faculties; while many senior academic chemists provided weaker support for such "modern" reforms than might otherwise have been expected. Generally, these anomalies can be explained through reference to generational factors, the university organization of the disciplines, and the peculiar course of German modernization.

Elisabeth Crawford, J.L. Heilbron, John G. May, Rebecca Ullrich

C.N.R.S, Paris, Office of History of Science, University of California, Berkeley

International leaders in physics and chemistry in the first third of the 20th century

The international reward system of science, well established around the turn of the century, produced a certain number of "winners" with high international reputation and visibility. We have constituted two data-bases made up of such "winners": on the one hand, the nominees and nominators for the Nobel prizes in physics and chemistry 1901-1933 ($N=$ca. 800) and, on the other hand, physicists and chemists elected to foreign membership in major academies of science (19 in all) in Europe and North America, 1880-1933 ($N=$ca. 400). We are now studying the characteristics of that part of the two populations whose identification with the international reward system was the strongest, i.e. those "winners" ($N=$ca. 100) who participated most actively in the Nobel system (and who were often prizewinners) and who were also elected to membership in a large number of academies of science outside their own countries.

To what extent did the group of most successful "winners" represent the international leadership of physics and chemistry and, perhaps, of science as a whole? We are addressing this problem through an examination of what constituted internationalism in science during three distinctive periods: before the First World War, during the War and its immediate aftermath, and in the interwar period. We have defined the term "internationalism" broadly so as to include both true international scientific activities (e.g. the International Association of Academies) and cross-national or bilateral ones (e.g. the grant programs of the Rockefeller Foundation or the Carnegie Institution, or the use of non-nationals on national scientific advisory boards). We distinguish between activities related to substantive or technical matters, like geophysical surveys or standardization of units and activities of honorific character that initially gave meaning to the term 'international scientific community'. Neither set of activities is necessarily characteristic of those who led physical science at the research front. We shall discuss the questions how far the "winners" led science and how far their international status was related to the production and diffusion of new knowledge.

Xb

Christian Hünemörder

Professor, Universität Hamburg, Deutschland (FRG)

DIE ENTWICKLUNG DER BIOLOGISCHEN HOCHSCHULDISZIPLINEN IM DEUTSCHEN KULTURRAUM VON 1800 BIS 1945

Die im deutschen Kulturgebiet sich seit 1800 regional sehr unterschiedlich entwickelnde Hochschulbiologie mit ihren zahlreichen Einzeldiszplinen (von Anthropologie bis Zytologie) war bisher nicht untersucht worden. Ziel eines vom 1.1.1981 bis 31.3.1984 von der "Stiftung Volkswagenwerk" geförderten größeren Forschungsvorhabens war es, unter Zuhilfenahme eines elektronischen Textverarbeitungssystems eine Gesamtübersicht in institutioneller, personeller und thematischer Hinsicht mit einer aus 2 Wissenschaftshistorikern und mehreren Doktoranden bestehenden Gruppe zu erarbeiten.
Der Vortrag möchte die Konzeption des aus mehreren verschiedenen Teilen (Historische Skizzen der Entwicklung der einzelnen Teildisziplinen, tabellarische Ubersichten z.B. über das wissenschaftliche Personal der Hochschulinstitute, Biologendatei in Form von Kurzbiographien) bestehenden "Handbuchs" vor der Drucklegung international vorstellen. Dabei erhoffen wir uns Anregungen und fördernde Kritik von den Bearbeitern vergleichbarer Forschungsprojekte.

Mary C. Rabbitt and Clifford M. Nelson
U.S. Geological Survey, 904 National Center, Reston, VA 22092

ORIGIN OF THE U.S. GEOLOGICAL SURVEY'S INTERDISCIPLINARY "MISSION APPROACH" TO SOLVING GEOLOGIC PROBLEMS

Scientific research before the beginning of the 20th century was largely a matter of individual enterprise, but by 1900 it was recognized that the solution of the large problems of science required organized efforts. The organization of scientific effort began in institutions established primarily for utilitarian purposes. Among the earliest were those established for the investigation of geologic problems because of the association of geology with the development of mineral resources, which became increasingly important as the industrial revolution progressed. The early national geological surveys founded in Great Britain (1835), Canada (1842) and Austria-Hungary (1849) were begun because of their importance to industry and mining. These organizations were predated by those in several Eastern States of the United States which established surveys in the 1820s and early 1830s to aid the development of agriculture and mining. In the United States, the Federal Government began national organization of geologic work after the Civil War and established the United States Geological Survey (USGS) in 1879.

The USGS was charged with responsibility for "classification of the public lands and examination of the geological structure, mineral resources, and products of the national domain." Its establishment, therefore, was clearly for utilitarian purposes. Clarence King, its first Director, realized that practical science could be used to further basic science, which he conceived to be the fundamental purpose of the organization. Thus, although he organized USGS work by geologic provinces, the first investigations--on the economic geology and technology of the most important mining districts--were designed to provide information of economic importance. Several disciplines became part of the effort. Geologic and topographic mapping were supplemented by chemical and physical investigations; petrography, then in its infancy, became an integral part of geologic field investigations; and mineralogy and paleontology assumed adjunct roles. In the search for data needed to solve geologic problems, the basic sciences were extended, and the basic data thus acquired became the foundation for further advances. This "mission approach" demonstrated its value not only in the solution of economic problems but also in the advancement of both practical and basic science. The development of the multidisciplinary approach in the USGS influenced the organization of the scientific work of the Federal Government, which remained largely utilitarian in purpose. The process also influenced private institutions, such as the Carnegie Institution of Washington, established for the purpose of furthering basic scientific research.

Frank N. Egerton

Professor of History of Science, University of Wisconsin-Parkside, Kenosha, WI 53141

THE ECOLOGY OF INSTITUTIONAL RESEARCH: A CASE HISTORY FROM THE GREAT LAKES

The U.S. Bureau of FIsheries established its Great Lakes Fishery Laboratory at Ann Arbor, Michigan in 1927, directed by John Van Oosten (1891-1966). In 1928 the Bureau sponsored a conference to develop a concensus concerning the important problems of fishery management on Lake Erie and to coordinate research on them. These were monitoring fishery yields, studying the life history of fish species, and studying the limnology of the lake.

Van Oosten hired 5 other fishery biologists, and the 6 of them did research in all three areas. His own doctoral research had been on age determination in lake herring and the use of this data for estimating fishing intensity. He and members of his staff conducted similar studies on other species, compiled and evaluated catch statistics obtained from commercial fisherman, and conducted limnological studies in western Lake Erie. Van Oosten also attempted to explain why the lake herring, which had been one of the most abundant commercial species in Lake Erie until 1925, fell that year to commercially insignificant numbers. During the Depression the budget of the laboratory was cut in half, and Van Oosten had to terminate some of his staff and some of the research programs. His conclusions in 1933 were that he and his staff had shown that the dangers to the fisheries from overfishing were great and from pollution were minor. Therefore, he cut the limnological research and never restarted it before his retirement from the directorship in 1949.

Ohio State University in 1896 established a field station for its biology faculty and students in western Lake Erie. Thomas Huxley Langlois (1898-1968), Chief of the Fish Section in the Ohio Division of Conservation and Natural Resources, also became Director of the University's Stone Laboratory at Put-in-Bay in 1938. Although the scope of research at the Stone Laboratory was very broad, its emphasis was more on limnology than fishery biology. In 1941 Langlois concluded that pollution from agriculture near western Lake Erie inhibited reproduction of lake herring and other species, and that habitat change rather than overfishing was causing the decline in their abundance.

Neither institution could easily test conclusions drawn at the other, so neither refuted the other's claims. Van Oosten and Langlois, for the rest of their lives, continued to argue their respective cases.

Glenn E. Bugos

University of Pennsylvania, Philadelphia, USA

"BORDERLAND SCIENCE: BIOPHYSICS IN THE U.S. NATIONAL RESEARCH
   COUNCIL, 1928-1942"

During the 1920s and 1930s, the National Research Council expanded its institutional presence in American science by actively promoting "borderland" or interdisciplinary science. The history of three NRC committees sponsoring research in biophysics illustrates the many problems in translating the rhetoric of borderland research into practice. The Committee on the Effects of Radiation Upon Living Organisms (1928-1938) was an attempt by biologists to exploit the technique of irradiation in studies of genetics, development and cytology. The NRC Subcommittee on Mitogenetic Radiation (1933-1936) used bacteriological and photoelectric techniques to resolve the controversy over mitogenetic radiation. In the Washington Biophysical Institute (1937-1942) physicists developed photochemical instruments for incorporation into medical and government laboratories. Tension, rather than cooperation, characterized relations between physicists and biologists on the administrative level of these NRC committees. As a result, the NRC biophysics programs emphasized service roles over theoretical promise to justify funding from philanthropic foundations and industry. The improvement of instrumentation and the quantification of biology became the principal basis for interdisciplinary cooperation. Finally, the administrators saw the future promise of borderland studies in the training of individual researchers rather than cooperative arrangements between specialists.

Stanley Wallen
Ph. D. candiate, New York University

Western Science and Eastern Religion: How compatible ?

There has been a great amount of discussion recently concerning the relationship between Quantum Mechanics and Eastern Mysticism. Writers such as Capra have attempted to portray Eastern Mysticism as the true precursor of Quantum Mechanics; indeed, that it provides the intellectual framework within which Quantum Mechanics can be understood. This paper will show that the similarities between these two fields of thought arise out of their common necessity to confront a well developed system that seems to be in good accord with our everyday perception of the world.
It is this denial of the reality of our common world view which produces the strong similarities in language and concepts which the two fields possess. But what is generally overlooked is that from this point on the two fields travel in opposite directions.
In Mysticism there is a real world that can be known beyond the illusory one: the path toward self-knowledge lies through self-negation. Quantum Mechanics however holds that there is no absolute reality and that our individual consciousness plays a large role in establishing such reality as does exist, for example, Wheeler's "No phenomenon is a real phenomenon until it has been observed." In short, the similarities of language and concepts in these two fields of thought have blinded us to the drastic differences in content.
This paper will then attempt to show that it is this dependence on a medium, language, that both fields regard as inadequate, and their common efforts to surmount its limitations, that hs led to their apparent similarities. The paper will also show that it is the failure to compare the goals of each of these two fields of thought that contributes to the common misperception that Eastern Mysticism and Quantum Mechanics are closely related.

Kjell Jonsson

Junior lecturer:Dept. of History of Ideas, Univ. of Umeå,
Sweden

IGNORABIMUS- a debate on the demarcation between science
and belief and its sociocultural implications.

In 1872 the well-known German electrophysiologist Emile
Du Bois-Reymond (DBR) told his collegues at the 45th
assembly of German natural scientists and doctors that
"was Materie und Kraft seien, und wie sie zu denken ver-
mögen, muss er ein für allemal zu dem viel schwerer ab-
zugebenden Wahlspruch sich entschliessen:'Ignorabimus'."
We will never know the essence of matter and force and
never solve the question of how consciousness occur in
connection with chemical or physical molecular processes
in the brain. DBR thereby demarcated natural knowledge
from metaphysics, science from belief, or Naturwissen-
schaft from Weltanschauung. This speech was met with an-
noyance from scientists who believed that modern science
constituted a complete world view hostile to an obsolete
Christendom. The zoologist Ernst Haeckel was one of them.
He found in Ignorabimus the "Ignoratis des unfahlbaren
Vaticans und der von ihm angeführten 'schwarzen Internationale'." But many theologians, scholars and idealistic
philosophers appreciated the distinct demarcation DBR
had made between the fields of science and belief.
The consequences of DBR's epistemological position and
the diffusion of the pragmatic interpretation of science
opened the way to a reintroduction of metaphysical idealism and an irrational thought style at the end of the
19th century. In a way DBR's Ignorabimus prepared the
way for a modern thought climate where we live without
a universal norm in a world of many truths: scientific,
artistic, philosophical, thological and so on. Each developed in its own community or thought collective.
The debate on Ignorabimus became part of the cultural
struggle in many countries. In Sweden DBR's statement
was adduced by established theologians, conservative critics and philosophers in their combating metaphysical materialism and the emergent radical counter-culture. In
the academic community the epistemological self-limi-
tation of DBR became hegemonial as professional ideology.
Radicals and socialists found that Ignorabimus discredi-
ted the idea of scientific and social progress.
The mind-body problem was impossible to solve according
to DBR. When almost every "modern" intellectual at the
turn of the century called him/herself a monist, it was
called a "dualistic" assumption, supporting the Cartesian
and Christian disintegration of the Western culture at
a time when it needed unification.

Brian G. Sullivan

Associate Professor, Western Kentucky University, Bowling Green, KY

THE PERSISTENT DUALISM OF SCIENCE AND RELIGION IN AMERICAN CATHOLIC THOUGHT

An intergrated view of science and theology has escaped the efforts of scholars to reduce the tension between these disciplines. Recent letters from American Catholic Bishops have not followed the lead of the theologians; they seem to continue the divergent account of the dualistic approach. This paper will review theological literature, pastoral letters, conciliar and post conciliar documents in an attempt to indicate the direction of a future unified account of science and religion.

George E. Webb

Assoc. Prof. of History, Tenn. Tech. Univ.

CREATION SCIENCE AND THE AMERICAN CONSTITUTION

Over the past two decades, the long-standing opposition in the United States to the teaching of evolution in the public schools has returned to its former significant level. Not since the height of anti-evolution during the 1920s, symbolized by the Scopes Trial in Tennessee, has there been so much public discussion concerning the place of evolution in the public school curriculum. Opponents of teaching this concept continue to be associated with fundamentalist religious groups in the nation, but the organized campaign to remove evolution from public schools shows a different focus from the earlier campaign of the 1920s. No longer characterizing themselves as anti-evolutionists, modern opponents of Darwin portray themselves as adherents to the concept of "creation science" or "scientific creationism," and claim to have found scientific support for their adherence to a literal reading of the first chapter of Genesis. In their attempt to minimize the teaching of evolution in public schools, creationists have resorted to legislative strategies designed to allow "equal time" for their views in the science classroom. Their success in several states has led to important court cases which have been decided on constitutional, rather than scientific grounds.

The basic question raised by creation science involves the First Amendment prohibition on establishment of religion. In several cases, creation science has been determined to be a religious, rather than a scientific world view. Having defined creation science as religion, courts have concluded that legislative support of this view, in the form of mandated presentation of creationism in science classes, represents an attempt to establish a particular religion, in direct conflict with the First Amendment. An examination of legislation and court decisions suggests that no state can implement creationism constitutionally. The focus of the debate will thus shift to the local level where the creationists have already enjoyed noticeable success.

Sven-Eric Liedman

Full professor, Department of the History of Science and Ideas, University of Göteborg, Sweden.

MANDARINS AND NON-MANDARINS IN THE GERMAN ACADEMIC INTELLIGENTSIA

Fritz K. Ringer's impressive work, The Decline of the German Mandarins: The German Academic Community, 1890-1933 has already reached the position of a classic inside intellectual history. Despite its brilliancy, Prof. Ringer's study has a certain weakness due to its functional type of analysis, according to which the attitudes of the German academic intelligentsia is seen as a function of their social position. Apparently, the differences between orthodox and radical "mandarins" are more impressive than the traits uniting them, which in fact seem to dwindle into a high esteem of Wissenschaft and Bildung. The uniting link is not so much attitudes and opinions as a limited range of controversial issues. It is not the position pro or con one issue that directly reveals the social position of a person or a group: it is the issue itself. This is universally true in all intellectual history, mandarin or not.

Secondly, the attitudes and opinions of the German Mandarins are not so unique as Ringer maintains. We can find a similar ideology in England and France, not to talk about Sweden at the same period. To the traditional academics, industrialization and capitalist development aroused confusion everywhere. Technological and economic development and socialism were important issues everywhere, whereas the attitudes to these issues varied very much inside all European "mandarin" groups.

This does not mean, that the situation in Germany and e.g. England was identical or even very similar. The difference between the German concept Wissenschaft and the English science illuminates the fact, that science and scientific institutions play quite another part in Germany as compared with England at the end of the 19th century, which I shall try to demonstrate in my paper. (A more extensive version of my paper will be published as an article in Comparative Studies in History and Society.)

G. Palló

A CASE FROM THE PERIPHERIES:
THE BACKGROUND OF THE "HUNGARIAN PHENOMENON"

The history of science, almost exclusively, deals with the events taking place in the most important scientific centres, although these are far from being identical with the scientific system in its entirety. The investigation of the scientific system, by necessity, includes the study of the relationships between centre and periphery, from the aspects of both theory and the history of science.

The paper wishes to contribute to meeting this requirement by conjuring up a characteristic case.

In the history of 20th century science the "Hungarian phenomenon" means the birth of a uniquely talented generation of Hungarian scientists in the period between 1885 and 1910, who scored momentous success abroad, mainly in the USA. That generation was represented by as outstanding scholars as Albert Szent-Györgyi, Eugen Wigner, John von Neumann, Leo Szilard, Edward Teller and others.

The names quoted are widely known but less known is the fact that, together with many more of their compatriots of similar calibre, they were Hungarians. Knowing it, the question inevitably arises: how was it possible that science in Hungary, represented by relatively few scholars and moving on the periphery, could attain such unbelievably high standard and how could the foremost personalities of Hungarian science occupy such predominant role in international science. In other words, whether the surprisingly large number of prominent scientists was in proportion to the general standard of science in Hungary, or not.

The answer to the question is in the negative. The standard of science in Hungary was far below the one suggested by the names quoted. One of the mysteries of the periphery lies just in the fact that, in spite of a relatively low general scientific level, immense values might be produced but, most likely, they are realised only in the centres.

Xe

Dr. Patricia Rife

Union for Experimenting Colleges & Universities

Emigration Dilemmas: Lise Meitner 1938-1939

Austrian physicist Lise Meitner faced serious repercussions concerning her decision to remain in Berlin at the Kaiser Wilhelm Institute for Chemistry after Hitler's <u>Anschluss</u> of Austria in 1938.

Drawing from primary documents of correspondence between Meitner, Bohr, the Rockefeller Foundation, and other sources, it is demonstrated that Meitner's mentor Max Planck played a considerable role in her decision to remain in Germany, and, as Hitler's persecution of Jews intensified, created a major "emigration dilemma" for Meitner.

Interviews with physicists and friends of Meitner's conducted by the author in Sweden highlight Meitner's year of adjustment in Scandinavia after her departure from Berlin in the summer of 1938. Meitner's circle of women friends are examined as a support network, and her relations with long-term colleague Otto Hahn analyzed in light of her personal and professional decisions, feeling of powerlessness, and frustrations as the "missing team member" within the 1938 discovery of nuclear fission.

Paul K Hoch and Edward Yoxen

Senior Fellow, STP, Aston University; and Lecturer,
University of Manchester

Schrödinger in transition between Britain and Ireland

This paper deals with the social and intellectual factors that eventually brought Erwin Schrödinger, one of the principal formulators of the new quantum mechanics, by 1940 to the newly founded Institute for Advanced Study in Dublin. We discuss his abandonment of his chair at Berlin in 1933; the events that initially brought him to Magdalen College, Oxford; his growing disenchantment with the attitudes and direction of physics in Oxford; his 1936 assumption of a chair at the University of Graz in Austria; growing conflict with the Nazi regime after the anschluss; and eventual departure from Austria on his road to Dublin. We consider his non-acceptance of other posts that were offered to him in this period in other countries, such as at Princeton in America and at Allabad in India; and also his inability (or unwillingness) to secure a senior post at what would have been thought of as the major contres of physics outside Central Europe. His intellectual environment at Dublin is compared with that at Oxford, as well as with others he might have potentially enjoyed at other centres.

Finally we discuss the effects his perapetetic existence from 1933-39 might have had on his growing disinchantment with the Copenhagen interpretation of quantum mechanics, his interventions in succeeding years in the debate about 'what is life', as well as his attempts to formulate a unified field theory of the electromagnetic and gravitational fields.

Considerable space is given to the imbalance between 'theoretical' and 'experimental' physics (as they have since become known) at Oxford in the 1930's, as well as between the arts and the sciences in the Oxford colleges of that period.

Xe

Kristie I. Macrakis

Harvard University, Graduate student

ROCKEFELLER FOUNDATION FUNDS REICH SCIENCE: THE DECISION TO FUND THE PHYSICS INSTITUTE OF THE KAISER WILHELM GESELLSCHAFT, 1934-1938

The initial question motivating the research on which this paper is based was a simple one: why did the Rockefeller Foundation fund the Kaiser Wilhelm Gesellschaft's(KWG) Institute of Physics during the Third Reich? In 1930 it had promised the Physics and Cell Physiology institutes of the KWG an appropriation of $655,000 for the buying of land, the construction of two research institutes, and the purchase of some equipment. By 1933 only the Cell Physiology Institute had been built, while the building of the Physics Institute had been delayed. Meanwhile, the National Socialists had siezed power in January of 1933 and the nature of the new regime had become well-publicized, especially with regard to its dismissal policies and to the Gleichschaltung of institutions with National Socialism. During the years 1934-35 the foundation faced a difficult decision.

We enter directly into the climax of the plot--the middle period in the relationship between the Foundation and the Society; we distinguish three phases in the period 1934-38: The first phase is the decision to fund the Physics Institute, 1934-35, where we follow the "essence" of the "decision-making" process--the Foundation's goals and objectives, its options, the consequences, and its final choice. The second phase is the public response to the decision -- the New York Times' exposé and Felix Frankfurter's response, 1936. The third phase is the follow-up visits made to Peter Debye's laborallory.

The officer's of the Foundation who knew what changes had occured in German Society were against the appropriation, but the executive committee and the legal counsel voted to go ahead with the project. Moreover, there was considerable pressure from Max Planck (the president of the KWG, 1930-36) to complete the project. The power structure of the Foundation is examined as well as the attempt to formulate a program and policy for Germany in response to the situation there during the thirties.

Lena Eskilsson

Junior lecturer, Dep of the History of Science and Ideas
University of Umeå, Sweden

Women in the academic world in Sweden 1870 - 1920.

At the beginning of the 1870:s women in Sweden got the right to enter the universities, but only to study medicine. From 1873 all other academic fields were opened to women, but they were not allowed to take degrees in theology and law.
Education was was an important issue to the women´s mowement. However,it was male politicians who took an interest in supporting the unmarried daughters of the middle class, who made the strongest demand for higher education for women.
Since women, it was argued, had fewer needs and no family to support, it would be more profitable for the government to employ well educated women than men in expanding public administration at the end of the 19th century.
From 1873 it took 50 years before the law was changed in such a way that women could also use their education in all employments, with the exceptions that they could not become ministers.
This is characteristic of the general view on women in higher education and science. It was more or less taken for granted that the reception of women in science and higher education was an exception. This was an alternative for those who could not devote their lives to the general roles of a woman as a wife and mother. The underlying assumption was that since so few women in any case would choose- this way of life, they would not form any threat to the male priviliges on the labour market.
Between 1875 and 1914 435 women in Sweden attained academic degrees, about 75% of them in Arts subjects and the rest in Medicine. Most of them became teachers (50%) and physicians in private practice (13%). Their choice of profession was belived to be in line with the view that women had "special skills" or "unique talents". Very few women were accepted for traditional kinds of scientific employment like university teaching and research.

Stanley, Autumn

Affiliated Scholar, Center for Research on Women, Stanford University

A Rose by Any Other Name: Omissions and Mis- and Dysclassifications in a List of 19th-Century American Women Patentees

As the Centennial of the Patent Office (1890) and the end of the century approached together, a women's-rights activist named Charlotte Smith persuaded the Patent Commissioner to compile a list of women receiving American patents since the Office opened. The list appeared from the Government Printing Office as <u>Women Inventors to Whom Patents Have Been Granted by the United States Government, 1790 to July 1, 1888</u>, with two later installments bringing it up to March 1, 1895. The significance of this unique source can scarcely be overestimated. Scholarly and nonscholarly researchers from Ida Tarbell to H.J. Mozans and the Women's Bureau of the U.S. Department of Labor have relied on it as gospel. Yet my research shows that the list is far from perfect-indeed, needs to be used with considerable caution. This paper takes a preliminary systematic look at some of its errors, omissions, mis-, and dysclassifications, and offers a possible explanation for some of them.

Sallie A. Watkins

Professor of Physics, University of Southern Colorado

A WOMAN'S PLACE IN EARLY TWENTIETH CENTURY PHYSICS--AN EXEMPLARY STUDY

Born in Vienna in 1878, Lise Meitner received the Doctor of Philosophy degree in physics from the University of Vienna in 1906. She went to Berlin in 1907 to continue her studies with Max Planck at the University of Berlin. In 1912, she was appointed his assistant. In 1917, she was given the task of establishing and heading the department of radioactive physics at the Kaiser Wilhelm Institute for Chemistry in Berlin-Dahlem. This paper discusses the Viennese educational system at the turn of the century, the barriers it presented to women, and the means Meitner (and others) used to gain access. It continues with an examination of the question of the employment of women in acadème in Germany (particularly in Prussia) in the early part of this century. Again, Meitner is put forward as an example.

Hoang Xuan Sinh

Professor of Mathematics, Hanoi Pedagogical University, Hanoi, VIETNAM

Women in Science in Vietnam

Since 1945, women have made impressive advances in certain areas of the sciences in Vietnam, and are receiving advanced degrees in increasing numbers. Of the five doctors of science in the Hanoi Biological Institute, for example, two are women, and a third is about to return home from abroad with her degree.

This paper catalogs the achievements of Vietnamese women in the sciences both before, but especially after, the overthrow of colonial regimes in the North and South of Vietnam. Statistics on the numbers of women in various branches of the sciences will be presented, as well as indications of women's presence at all levels of the Vietnamese scientific establishment.

The accomplishments of several individual women scientists will be catalogued, as well as the scientific and agitational activities of the Vietnamese Union of Women Scientific Workers. Stress will be put on the progress made since 1945, and on attempts to change and improve the educational system. Predictions and projections for the future will be included.

Carita Peltonen

Institution of Organic Chemistry, Åbo Akademi, Åbo, Finland

WOMEN IN SCIENCE AND TECHNOLOGY - SOME ASPECTS TO IMPROVE THEIR SITUATION WITHIN UNIVERSITIES IN FINLAND

There are not very many women doing research work within science and technology at the universities in Finland today. Has this anything to do with how women look at their position within the university? Female students do not see too many role models during their studies and as they see the female researchers have many problems, they perhaps consider carefully before they enter the same branch. Perhaps they know that they do not have the same possibility as their male colleagues to do research work fifteen hours a day.

Girls' and women's world is not the same as a male scientist's world. Young people think science and technology are the most masculine fields. As the university is a male community, men are the majority, men are more often encouraged to continue with postgraduate studies than women with more talents. How could the image of science and scientists be changed? Can the subjects be adapted to the interests of women? Even today a common opinion within the university is that a female scientist should stay unmarried in order to get through the demanding and rewarding research work.

Today the scientific community is isolated from the people. Science as a social institution is so dehumanized that women feel ambivalent and sick. This has to be changed and women are needed in changing the process. Even in science and technology much more human aspects are needed because what we have to do is to protect life and Nature - not destroy it!

Olodo Kokou

Professeur; President, Académie d'Histoire des Sciences, "F.A.H.S.T."

La tradition des idées des sciences techniques à l'évolution des mathématiques traditionnelles

Dans le règne animal on rencontre l'emploi d'instruments rudimentaires par lesquels on entend un fragment de matières naturelles comme le bois, l'os, les cornes ou les pierres utilisés par les premiers hommes pour atteindre un but. La technique est aussi ancienne que l'homme; l'outil est un dispositif adapté à ses fins et propre à une utilisation réitérée, et implique nécessairement là où il paraît l'existence de l'homme industrieux. La tradition des idées de créativité technique a engendré la pensée des sciences de culture des techniques primitives des premiers hommes industrieux à l'innovation du savoir et du "savoir faire" des sciences de l'homme dans la nature: celles des sciences techniques de constructions, des habitations, des forges, l'invention traditionnelle des formes et multi-formes de géométries, les arts graphiques, porteurs de la représentativité d'éléments cosmogoniques de l'univers, et des calendriers traditionnels du savoir des nombres numériques à la calculabilité évolutive des mathématiques sociales.

Waldemar ROLBIECKI, Warsaw, Poland.

Institut of the History of Science, Education and Technology of the Polish Academy of Sciences.

Centralization of the Management of Scientific Activity in the European Socialist Countries.

1. After World War II a number of European countries – Poland, a part of Germany, Czechoslovakia, Hungary, Rumania, Jugoslavia, Bulgaria and Albania – entered onto the path of socialist development. This also resulted in organizational changes in science in these countries. The guiding principle behind these reforms was the <u>utilitarian</u> idea of closely tying all of national science to the service of its own society /autotelic conceptions of science were condemned and rejected/, which also ment the <u>planned</u> utilization and further development of the nation's "scientific potential". These ideas were the justification for the central management of science in each of these countries.

2. Owing to different local conditions, national traditions and inventiveness of the reformers, initially in each of these countries these reforms were carried out <u>independently</u> of one another, benefitting as each country saw fit from the experiences of the oldest socialist state, the USSR. Around 1948, however, as a result of an offensive by the supporters of dogmatism and the personal intervention of Stalin, in the majority of the above countries there took place a <u>unification</u> of development schemas, also in the organization of science. Soviet models became obligatory. One of these is the model of the <u>national academy of sciences</u>. This idea derives from the 17th century Parisian Académie des sciences. Further stages in the development of this model are the 18th century academies of science in Berlin and St.Petersburg and its mature form – the Academy of Sciences of the USSR. Such academy is an elitist, self-coopting corporation of scientists with lifelong tenure. The executive committe of this body is entrusted with autority over the national network of basic research institutes as well as certin coordinating functions in relation to other "organizational sections' of science. /e.g. schools of higher education/.

3. Such centralization had and has a number of <u>advantages</u>. Above all, it contributed to a rapid growth of "scientific potential" in all of the above countries. It also creates <u>difficulties</u> and <u>dangers</u>, however. One of the most serious seems to be the formation of a collective and therefore semi-anonymous monopoly of scientific authority which has at its command administrative sanctions, which in turn hampers spontaneous and controversial scientific activity that is the driving force in the development of science itself. These difficulties and dangers have given rise to appearance of various <u>anti-centralist trends</u> in the scientific circles of these countries.

Paerle KORNBAUM - Annie TANDE

Rédactrices scientifiques - CDSH/CNRS, Paris, France.

LA DOCUMENTATION AUTOMATIQUE EN HISTOIRE DES SCIENCES ET DES TECHNIQUES : PRESENTATION DU FICHIER FRANCIS.

Le Centre de documentation Sciences humaines (CDSH) du CNRS alimente et gère depuis 1972 vingt trois banques de données bibliographiques dans le domaine des sciences humaines et sociales. Celles-ci sont regroupées dans le fichier FRANCIS (Fichier de recherches bibliographiques automatisées sur les nouveautés, la communication et l'information en Sciences humaines et sociales).
L'une de ces banques de données est la banque *Histoire des sciences et des techniques*.
La littérature scientifique mondiale y est analysée (essentiellement des périodiques). Actuellement, cette banque de données contient environ 60.000 références bibliographiques, avec un accroissement annuel d'environ 4000 références.
Différents produits et services sont offerts à partir de cette banque de données *Histoire des sciences et des techniques*.
 - **Une publication bibliographique trimestrielle**. Le bulletin signalétique *Histoire des sciences et des techniques*.
 - **La diffusion sélective de l'information (DSI)** sous forme de bibliographies :
. *personnalisées* (sujet défini par l'utilisateur) ou *standard* (thèmes sélectionnés par le CDSH).
. *périodiques* (année en cours) **ou** *rétrospectives* (depuis 1972)
 - **L'interrogation en conversationnel**
L'utilisateur, équipé d'un terminal, peut interroger lui-même la banque de données soit par le Centre de calcul du CNRS (CIRCE), soit par le Centre serveur QUESTEL (diffusé aux Etats-Unis par QUESTEL INC. Washington).
Le CDSH travaille actuellement avec de nombreux partenaires, procédure qui se développe de plus en plus actuellement. Dans cette optique des collaborations sont souhaitées.

INDEX

Ahonen, Guy: Hc 7.A
Aiton, Eric John: Md 1.P
Aldrich, Michele: Ga 3.A
Alvarez, Carlos: Mf 6.A
Applebaum, Wilbur: Ac 2.P
Ausejo, Elena: Te 5.A
Azmanov, Iskren: Bh 6.P

Baader, Gerhard: Bh 6.P
Badash, Lawrence: Pp 6.P
Bag, AK: Mb 2.A
Balan, Florin: Sc 7.A
Balan, Stefan: Sc 7.A
Barmark, Jan: Hd 7.P
Battimelli, Giovanni: Td 7.A
Baumer, Anne: Bc 2.P
Beatty, John: Me 7.P
Becher, Harvey W: Xb 6.P
Beer, John J: Sb 5.A
Benguigui, Isaac: Bc 2.P
Benis-Sinaceur, Hourya: Mg 8.A
Bennett, John: Qd 6.A
Berggren, John L.: Qa 3.A
Berkel, Klaas van: Pa 2.A
Berry, William: Bi 5.A
Berry, William: Gb 7.P
Beukers, Harm: Ce 5.P
Bevilacqua, Fabio: Pj 7.A
Bhardwaj, HC: Tb 3.A
Biernacki, Andrzej: Qd 6.A

Billo, Saleh: Gb 7.P
Blay, Michel: Qb 1.A
Blondel, Christine: Pf 5.A
Blondel-Megrelis, Marika: Ce 5.P
Botelho, Antonio Jose J: Sa 8.A
Brackenridge, J Bruce: Pa 2.A
Braun, Ernst: Pg 5.P
Briggs, J Morton: Xa 3.A
Brown, Laurie M: Pm 5.A
Buchs, Mina: Bc 2.P
Bugos, Glenn E: Xc 7.A
Burch, Christopher: Pa 2.A
Burch, Christopher: Pe 7.P
Burian, Richard: Bf 5.P

Cabral, Regis: Pd 6.A
Camacho, Luis A: Sc 7.A
Carandell, Juan: Ab 2.A
Cassidy, David C: Ph 6.A
Chalhoub, S: Ab 2.A
Chapin, Seymour: Ad 3.A
Chappell, John: Hc 7.A
Chappell, John: Pn 5.P
Chen, Joseph Cheng-Yih: Ma 1.P
Chen, Joseph Cheng-Yih: Tb 3.A
Chen, Ke-jian: Qe 6.P
Chevalley, Catherine: Pk 7.P
Chhabra, JG: Ae 8.A
Chigira, Eiji: Mh 5.A
Chriss, Michael: Ac 2.P

The number and letter following the scientific section code indicates the date and time, morning (A) or afternoon (P), the paper is to be given.

Cifoletti, Giovanna C: Qb 1.A
Cohen, H Floris: Qb 1.A
Cooke, Roger L: Mf 6.A
Cooper, Gail: Te 5.A
Cowan, Ruth Schwartz: Bg 6.A
Crawford, Elisabeth: Xb 6.P
Crosland, Maurice: Cb 1.P
Cross, Stephen J: He 8.A
Crowe, Michael J: Ae 8.A
Cushing, James T: Pn 5.P

D'Agostino, Salvo Salvatore: Pf 5.A
Dale, Andrew: Pb 1.A
Darden, Lindley: Bf 5.P
Darrigol, Olivier: Pm 5.A
Das, Sachidanand: Ma 1.P
Daston, Lorraine: Me 7.P
Davies, Susan: Po 6.A
Davis, Gregory: Tg 6.A
Dawson, Virginia: Bb 3.A
Day, Anne: Pd 6.A
Debarbat, Suzanne: Ad 3.A
Debru, Armelle: Ba 2.A
Debru, Claude: Bg 6.A
Delaporte, Francois: Bl 6.P
Diederich, Werner: Qc 1.P
Dobbs, Betty Jo Teeter: Ca 1.A
Doncel, Manuel G: Pm 5.A
Dou, Albert: Mc 1.A
Drucker, Thomas: Qe 6.P
Duchesne, Raymond: Tg 6.A
Dugan, Kathleen: Sc 7.A
Duncan, Alistair M: Cc 2.A

Echeverria, Javier: Md 1.P
Egerton, Frank N: Xc 7.A
Eisele, Carolyn: Qd 6.A
Escoffier, Jeffrey Paul: Se 8.A
Eskilsson, Lena: Xf 6.P
Etzkowitz, Henry: Te 5.A
Evans, James: Pc 2.P

Feher, Marta: Qc 1.P
Feldman, Theodore S: Pb 1.A
Felix, Annette: Bc 2.P
Fernald, Anne: He 8.A
Ferraz, Antonio: Qf 7.A
Field, JV: Mc 1.A
Fischer, Karl: Ae 8.A
Fisher, Nicholas: Cb 1.P
FitzGerald, Desmond J: Qc 1.P
Flegg, Graham: Md 1.P
Fleming, James R: Ga 3.A
Florance, ET: Cc 2.A
Fournier, Marian: Bb 3.A
Frangsmyr, Tore: Gb 7.P
Freudenthal, Gad: Ca 1.A
Fujii, Kiyohisa: Cc 2.A
Fullmer, June Z: Pb 1.A
Furukawa, Yasu: Cd 7.P

Gabbey, Alan: Pb 1.A
Galdabini, Silvana: Pg 5.P
Galison, Peter: Pn 5.P
Gao, Dasheng: Sc 7.A
Garcia-Diego, Jose A: Tc 2.P
Garciadiego, Alejandro R: Mf 6.A
Ge, Ge: Pl 8.A
Gibson, Jane Mork: Tc 2.P
Gigerenzer, Gerd: Me 7.P
Gilain, Christian: Mc 1.A
Gingerich, Owen: Ac 2.P
Gingerich, Owen: Md 1.P
Gingras, Yves: Po 6.A
Giuculescu, Alexandru: Mg 8.A
Giuculescu, Alexandru: Qg 7.P
Giuliani, Giuseppe: Pg 5.P
Glas, Eduard: Mc 1.A
Goddu, Andre: Qa 3.A
Gokalp, I: Td 7.A
Goldberg, Stanley: Pn 5.P
Gomez Pin, Victor: Mg 8.A
Gooding, David C: Pc 2.P
Goodstein, Judith R: Po 6.A

Gorman, Mel: Sb 5.A
Goucher, Candice L: Sc 7.A
Greene, John C: Bj 5.P
Gribanov, Dmitry: Pn 5.P
Guerrino, Antonio Alberto: Ba 2.A
Guillerme, Andre: Tc 2.P

Hacker, Barton: Tf 5.P
Hakfoort, C: Pc 2.P
Hakkarainen, Heikki J: Hb 6.P
Hall, Howard: Se 8.A
Hall, Robert Edmund: Qa 3.A
Harman, Peter M: Pf 5.A
Hashimoto, Keizo: Aa 1.A
Heilbron, JL: Xb 6.P
Hempstead, Colin A: Pg 5.P
Hessley, Rita K: Pe 7.P
Hill, David K: Qb 1.A
Hiromasa, Naohiko: Po 6.A
Ho, Peng Yoke: Aa 1.A
Hoch, Paul K: Pm 5.A
Hoch, Paul K: Xe 5.P
Hoering, Walter: Qf 7.A
Hollinger, David A: Pe 7.P
Hoppe, Brigitte: Be 5.A
Hormigon, Mariano: Mf 6.A
Hornix, Willem J: Cd 7.P
Howse, H Derek: Ad 3.A
Hoyer, Ulrich: Pj 7.A
Hoyer, Ulrich: Pk 7.P
Hu, Naichang: Bl 6.P
Hubenstorf, Michael: Bh 6.P
Hunemorder, Christian: Xc 7.A
Hurwic, Jozef: Pp 6.P

Infante, Jose M: Hc 7.A
Ionescu-Pallas, Nicholas: Pf 5.A
Ionescu-Pallas, Nicholas: Qf 7.A
Isaksson, Eva: Pi 6.P
Israel, Giorgio:

Jain, Suresh C: Sb 5.A
James, Frank AJL: Xb 6.P

James, Matthew J: Bi 5.A
Jaquel, Roger: Xa 3.A
Jia, Sheng: Pi 6.P
Jimenez-Olivares, Ernestina: Ha 3.A
Johannisson, Karin: Ha 3.A
Johansson, Sif: Bg 6.A
Johnson, Jeffrey A: Xb 6.P
Johnston, Ron: Po 6.A
Jonsson, Kjell: Xd 7.P

Kaiser, Walter: Ce 5.P
Kamlah, Andreas: Me 7.P
Kauffman, George B: Pp 6.P
Keel, Othmar F: Bc 2.P
Keller, Alexander G: Tc 2.P
Kellogg, Mary: Pa 2.A
Kim, Dong-Hyun: Hc 7.A
Kim, Yong Woon: Qg 7.P
Kipnis, Naum S: Bc 2.P
Klein, Martin J: Pj 7.A
Knoefel, Peter: Bd 1.P
Knorr, Wilbur: Ma 1.P
Knudsen, Ole: Pj 7.A
Ko, Tsun: Tb 3.A
Kokou, Olodo: Xg 8.A
Kondratas, Ramunas: Bf 5.P
Kornbaum, Paerle: Xh 5.A
Kox, A J: Ph 6.A
Kruger, Lorenz: Me 7.P
Kumar, Deepak: Sb 5.A

Langeler, Jan: Be 5.A
Larsen, Mogens Esrom: Ma 1.P
Lashchyk, Eugene: Pe 7.P
Laszlo, Alejandra: Bg 6.A
Laurent, John: Bk 6.A
LeGrand, Homer: Gb 7.P
Leary, David E: Hd 7.P
Leikola, Anto: Bb 3.A
Leopold, Joan: Bi 5.A
Leopold, Joan: Hb 6.P
Leviton, Alan: Ga 3.A

Li, Pei-shan: Bh 6.P
Li, Zhaehua: Mg 8.A
Liedman, Sven-Eric: Xe 5.P
Lih, Ko-Wei: Mh 5.A
Lindberg, Bo S: Xa 3.A
Liu, Kwang-Ting: Ca 1.A
Lomax, Elizabeth: Be 5.A
Losee, John: Qg 7.P
Love, Rosaleen: Bg 6.A
Lowood, Henry: Bb 3.A
Lu, Jingyan: Qa 3.A
Lu, Jingyan: Ta 2.P
Lundgren, Anders: Cb 1.P
Lung, Tsuen-ni: Ta 2.P

MacKinnon, Edward: Pn 5.P
Macrakis, Kristie: Xe 5.P
Mark, Robert: Qa 3.A
Maghout, Khalil: Ab 2.A
Maieru, Luigi: Mc 1.a
Martinez-Contreras, Jorge: Bj
Martinez-Gazquez, Jose: Ab 2.A
Mason, Stephen F: Be 5.A
Mason, Stephen F: Qc 1.P
Mathur, Krishan: Ba 2.A
Matsuo, Yukitoshi: Cb 1.P
Matsuo, Yukitoshi: Qe 6.P
Matthews, Glenna: Bk 6.A
Mauskopf, Seymour H: Cc 2.A
May, John G: Xb 6.P
Mayor-Mora, Alberto: Sa 8.A
McClellan, James: Xa 3.A
McLaughlin, Peter: Bb 3.A
Mei, Jianjun: Tb 3.A
Mei, Rongzhao: Ma 1.P
Meinel, Christoph: Ca 1.A
Melhado, Evan M: Bl 6.P
Mendoza, Celina Ana Lertora: Ab 2.A
Meyenn, Karl von: Pm 5.A
Michel, John L: Pk 7.P
Millman, Arthur: Bi 5.A
Miloshev, Boris Yossifov: Sc 7.A

Mintz, Lee: Ha 3.A
Mokrzecki, Lech: Qc 1.P
Mora-Charles, Maria Sol de: Md 1.P
Morris, Robert J: Cb 1.P
Moulin, Anne Marie: Bg 6.A

Nagaoka, Ryosuke: Mc 1.A
Nakagawa, Yasuo: Pd 6.A
Narayan, VA: Sb 5.A
Nasr, Zuraya Monroy: Ha 3.A
Nelson, Clifford M: Xc 7.A
Neustadt, Mark: Qc 1.P
Newcomb, Sally: Cc 2.A
Newman, William: Ca 1.A
Ngoc, Nguyen Dinh: Mh 5.A

Ogawa, Tsuyoshi: Pl 8.A
Ohbayashi, Masayuki: Bg 6.A
Olausson, Lennart: Hb 6.P
Oosterhoff, Jan: Po 6.A
Ortiz, Eduardo L: Mf 6.A

Padian, Kevin: Bj 5.P
Pallo, Gabor: Xe 5.P
Palm, Lodewijk: Bd 1.P
Pancaldi, Giuliano: Bi 5.A
Papp, Desiderio: Bi 5.A
Papp, Desiderio: Ph 6.A
Peltonen, Carita: Xf 6.P
Perlman, James S: Pa 2.A
Petit, Annie: Qf 7.A
Pittman, Walter: Ga 3.A
Polanco, Xavier: Sc 7.A
Polikarov, Azaria: Qg 7.P
Porter, Theodore M: Me 7.P
Pressman, Jack D: Hd 7.P
Proctor, Robert: Sd 7.P
Proverbio, Edoardo: Ad 3.A
Puig, Roser: Ab 2.A
Pyle, Andrew John: Qc 1.P

Qian, Wen-yuan: Qe 6.P

Rabbitt, Mary C: Xc 7.A
Ramirez, Santiago: Mf 6.A
Rao, Ranganatha S: Tc 2.P
Rasch, Manfred: Sd 7.P
Reeves, Barbara J: Pi 6.P
Rhees, David: Cd 7.P
Rickey, V Frederick: Mf 6.A
Rife, Patricia: Xe 5.P
Rigden, John S: Pl 8.A
Rivera, Zely E: Sa 8.A
Roberts, Lissa: Cb 1.P
Robotti, Nadia: Pk 7.P
Roe, Shirley A: Bb 3.A
Rolbiecki, Waldemar: Xg 8.A
Romanovskaya, Tatiana: Pp 6.P
Rossi, Arcangelo: Qe 6.P
Roy, Mira: Qa 3.A

Saha, Margaret: Bf 5.P
Said, Hakim Mohammed: Qa 3.A
Samarescu, Dan Florian: Pf 5.A
Sangwan, Satpal: Sb 5.A
Saran, S: Ta 2.P
Schmitz, Rudolf: Bl 6.P
Schoijet, Mauricio: Bh 6.P
Schoijet, Mauricio: Sa 8.A
Schubert, Helmut: Pg 5.P
Sebestik, Jan: Qe 6.P
Secord, James: Bj 5.P
Servin-Massieu, Manuel: Sa 8.A
Shames, Morris: Pe 7.P
Shamin, Alexei: Bd 1.P
Shamin, Alexei: Be 5.A
Sharma, Shakti Dhara: Aa 1.A
Sharma, Shakti Dhara: Ae 8.A
Sharma, Shakti Dhara: Ma 1.P
Shea, William: Pb 1.A
Shen, Kangshen: Mh 5.A
Shimodaira, Kazuo: Mg 5.A
Sime, Ruth L: Pp 6.P
Singh, Ravindra N: Ta 2.P
Sinh, Hoang Xuan: Xf 6.P

Sleeswyk, Andre: Tg 6.A
Smith, CL: Bi 5.A
Smolenov, Hristo: Pl 8.A
Snelders, Henricus: Ce 5.P
Sofonea, Liviu: Pf 5.A
Sofonea, Liviu: Qf 7.A
Sokal, Michael M: Hd 7.P
Song, Sang-yong: Pd 6.A
Sorbom, Per: He 8.A
Sorlin, Sverker: Tf 5.P
Sperling, Norman: Ae 8.A
Srinivasan, Raman: Sb 5.A
Stanley, Autumn: Xf 6.P
Stranges, Anthony: Sd 7.P
Sturchio, Jeffrey L: Cd 7.P
Sugiyama, Shigeo: Bl 6.P
Sullivan, Brian G: Xd 7.P
Sun, Fang-Toh: Ta 2.P
Sun, Suyun: Tb 3.A
Swijtink, Zeno G: Bf 5.P
Sylla, Edith: Qa 3.A

Tagawa, Masa-yoshi: Mh 5.A
Tande, Annie: Xh 5.A
Taylor, Kenneth L: Ga 3.A
Teichmann, Jurgen: Pg 5.P
Theodorides, Jean: Bl 6.P
Tian, Changhu: Tb 3.A
Todd, Edmund N: Tf 5.P
Tsuji, Tetsuo: Mb 2.A
Tsuneishi, Kei-ichi: Sd 7.P
Tuman, Vladimir: Aa 1.A

Ulansey, David: Aa 1.A
Ullrich, Rebecca: Xb 6.P
Unno, Kazutaka: Xa 3.A

Van Egmond, Warren: Md 1.P
Van Keuren, David: Bk 6.A
Vidal, Fernando: Hd 7.P
Viladrich, Merce: Ab 2.A
Vincent-Bensaude, Bernadette: Pi 6.P

Waff, Craig B: Ad 3.A
Wallace, William A: Qb 1.A
Wallace, William J: Tg 6.A
Wallen, Stanley: Xd 7.P
Wang, Jinguang: Pc 2.P
Watkins, Sallie: Xf 6.P
Weart, Spencer: Pg 5.P
Webb, George: Xd 7.P
Weiss, Burghard: Pc 2.P
Weiss, Sheila: Bh 6.P
Wells, Ronald A: Aa 1.A
Welther, Barbara L: Ae 8.A
Westfall, Richard S: Ac 2.P
Westman, Robert: Ac 2.P
Westman, Robert: Md 1.P
Widen, Solveig: He 8.A
Widmalm, Sven: Xa 3.A

Williams, Bernard O: Td 7.A
Williams, Mari EW: Tg 6.A
Winsor, Mary P: Bj 5.P
Wisan, Winifred L: Ac 2.P
Wu, Yi-yi: Pc 2.P

Xie, Huan Zhang: Ba 2.A
Xu, Liang-ying: Qg 7.P

Yagi, Eri: Pj 7.A
Yamazaki, Eizo: Pc 2.P
Yao, Shu-ping: Se 8.A
Yeo, Richard: Xb 6.P
Yoshida, Akira: Ce 5.P
Yoxen, Edward: Xe 5.P

Zhang, Yunming: Tb 3.A
Zhang, Zhong-jing: Ta 2.P